METHODS OF

BOOK DESIGN

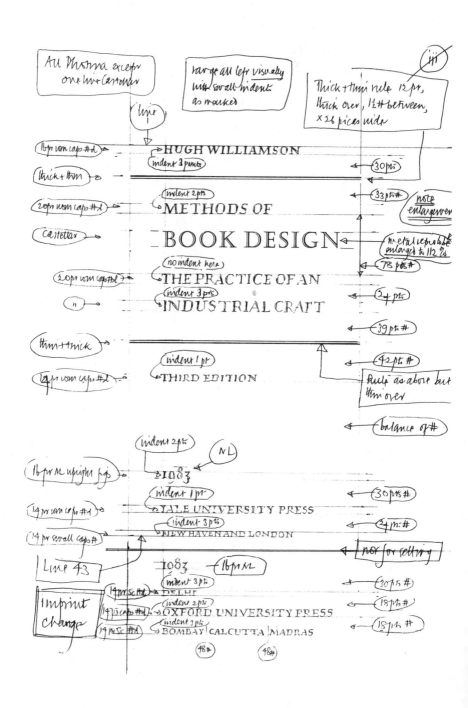

FRONTISPIECE. Layout for the title-page opposite.

HUGH WILLIAMSON

METHODS OF

BOOK DESIGN

THE PRACTICE OF AN

INDUSTRIAL CRAFT

THIRD EDITION

YALE UNIVERSITY PRESS

NEW HAVEN AND LONDON

Copyright © 1983 by Yale University
Second printing 1985

Designed by the author, set in Lasercomp Photina,
and printed in Great Britain at the Alden Press, Oxford

Library of Congress cataloging in publication data
Williamson, Hugh Albert Fordyce, 1918–
Methods of book design
Bibliography: p. Includes index
1: book design. I: title.
Z116.A3W5 1983 686 83–3610

ISBN 0–300–02663–3
ISBN 0–300–03035–5 (pbk.)

Contents

Illustrations

Illustration numbers are those of the sections in which the illustrations appear. When there are more illustrations than one in a section, the order of illustrations within the section is indicated by a lower-case italic letter.

Preface to the third edition

M*ethods of book design* was first published in 1956, nine years after I completed a short course at the London School of Printing (as it was then) and joined the London publishing business of Oxford University Press. Book production was comparatively simple in those days, though it may have seemed so only to a beginner. A year of formal printing education, a few years of practical experience of book design and production in a great publishing organization, and many absorbing hours in the Printing Library of the St Bride Institute nearby provided foundation enough for a text which was to continue to sell in no great numbers but very steadily for twenty years. The first edition even included a section about photo-composition and a number of passages which referred to photo-composition without mentioning it; examples included the difference in design between capitals and small capitals, and the effect of optical scale on letter-design. Several chapters were vigorously polished by specialists who knew a great deal more than I did. Publishing and printing friends at Oxford University Press joined in, beyond the call of duty. The completed edition was born of a team operation.

Ten years later I brought the text up to date for its second edition, published in 1966. An appendix listing composition founts was replaced by an analysis of type-face popularity. There was no other major change.

In 1975 Oxford University Press, which had published the first two editions, indicated that stock was running low and that a third edition should be prepared within the year. I was flattered and a little dismayed. I would have thought somebody younger would have written a different book to take the place of mine by then. I had left publishing and returned to the printing industry. I had not designed a hundred assorted editions a year since the 1950s. Above all, the balance of book production was tilting towards photo-composition and offset printing, and other radical changes were on the way to enforce rewriting rather than revision. I started at once, but progress was slow, partly because I was in my fifties and less energetic than I had been twenty years before, partly because I had difficulty in deciding how best to write about a scene which was changing so rapidly, and partly because as a printer I was deeply involved in technical and economic change. I spent several years, on and off, preparing the third edition.

By the time it was ready, it was too late. Economic recession had settled in like a cold winter, and changes in editorial policy excluded my book from the famous list in which it had twice appeared. Accelerating change in book-production technique compelled me to rewrite much of the rewritten version. This has not been easily done; I have had to examine metal typography closely and at length, for example, in order to explain photo-composition to readers who may need to know nothing else about metal type. There are areas of book production — the latest type-designs, for example, the electronic scanning of half-tone subjects, the belt-fed Cameron press, and the burst or notched techniques of unsewn binding — of which I have heard or read but into which I have never entered. Lacking opportunity to study these innovations, I have based my text for the most part on information acquired through my own experience, trying to demonstrate an approach to innovations which others may follow towards innovations not yet begun. I have spent a long time on this, and if Yale University Press had not accepted the book for publication before revision began in earnest I might never have finished it.

Since I started work on the new edition, metal type has gradually disappeared from the Alden Press, at which I work, and has been replaced by a computer-centred Lasercomp system of composition. I have been intimately involved in this change, compiling and designing a type-specimen book for it, selecting type-faces for installation, and regulating spacing parameters among other items, and I have learned a good deal about such systems. In my view they are at some points in advance of the metal type-setting systems they have replaced, and at other points they have not yet caught up. The same is true of offset printing; photo-composition and offset still have scope for improvement before they match all the advantages of metal type and letterpress printing. This is one of the reasons for the number of pages I have devoted to techniques no longer in use in most bookwork printing houses.

The plan of the third edition is much the same as that of the first two. Paragraphs and even pages survive from the earlier editions, but otherwise this is a new version of the original work. The illustration of books such as this one has greatly developed in recent years, and my text might indeed have looked more attractive if it had been embellished and explained by many more illustrations which would have been easy to find; but the students and beginners who pay for the book themselves would have had difficulty in affording it. Reproductions of printed pages are now easier to find elsewhere, and examining books is more instructive than looking at pictures of them.

As in earlier editions, chapters are divided into sections for ease of reference, but few sections if any deal comprehensively with their subjects; there are likely to be other references to the matter elsewhere, indicated by the index. In this edition sections are numbered serially by chapters. Illustrations are not serially numbered; each takes its number from the section in which it appears, with the addition of letters when more than one illustration appears in a section, enabling the reader to trace illustrations as well as sections through the headlines.

Italic type is used in this edition to distinguish a word or phrase from its neighbours, usually to show that it will be defined, and sometimes for another purpose identifiable by context, but in no instance with the intention of emphasis. The number of a year in parentheses indicates reference to a title described in full at the end of the book under the heading *Publications*, a section which replaces all the lists in previous editions but in abbreviated form.

Printers have been developing the vocabulary of their trade for centuries, and printing designers who understand the language do well to use it. But it was not methodically compiled; there are some terms (such as *beard*) to which different authorities assign different meanings; there are words (such as *folio*) which may be correctly used in two or more distinct senses; and there are items (such as *ascenders and descenders*) which seem to have generated no generally used collective name. Electronic engineers and machinery manufacturers have debased parts of this vocabulary by appropriating old terms for new techniques; *kern* becomes a verb which does not refer to overhanging strokes, and *reverse leading* is possible only in the absence of lead. Book designers must still speak the printer's language when they instruct him in printing matters, and like anybody else who has the privilege of giving instructions designers have a duty to make their instructions clear and unambiguous. In writing this book about the making of books I have done the best I can with words familiar to me after many years among printers, and where there seemed to be no appropriate word in general use I have invented one (such as *extender*); I cannot expect everybody to agree than my vocabulary is correct, but I hope very much that my meaning is always clear.

The English language, like the language of printers, has not yet developed all the words we now need. No single word, for instance, represents a male or female person, except perhaps the irritating 'one' — one can hardly rely on *he/she* or *his/her*. Accordingly I have throughout the book referred to the designer as though all designers were men; Beatrice Warde, at a time when very few women were typographers, expected nothing else in the first edition of 1956. A generation later I trust I shall not be suspected of not knowing that there may be as many women as men in book design in Britain and the United States.

The historical introduction to the first two editions is now omitted, and part of the preface to the first edition has been written into a new introduction. Changes in book-production technique are reflected in changes among the chapter titles. Appendixes dealing with format and with colour analysis and synthesis have been for the most part absorbed into the main text and omitted as separate items.

Nobody should suppose that this book or any other contains all the information any book designer is likely to need. The first edition represented a proportion of the opinions and experiences I accumulated in the first eight years after my release from military service in 1946. This edition also includes other ideas evolved in more than twenty-five further years of practice. But my work has always been that of an administrator of the publisher's and the printer's part in book production, and never that of a specialist book designer. No doubt there will be readers who quite early in their careers know very much more about various aspects of book design than I do after more than thirty years in my trade. To these and all other readers I suggest that what we all need most is not an encyclopedia of printing or a comprehensive work of reference, which I have not attempted to provide, but a common purpose to maintain a continuing, methodical, and critical inquiry into book design, and into everything of technical, literary, aesthetic, and economic substance and value we can observe in it. If the chapters that follow indicate a method of gathering the necessary information and of applying it effectively to the practice of book design, they have succeeded in their purpose.

Oxford, 1983   H.W.

METHODS OF

BOOK DESIGN

CHAPTER I

Introduction

F OR THE PURPOSE of the chapters that follow, I define *book design* as the planning which determines the physical characteristics of the book, including particularly its dimensions, its general appearance to the eye, its structure and mechanism, and its durability. Planning part of a book — the arrangement of type and illustration, for example — is no more than part of book design, which is here intended to mean the planning of the whole book. *Book production* may be defined as the other processes by which typescript is translated into edition, including such essential procedures as working out estimates and deducing published prices, buying materials and services from suppliers, keeping track of progress and of specification changes, and controlling expenditure.

Only the industrially produced book is discussed in these pages, though in terms which may at some points apply to the hand-made printed book, to periodicals, and to other kinds of printing. The theme of this book is an intricate form of industrial design, devoted to the service not only of author and reader but of commerce. The printed book appears in innumerable forms, some differing radically and others in detail from the majority. Books of entertainment for young children, school-books, books concerned with the arts and sciences, and works of reference, for instance, all pose their own problems of design and offer their own opportunities for the exercise of skill. To delve into these would be pleasant and might be instructive, but the preparation of many other special kinds of book would remain undiscussed. The whole craft of book design cannot be described in a single text; better to attempt a description of the principal operations of the non-specialist book designer.

The term *book design* is used here because it is familiar; my subject would be more accurately entitled *edition design*. The task is to design not a single book but a whole edition, the last copy of which will be no less well-made than the first, and which will lend itself to whatever reprints may later be ordered. From time to time the design of an edition already produced may have to be modified for the production of a new edition of the same work or for that of a different publication. Few book designers deal with a single edition at a time for long; most have to plan a stream of editions, some intended to move more rapidly through production than others, and each appearing and reappearing on the designer's desk in various forms long after his plan for it has been completed, and probably while he is designing a subsequent publication. And probably few book designers — as distinct from typographic designers — spend all their time, or even most of it, on design. The control of book production is a matter of timing and cost as well as quality, and some share of departmental and general management may have to be allowed priority. Time has to be found for careful and imaginative work, but all too often the best design is that which is ready when it is needed.

The craft of book design consists of more than planning. The efficacy of the plan has to be verified at various stages by the designer himself. Examples of the kind of evidence on which he relies for this verification include the estimate of production cost and the proofs of the type-setting. Instructions may have been misunderstood or neglected; or if they have not brought about the intended effect, they may have to be amended. The designer examines proofs for the same reason that the author reads them: each knows his own intentions better than anybody else, neither has other means of making sure they are effectively implemented at every point, and both rely on accuracy and consistency to deliver the result they seek.

The aesthetics of book design are not easily described, if indeed they can be properly described at all. Taste and fashion have their influence here, and the tendency is towards continual slight change. To lay down rules of style would be easy enough — we need only consider how things were done yesterday, or how they are done today, or how we prefer to do them ourselves, and to elevate these practices or preferences to the status of dogma. Principles of design which apply to all kinds of book, which will continue to apply for more than a year or so, and which need to be stated, are less easily found. The main points about legibility can be asserted; but when deciding what kind of pattern the printed image is to present, the book designer will do better to rely on his own ideas about the nature of the text before him than on those of writers discussing the theory of his craft. The conflict between centred and eccentric styles of display typography, for instance, will not be terminated by theory, nor can any text describe or illustrate the innumerable valid styles of typographic arrangement. An hour spent on planning a flat opening and a durable spine may be worth two hours of searching for an original and striking arrangement for a title-page. Printing is a mass-production technique of communication, and effective communication and economical production are worth more than patterns. For these reasons the aim of this book is to show something of the possibilities of industrial book design rather than to say what use should be made of them.

An edition faultless in design, in materials, and in workmanship is all too rare an achievement. Those who have shown themselves capable of such a triumph are entitled to move with certainty among such subtleties as the arrangement of printing types not merely in a legible but in a graceful style; the appreciation of the minutiae of type design and of presswork; the aesthetics of book illustration; and the inchoate but attractive opportunities of art, atmosphere, and allusion in typography. There are fields of experiment to be explored too; books will not keep their present form for ever, as electronic revolutionaries have been rather stridently declaring for some time, but their extended and elaborate messages will have to be published somehow, and book designers will be among those who evolve the method. The most welcome innovations are likely to be those which have been shaped by the traditions and development of the printed book throughout more than five centuries, and by those conventions of good printing which are still current today.

Industrial book production is technical, complex, diverse, and continually developing. No designer should suppose himself to be intimately familiar with all the materials, techniques, and machinery that may be available for the manufacture of his editions. He is always likely to encounter something new to him even in a process long and often used, something unexpected in the operations of a different supplier or in the properties of a different material, and something unforeseen in the evolution of a new technique. Then he will do well to identify if he can, in each new element of his book production, those characteristics which affect his plans. These may be termed the *design factor*, the cardinal point of selection and comparison by any designer, the aspect which deserves to be probed with all the technical knowledge at the designer's disposal. An example is half-tone printing by web offset, the quality of which should not be relied on until the specific printer's previous performance of such tasks has been assessed. Some printers can reproduce monochrome or colour half-tones very well indeed by web offset, others seem to be incapable of a tolerable result; an edition which includes half-tones intended to be of good quality, about to be produced by a printer whose half-tone abilities (the design factor) have not been assessed, is like an accident approaching the place at which it is likely to happen. A designer capable of identifying and observing those factors in book production elements new to him can find his way along unfamiliar paths without relying too much on publications about his craft.

The available qualities of workmanship and materials for book production range from excellent to intolerably bad. Within this range the conscientious designer seeks the best that can be afforded for each edition. The proprietor of an edition — usually a publishing organization — does not always ask for the best and may not always be willing to afford it. Publishing colleagues of the designer are likely to judge book production primarily by its economy and punctuality. In terms of sales, the public response to excellence is not often identifiable. The designer who seeks to set and to maintain standards higher than those of the proprietor acts all but alone, in the service of authors, readers, and books, but in doing so he is likely to gain for the proprietor a commercial advantage. Every book has to be good enough in every way for its purpose, for its price, and for the good repute of its publisher; it may have to compete with a better-made book which is more apt to catch the attention of a buyer. By continuing selection, designers collectively foster good workmanship and materials, which will not be available for long if demand is allowed to wane. The best chance of success lies in planning all editions for economical production, as in other forms of industrial design, and in making the most effective use of permissible expenditure.

Ideally a book designer should be something of a published writer, substantially a general reader, a committed member of the book-buying public, and even a little of a bibliophile and a bibliographer — in short, a bookman — as well as a publisher or a printer or both. But most people have in their nature enough of the designer and craftsman, and enough ability to recognise evidence of the difference between good work and bad in any field, to design books well if they will only try. The general public is far from blind to that difference, without claiming any great perceptive talent. An abiding enthusiasm for books — primarily to read, but also to look at — encourages the imagination. Knowledge and skill and confidence grow with practice; each completed edition leads the way to better work next time.

The nature of the task has never been more eloquently commended than by Daniel Berkeley Updike, in the closing words of his great work on *Printing types*, first published in 1922 —

The practice of typography, if it be followed faithfully, is hard work — full of detail, full of petty restrictions, full of drudgery, and not greatly rewarded as men now count rewards. There are times when we need to bring to it all the history and art and feeling that we have, to make it bearable. But in the light of history, and of art, and of knowledge, and of man's achievement, it is as interesting a work as exists — a broad and humanizing employment which can indeed be followed merely as a trade, but which if perfected into an art, or even broadened into a profession, will perpetually open new horizons to our eyes and new opportunities to our hands.

CHAPTER 2

The typescript

THE TYPOGRAPHICAL COMPOSITION of books should above all be correct, consistent, and clear. Textual defects cannot be redeemed by triumphs in other aspects of book production. Quality of a high order depends on skilful and diligent sub-editorial work, and a designer who aims at such quality will succeed only if he holds himself responsible for work of this kind whether or not he does it himself. If for example his pages, evenly composed and clearly printed on good paper, are marred by a conspicuous excess of capital letters and by a less obvious number of spelling mistakes, the edition has been spoilt by his own neglect.

Good printing in this sense begins with the copy. *Copy* is that version of the text the printer copies when he sets type. Writers may have to be permitted to act the amateur in preparing copy; their duty and talent are not spelling and punctuation but the assembly and presentation of ideas. The publisher, in his contract with the author, normally claims control of book production, which by custom includes the regulation of textual apparatus. Whatever the state of the typescript when it arrives from the author, by the time it enters the printer's composing-room it needs to be in such order that a textually sound edition is likely to emerge from it, and that alterations to type will be kept to a minimum. Alterations are not only expensive; they are all too likely to lead to additional inaccuracies and other defects of quality.

§ 2-1 Author and typescript

The more the author can be induced to do by way of preparing good copy, and the less amendment publisher and printer impose on the typescript as a result, the more directly the author will address his reader from the pages of his book. Most authors begin to compile typescript without much study of the requirements of copy for publication. The publisher will do well to intervene before the typescript is begun, if he has the chance, and to explain these requirements. Publications about copy preparation, for the guidance of prospective authors, are available or can be compiled, and are likely to cover the items common to most projects. But an author impatient to begin his own work may not be eager to spend much time studying the requirements of book-production organizations.

All copy intended for a printer should be typewritten, partly for clarity, and partly to enable the number of characters in the text to be calculated by averaging (§ 18-1). Each typewritten character and space should occupy the same lateral space as the others; typewriters with varying character widths, in imitation of printing types, make calculation difficult.

Well-prepared typescript will not only start by being clear, but will also be durable enough to remain clear throughout the repeated handling it will endure, folio by folio, in publishing house and printing office. In this chapter, a *folio* means a single-sided leaf of copy; elsewhere it may also mean the size of a book (§ 3-1) or the number of a printed page. The sheet must not be too wide or too deep for the copy-holding bracket beside the keyboard operator; the European standard size A4 (297×210 millimetres, $11\frac{3}{4} \times 8\frac{1}{4}$ inches) is big enough, and one of the larger photocopying leaves is unlikely to be too big. Flimsy paper may be an advantage in airmail costs, but is awkward to handle and mark, and likely to be torn and crumpled in handling.

At the head of the folio, a margin equivalent to two or three lines of typing will allow the copy-holder to grip the paper without concealing the first line. On the left, a margin equivalent to about fifteen typed characters will provide room for instructions written by sub-editor and designer. At the tail of the sheet, a margin deep enough for the typewriter to grip while the last line on the folio is being typed will ensure that the last line is as straight as the first. On the right, some kind of margin is needed to prevent characters from falling off the paper unnoticed by the typist.

A deep enough space between the lines will accommodate any line of typing or handwriting that may be added by way of amendment. All lines of typescript, without exception, should be separated from their neighbours by this amount of space. Sometimes small type is unnecessarily indicated by the typist's use of close spacing; all too often, such passages need the most extensive sub-editorial markings, for which there is not enough room. Prose extracts quoted from another work, when they make say four or more lines of typescript and are likely to be set in smaller type or in some other style distinct from that of the main text, should start a new line and should be indented at the left with space above and below, but spaced between the lines in the same way as the main text. Quotation marks are unnecessary round such extracts.

The typing should be black and sharp, so that the keyboard operator can read it without error at arm's length. Amendments are best placed between the lines, where the keyboard operator will be looking, not in the margins; if they are not typed, they can at least be clearly written in black. As the folio will be fixed upright in a copy-holder, additions should always be horizontal, and all the copy should appear on one side of the paper only. All sheets should be the same size, even if a few lines of addition have to appear on a single folio, to facilitate the handling and safeguarding of the copy.

The typist might be invited to study a few well-printed pages before beginning, and to emulate the way in which the type is set out. An indent of three or four characters is ample to indicate the beginning of a new paragraph, and there should be no extra space between paragraphs; at the end of a sentence, a single space should be enough.

Imitation of the printed page need not involve the typist in the laborious centring of headings and sub-headings. These may be aligned at the left, and separated from the adjacent text by extra space. When there is more than one kind of heading, the grade of each should be indicated in the margin by one of a series of letters, circled to show that it is not to be set. Capitals should be used in headings only where they are essential, as in the text, since the designer may decide to set them without extra capitals. Underlining is best restricted to words for which italic type is obligatory; it is for the designer to decide whether italic is to be used for headings. Except by agreement with the publisher or in accordance with well-established convention, as in volume numbers when journals are referred to in learned works, the wavy underline which indicates bold type is better avoided by the author. The same applies to the double underline for small capitals, which the designer may decide not to use. Any kind of underlining is difficult to delete tidily during sub-editing. If headings which make more than one line are divided into lines at a natural break in the wording, designer and printer may find a clear and natural style of heading will emerge as a result of following the copy on this point; but shorter headings, less than a line long, are usually more effective. There is no need to number sub-headings if the text includes no reference to these numbers.

Each folio should be so numbered that its position in the book is unmistakable. Numbering by chapters will do if the chapter number precedes the folio number; the first folio of chapter 12, for example, would be numbered 12-1 or 12/1. If an inadequately numbered typescript is dropped and scattered at any stage, reassembly may introduce mistakes as well as delay. The typescript of the text is often the framework into which illustrations, tables, and other such items are fitted. Marginal notes, such as 'figure 3 near here', indicate the approximate position of each. Text references to exact positions ('above', 'below', 'on this page') may have to be altered during *make-up* (§ 4-8).

When the composition of the printed page will be more complex than usual, or when the number of pages is crucial, the publisher may be able to arrange for the typescript to be set out in the style of his plan for composition in type, perhaps by providing specimen folios of typescript. This will reduce otherwise extensive marking of the copy which may make it difficult to follow accurately, and may enable the author to tailor his text to the space available while he compiles it.

The designer may also have an opportunity to commend simplicity in the presentation of text. On the whole, continuous reading is easier when it is not continuously interrupted by sub-headings, paragraph numbers, *indents* (§ 7-4) and so on. The text may indeed have been systematically assembled by its author, but the structure may not benefit from this typographic emphasis. Paragraph and other item numbers are better omitted when they are not useful to the reader.

In the copy as in the edition, any word or phrase spelt, capitalized, hyphenated, italicized, or presented in any particular way should be typed in exactly the same way throughout the work, whenever its meaning is exactly the same. An observant reader may otherwise suppose that inconsistency in presentation indicates some change in meaning. If while he prepares his copy the author compiles a list of his preferences about such details of style, the publisher is likely to accept and apply such preferences, unless he believes a reader might think some of them incorrect.

In the same way, an author whose work requires unusual letters, signs, accents, or figures can help publisher and printer to get them right by listing at least the first appearance of each with its folio number, and by adding printed or carefully drawn examples of the character he means to designate. Accents and letters for the major languages of the Western world may be taken for granted; mathematics, Greek, phonetics, and other special alphabets may be mentioned collectively, as the necessary equipment is likely to be available in the office of an established printer of such works.

Correct type composition may be defined as text which in all its details presents what the reader is entitled to expect. Oddities may be permissible — by the time the reader has swallowed one or two of them he may come to tolerate the taste — but no reader is likely to enjoy what he believes to be a mistake. The poet e. e. cummings spelt even his own name without capitals. T. E. Lawrence preferred inconsistency in the transliteration of Arabic names. Bernard Shaw used *letterspacing* (§ 4-5) for emphasis in order to avoid italic, and shunned the apostrophe. Nobody seems to have complained, but hardly anybody seems to have followed such examples either. When the publisher has contractually claimed control over book production, he is entitled to decide what is correct. Bernard Shaw paid for and controlled his own book production.

Copy which is consistent in detail, and not obviously incorrect, deserves to be followed closely in detail by publisher and printer. Such typescript is all too rare.

§ 2-2 The sub-editor's task

Sub-editorial work on the typescript may be carried out by publisher or printer or both. There are also editions which appear to have received no attention of this kind at all. The nature not only of the text but of the organizations which print and publish it determine the pattern.

Many printers now require most copy to be followed exactly by their keyboard operators, compositors, and readers. If on arrival from the publisher the typescript appears to be inconsistent, incorrect, or obscure, a conscientious printer has it amended until it is capable of easy translation into a textually sound edition. The best guides in this task are the evident preferences of author and publisher, and the usage of the printing trade. Such a printer may well be more meticulous and knowledgeable than most authors and publishers about such matters as punctuation, and may regard convention in this area as a part of good printing on which he is entitled to rely for his reputation.

After some years of handling typescript of various kinds, printers tend to accumulate a body of precedent and custom by which doubtful matters of style can be resolved. *Hart's rules for compositors and readers at the University Press, Oxford* (1978) provides an authoritative example of a printer's house style. Such legislation co-ordinates the work of different people within the same organization, and becomes habit as well as law. The publisher who seeks to change such customs by imposing his own house style throughout may jeopardize the accuracy of composition. He would do better to perfect the copy and require the printer to follow it in detail. If he merely indicates some preferred spellings and other details of style, he may have to pay the printer for going through the whole typescript in search of points at which these preferences are to be applied.

A printer's sub-editor will not as a rule change the wording of the text unless there is an obvious mistake in the copy; and even if a sentence appears to be nonsense, he will probably do no more than politely query its construction. Any printer is contractually entitled to follow copy exactly, unless he has specifically undertaken to correct it. The conscientious printer who corrects as a matter of course is no longer typical; when most of his customers no longer intend to pay for this service, they can no longer expect it to be provided.

The publisher's sub-editor may well do more, in the exercise of editorial privilege. Whatever changes are made in the text, the copy should show what has been done to it and by whom, for the guidance of the printer, and to provide evidence for any subsequent inquiry into alterations or their cost. If author, publisher, and printer each mark the copy in a different colour, there will be little confusion. The designer's colour could be that of his employer, though an independent designer might introduce a fourth colour. The printer should then work from typescript showing the different colours, not from a carbon copy or photocopy; and his copy should be the clearest available.

Any author with access to a word processor can now provide electronic copy in the form of magnetic tape or disc which a printer with compatible equipment can use to drive a phototypesetter. The printer usually has to edit such copy on a *visual display unit* (§ 5-10) in order to insert typographic commands. The text itself is better left alone until the proof can be edited, unless a printout of the text has been supplied for sub-editing. Editing and sub-editing are best done in writing for the author to see; keyboard sub-editing can be examined only by meticulous comparison between draft and proof, and may appear to the author to be clandestine. All sub-editors should beware of an excess of zeal; Ronald Mansbridge (1980) has warned them that nothing is more exasperating even to the most experienced of authors.

The marks made on the copy must be clear and self-explanatory. If handwriting is inadequately legible, it is likely to be misread. Standard proof-reading marks, such as those of the British Standards Institution (1976), are useful for amendments to copy and should be familiar to all book printers. The purpose of any non-standard mark may need explanation to the printer if it is to be carried out correctly.

Any instruction which applies to the text as a whole, rather than to a specific folio, should appear in a separate list of such instructions. The marks on any folio may apply to some subsequent folios, but all should have a limited and local application. More than one person may be at work on the book at any stage; if copy is divided into two or more batches, or 'takes', any directions on the earlier folios may not be observed during work on the later folios.

The author's marks on copy tend to amend the meaning of the text. The publisher's marks may change its arrangement and the construction of sentence or paragraph. A good printer may be left to add any further typographical notes his composing-room may need, and to attend to consistency in style, in accordance with any requirements of author and publisher he may have observed.

Any kind of ambiguity in copy may cause queries to be raised, or type to be wrongly set. When copy presents problems which can be solved for the printer only by author or publisher, production comes to a stop while the answer is sought. In a busy printing-office the time so lost cannot always be made up; production of the edition slows perceptibly and sometimes inopportunely. The printer himself may wish his staff to get on with production rather than to raise such queries, which may have been put up as an excuse for delay arising from another cause. If author and publisher favour continuation of progress whatever the textual problems, they have a duty to make their intentions unmistakable in the copy.

§ 2-3 Designer and copy

The designer's aim in marking copy should be to render his design intentions clear at any point where doubt could possibly arise. Even when separate design instructions have been sent to the printer, these instructions may be clarified and emphasized by marks on the copy. Any item not covered by such instructions will need to be marked up by the designer. A well-designed book is well designed in all its parts, however difficult, and if such intricacies as tabular setting are left to the printer, the designer may be responsible for local failure he has not tried to avert.

The designer who works for a publisher may expect to be allowed at least some sub-editorial rights, with particular regard to the wording and arrangement of *preliminary pages* (chapter 8), headings, and other *display* (chapter 9). While he will do well to leave the text itself to others, he may with editorial consent improve the appearance of the text page, for example by reducing an excess of capitals, punctuation, or superfluous item numbering.

Industrial book design is not an art, devoted to the design of new kinds of graphic image. It is, rather, a kind of translation, generating an edition from copy. The form of the edition should grow from the nature of the text, should express something of the way in which the text is written, and should suit the kind of reader to whom it is addressed. If that form is also likely to prove acceptable to booksellers and librarians who distribute works of that kind, the book has its best opportunity of reaching its potential reader. The designer has obligations to author, publisher, bookseller and reader.

CHAPTER 3

Format and margins

THE TERM *format* has long been used to mean the shape and size of a book in particular, as well as its style of presentation in general. In recent years the word has been appropriated to mean a typographic specification for which a computer can be programmed. In this book I use the word *format* to mean the dimensions of the leaf of paper on which each printed page appears. Elsewhere a leaf is often called a *page*; here, a *page* is the printed area on one side of a leaf, or the material from which it was printed if that material is separable from other pages.

Format has to be settled at the outset of design planning. If it is not, much of the subsequent work may have to be altered, or will have to be provisional. Bruce Rogers (1936) was an eminent and experienced book designer when he started work on the Oxford Lectern Bible of 1935, but he seems to have done so without adequate information.

As is my usual procedure I first tried to visualize and plan a page which seemed to me the ideal one for such a purpose, disregarding for a time the limitations that would afterwards have to be met somehow. . . . I decided that the proposed Bible should, if possible, be set in a larger type than had recently been used, . . . the first step was to have several lines of Bible text set in the regular 22-point Centaur to an approximate column width. I printed enough of these proofs to paste up a large folio page with two columns of type, leaving an ample space between columns and adequate margins all round. . . . My pasted-up pages were admired; but, alas, in size they were even more impractical than I had guessed, for the size of the leaf, I now learned for the first time, must not exceed the standard dimensions of folio Bibles used on the brass lecterns of most English churches — and this size was 12×16 inches. As $1\frac{1}{2}$ inches was the minimum for the back or binding margins my ideal type page left only about an inch of margin round its other three sides — plainly impossible to contemplate. But Mr Milford and Mr Foss still preferred Centaur to either of the other types; so . . . I redrew all the ascending and descending letters, shortening their stems and compressing laterally the rounded loops . . . in the end thirteen new lower-case letters were cut. . . . All this time, however, I had the uncomfortable feeling that I should never really like this Bible, or be proud of it, on such a small size leaf with margins that, although as ample as is usual in folio Bibles, were still to my eye inadequate.

The 1935 Bible is generally considered to be a masterpiece, but its designer, committed to a type too large for the leaf, was disappointed. He would have done better to suit the leaf to the lectern before anything else. With any other book he would have done well to bear in mind the intervals between shelves, and the maximum sheet size of the press, at the same stage.

In much the same way, the format of any edition may have been fixed before the designer begins his work. Any series of books is likely to be produced in the same format throughout. Other books, not part of a series, may fall naturally into a classification in which the publisher specializes and for which he has previously preferred a certain format. The designer may be able to argue a case for change, but will often find that a final decision about format is not his to take.

§ 3-1 Format calculation and description

When a standard size of sheet and a standard format will suit an edition, there is no need for much calculation. Non-standard sheets and formats offer opportunities as well as pitfalls, and should be calculated to suit printing and binding methods as well as the content of the book and its reader.

The derivation of folded size from a given sheet is outlined below, with particular reference to standard formats, but for general application. To derive the dimensions of the cut leaf from those of the folded *section* (§ 16-2), the folded height is reduced by 6 millimetres or $\frac{1}{4}$ of an inch, to allow for cutting at *head* and *tail* (§ 16-2), and the folded width is reduced by 3 millimetres or $\frac{1}{8}$ of an inch to allow for a single cut at the *fore-edge* (§ 16-2). To derive sheet size from format, the same allowances for cutting are added to the dimensions of the cut leaf, doubled when any illustrations are to *bleed* (§ 14-3), and the totals are multiplied as indicated below.

A diversity of sheet and leaf sizes is as old as printing itself. The approximate measurements of each size of sheet came to be designated by one of a family of picturesque names. In 1951, for example, the British Standards Institution (BSI) standardized *foolscap, pinched post, imperial, ledger super royal,* and *double elephant* among its paper sizes for various purposes, but these were uncommon in book production even then.

In order to describe the format of a book, printers of the hand-press era used to state the sheet size by name in this way, and indicate the number of pages to be printed on each side of it. The full size of the sheet, too great to be used as a leaf in a book, was the *broadside*; this old term, no longer widely used in bookwork, here refers to the old named sheet sizes themselves. Two pages printed on each side are *folio* pages. Four *quarto* (4to) pages appear on each side of the sheet, or eight *octavo* (8vo), or sixteen *sexto-decimo* (16mo) and so on. If the folding methods are available, pages can be printed six or twelve each side of the broadside (*sexto* and *duodecimo,* § 12-1).

[13]

The folded but uncut size of the section is usually derived from that of the broadside by dividing the longer side of the broadside by 2 (folio, 4to, or 6to), by 4 (8vo or 16mo), or by 8 (32mo), and the shorter side by 2 (4to or 8vo), by 3 (6to or 12mo), or by 4 (16mo or 32mo).

A 6to in these terms makes a *landscape* format only — wider than its depth — when printed on a broadside, but on a sheet twice that size or more a 6to could be *portrait*. The factors by which the shorter or longer side of the broadside are divided can be exchanged. A folio format, for instance, would still be a folio if the shorter side of the broadside were divided. The resulting formats would in some sizes be all but useless, but in others an unusual but useful proportion could be derived.

The wooden hand-press, whose limited size determined the dimensions of the broadsides, could apply even pressure to quite a small sheet only; the broadsides standardized in 1951 ranged from *foolscap* ($13\frac{1}{2} \times 17$ inches) to *royal* (20×25 inches). The cylinder press, which superseded the hand-press, is not so limited; sheet sizes commonly printed today, four times the area of the broadside, are known as *quad* sheets (*quad royal*, for example) and *double-quad* sheets (such as *double quad crown* or *eight-crown*) are also used.

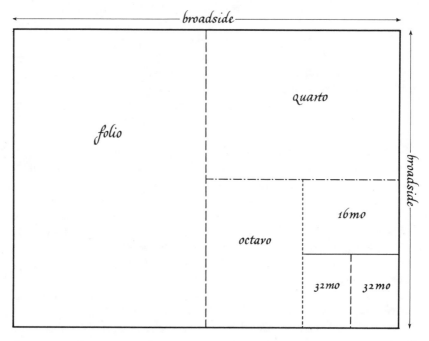

Figure 3-1*a*. The proportion between the uncut leaf and the broadside. The diagram shows the greater width of the 4to and 16mo leaves in comparison with folio and 8vo.

The method of description, however, still refers to the number of pages printed on each side of the broadside, not on that of a double, quad, or double-quad sheet.

In British Standard 1413 of 1970, four traditional broadsides were standardized in metric units (table 3-1). The traditional royal format was reduced, and two paperback formats — 181 × 111 and 178 × 111 millimetres, 7.13 × 4.37 and 7.01 × 4.37 inches — were stated to be in general use. In the table below, the four standard sheet sizes and formats are shown first in millimetres and then in inches.

TABLE 3-1 *British standard sheet sizes and formats*

	quad sheet	quarto	octavo	
crown	768 × 1008	246 × 189	186 × 123	*millimetres*
large crown	816 × 1056	258 × 201	198 × 129	
demy	888 × 1128	276 × 219	216 × 138	
royal	960 × 1272	312 × 237	234 × 156	
crown	30.24 × 39.69	9.69 × 7.44	7.32 × 4.84	*inches*
large crown	32.13 × 41.58	10.16 × 7.91	7.80 × 5.08	
demy	34.96 × 44.41	10.87 × 8.62	8.50 × 5.43	
royal	37.80 × 50.08	12.28 × 9.33	9.21 × 6.14	

The quarto and octavo sizes above are shorter and narrower than one-quarter or one-eighth of the quad sheet after being cut at head, tail, and fore-edge.

In the United States there is no equivalent to British Standard 1413, but Lee (1979) explains the American system in detail. The dimensions of sheets held in paper merchants' stock and of the equivalent of the octavo cut pages derived from them are listed in table 3-2, in inches as shown by Lee and in millimetres. The first two sizes are short-grain (§ 15-3) and the last four long-grain; the machine-direction of the sheet runs along the short or long dimension respectively. These sizes, and other measurements used in the States, apply to book papers only; cover papers are measured by different methods.

Whatever the cut page dimensions in table 3-2, the grain of the paper is parallel with the spine, as it should be; if Lee had added the equivalent of quarto sizes (an equivalent because such terms seem hardly ever to be used in the States), they would all have been cross-grain.

The term *quad* is not generally applied in the States to printing presses or sheet dimensions. The two sizes described in this table as quad are British in origin — 35 × 45 inches is quad demy, and 38 × 50 is the twentieth-century version of quad royal, both before metrication. The 45 × 68 sheet is the same as that standardized in the 1950s for the London publishers William Heinemann Ltd when their Windmill Press at Kingswood, Surrey, had a big letterpress perfector (§ 12-4), a Linotype Miehle, with a maximum sheet capacity of $45 \times 68\frac{1}{2}$ inches on which they used to print 128 pages 'short demy octavo' at one pass through the press.

TABLE 3-2 *American stock sheet sizes and formats*

	inches		*millimetres*	
quad sheet	*cut page*		*quad sheet*	*cut page*
35×45	$5\frac{1}{2} \times 8\frac{1}{2}$		889×1143	140×216
38×50	$6\frac{1}{8} \times 9\frac{1}{4}$		965×1270	156×235
double quad	*cut page*		*double quad*	*cut page*
41×61	$5 \ \times 7\frac{3}{8}$		1041×1549	127×187
44×66	$5\frac{3}{8} \times 8$		1118×1676	137×203
45×68	$5\frac{1}{2} \times 8\frac{1}{4}$		1143×1727	140×210
46×69	$5\frac{5}{8} \times 8\frac{3}{8}$		1168×1753	143×213

The International Organization for Standardization (ISO) has established a different kind of standard size for paper. The ISO standards refer to the size of the finished work — in bookwork terms, that of the format. The bookwork section of the ISO standard (ISO/R216) presents A sizes: B and subsequent letters indicate posters, envelopes, and other uses. The A sizes are based on a rectangle (A0) almost exactly one square metre in area, and whatever the subdivisions of this rectangle by halving one side, the sides almost exactly retain the proportions $1 : \sqrt{2}$. By contrast, the British standard sheet sizes differ from each other in the proportion between shorter and longer sides, and successive folds produce leaves wider or narrower in proportion (figure 3-1a). In the ISO series, the number after the A indicates the number of times the basic rectangle has been halved — hence A0, for the 'broadside' before subdivision. A0 is 841×1189 millimetres. The formats likely to be useful in book production are A4, 297×210 millimetres, 11.69×8.27 inches, and A5, 210×148 millimetres, 8.27×5.83 inches. The standard quad sheet from which these sizes can be produced allows for cutting and is known as RA0, 860×1220 millimetres, 33.86×48.03 inches. In Britain these A sizes have so far gained more ground in the periodical field than in bookwork.

Systems of named sizes usually indicate sizes of paper likely to be available from stock, but apart from this are a matter of nomenclature only. The names are briefer than description by measurement in millimetres or in inches when they include fractions, and when these names are used disaster is less likely to result from a typing error on a paper order. But some designers prefer to rely on measurements.

Book formats may be based on any size of sheet which can be made, printed, and bound; the names of some sizes indicate only that they are or have been in general use. Non-standard sizes are best defined in terms of the dimensions of sheet and format. In Britain the height of the leaf is usually stated before the width, in the United States width usually precedes height.

---

to buy larger founts. With each size of type spacing material must be ordered.

If your plans for printing are not yet formed and you wish to begin on an experimental basis I would suggest you purchase two complete card founts each of say 12, 18 and 24 point roman and one complete card fount each of 12, 18 and 24 point italic – all of the one face you have chosen to live with.

The type suppliers' lists and catalogues which are mentioned in Chapter 9 will not be too helpful unless you know what a few lines of Bembo or Plantin or Bodoni may look like on the page. It is the practice of some publishers to state the name of the typeface in which their books are printed and such information is usually to be found on the *verso* of the *title-page*. You may find this helpful in getting to know more about type designs and identifying them. Also, a number of book-printers issue type specimen books and some of these, like Mackays', can be bought (see page 120).

You already have an idea of what Ehrhardt (Mono-type series 453) looks like from reading the text of this book but you may not realize that there is spacing between the lines of type. Line spacing is known as *leading*. The purpose of additional spacing between the lines is to increase the legibility of the typeface – it helps your eye to keep to the line it is reading and to find with ease the beginning of the next line below. Short lines are of course easier to read than long ones, but this can go too far and cause frequent breaks in words. The page with many hyphens at the ends of lines begins to look unsightly. You will find that good design depends on a delicate balance of typographical factors.

At the end of this chapter will be found a few speci-

38

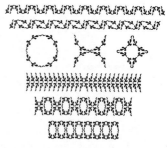

economic and aesthetic considerations. The typographical craftsman should have sufficient experience and skill in his hands to cut with some confidence an alteration on the face of a type. For instance the removal of the 'accent' on this flower (🌸) may be desirable in certain instances although it would not justify the cutting of special matrices.

[45]

A page from A Suite of Fleurons *showing Monotype flower 468-9 in use*

Figure 3-1*b*. John Ryder's *Printing for Pleasure*, reduced to 68 per cent, is an example of demy 12mo, printed with 48 pages on each side of a sheet 35 × 45 inches (in 1976).

Methods of manufacture, reading habits, and shelves are all adapted to the traditional shape of books, an upright oblong. Books may however be square, or wider than tall; whatever other details are specified, a leaf wider than its depth should be termed *landscape*. The practice of stating height before width is not invariably recognised, and an additional precaution against misunderstanding is to describe a landscape crown quarto, for instance, as '186 millimetres *tall* × 249'. The small difference in dimensions between a landscape and a portrait format derived from the same sheet is due to the application of two cuts to the shorter side which has become the height of the leaf.

§ 3-2 Format and function

Some of the factors which govern the selection of format are mentioned in this chapter, but this selection affects, and is affected by, other factors such as width of typographic line, type size, and illustration area, discussed in subsequent chapters. The order of events varies from edition to edition, and each designer has to find his own way to his own final design, preferably with the example of Bruce Rogers in mind to show what can happen when one factor is fixed in the wrong order.

One first step is to find out the approximate length of the typescript, by some form of *casting-off* (§ 18-1), however simple and brief. The number of printed pages in a book may have some influence on its format. A particularly long book can be more economically produced on larger and fewer pages, and may also command a higher price, than a thicker book in a smaller format. A book too thick for its format is ungainly to handle and unattractive to see. When an edition is to be a large one, an *even working* (§ 18-1) may become economically desirable enough to influence format.

If illustrations are to be a feature of the book, a large format is likely to display them adequately. Verse tends to need depth in the page, rather than width; a line of iambic pentameters, for example, contains ten syllables, but in prose ten words of average length make a line of moderate width. The extra width of quarto pages renders them suitable for text set in double column, but in single-column setting may invite rather too long a line of text, unless the type is large.

The best size is that which will best present the book's contents to its reader; if this size has to be modified to suit production processes, the book trade, or anything else, the reader may be the loser.

The book has to be sold in order to reach its readers. There are certain conventions about size which, like other conventions in book design, are best observed if there is no good reason for discarding them. The conventions are neither many nor rigid, but deserve attention because booksellers, librarians, and the book-buying public may hesitate before buying a book of unusual shape. An unnecessarily large format, for example, might prejudice the sale of any book, if it would appear to be too big for the appropriate shelf; an unusually small leaf might suggest that the text printed on it was of little importance.

A book of the right dimensions to fit on a church lectern would not fit the shelves of most households, nor could it very well be carried about; lecterns apart, books of such size are apt to be inconvenient both in and out of use. When they are laid open on a table, the head of the page is uncomfortably distant from the eyes of a seated reader.

Quartos are more manageable. Except in the smallest sizes, they are rather too big to be conveniently portable, but only the largest will fail to fit into shelves with a vertical capacity of 12 inches or 300 millimetres.

Octavo books are the most manageable. They can be held on the knee for reading; below about royal octavo they can be carried about quite easily, and they fit into most bookshelves. Crown octavo and smaller sizes are useful for books which are likely to be carried in pocket or handbag and which may have to be held in the hand for reading.

For holding and carrying, 16mos would be best; little books have in the past been the mainstay of publishing houses such as that of the Elzevirs. But the reader of today requires a larger type than can be accommodated on the narrow pages of sizes below demy 16mo. The larger 16mos may be found useful as a squarer alternative to the smaller 8vos.

Landscape formats may be useful for illustrated works, but are unlikely to supersede the portrait shape for general use. The book is awkward to hold, and is less strong in the spine and in tenacity between case and text (§§ 16-6 and 16-7) than a book sewn along the longer edge.

In the four most popular pre-metric sheet sizes in Britain — crown, large crown, demy, and royal — the shorter side as a proportion of the longer rose from about 75 per cent in crown to about 80 per cent in royal, thus becoming wider in the larger sheet. In each sheet size, the proportions of the quarto leaf were the same as in the sheet. Those of the octavo leaves became proportionately narrower as the sheet became larger, from 67 per cent to 63 per cent. In the metric versions of these sheets, except in royal, the proportions are much the same. The shorter side of the metric royal sheet is only 75 per cent of the longer; the quarto leaf is accordingly narrower in proportion, and octavo wider, than in other sheet sizes. All A sizes have the same proportion of 71 per cent, so that A5 is wider than octavos and A4 narrower than quartos.

§ 3-3 Format and production

The standard sheet sizes can be caused to produce leaves in a variety of proportions, large and small, wide and narrow. If one of these does not appear to suit a projected edition, a non-standard sheet may be selected or devised. Whether the sheet is standard or not, the most economical use of printing machinery is that which occupies the highest possible proportion of its maximum printing area. When editions are printed on a web of paper (§ 12-6), instead of on sheets, the width of the web may be varied from edition to edition, though not without cost to short editions; the length of the *cut-off* (§ 12-6) depends on cylinder diameter and is capable of variation only when cylinders can be changed. Web printing accordingly restricts format selection.

Caution is needed in planning based on a sheet which approximates in either dimension to the maximum capacity of a specific press. On any press, the maximum dimensions of the printing image are significantly smaller than those of the maximum sheet. Illustrations which run off the edge of the cut leaf (or *bleed*, § 14-3) require a larger sheet than usual for the same leaf size, and unsewn books (§ 16-4) also need additional sheet size for the extra cut at the spine, unless a reduced leaf width is intended.

When more than a *tonne* of paper (1000 kilograms or 2205 pounds) will be needed for an edition, there is a possibility in some countries that paper can be made to order for it. In other countries, a *making order* (as distinct from a *stock order*, for which paper is likely to be held in stock by mill or merchant) may have to be substantially bigger. A making order offers the designer an opportunity of stipulating his own specifications for the paper. A reprint of the edition may however call for much less paper, and there may then be no possibility of matching the paper specification of the first printing. An uncommon size of sheet may be impossible to match in a small quantity, and whatever larger sheets may be available will have to be cut down to size, increasing the cost of the reprint. The most convenient sheet size and quality will usually be one which is readily available from stock.

Binding machinery has limitations which deserve study during the planning of any edition designed to be unusually thick or thin, large or small. Different printers and binders have different equipment, and a supplier can sometimes be found for a specification impracticable or uneconomic elsewhere.

Taste as well as reason should attend the choice of a book's shape. Some designers like their books a little narrower than usual, others prefer them squarer; there are even those who assert that one proportion of width to height is inherently better than all others. On the whole, leading designers who wish to produce a particularly elegant book seem to favour a slight narrowing of the leaf from the standard proportion of octavo books. Utility and aesthetics are closely linked, and this preference for a narrow leaf may be caused by the advantages of a narrow measure (§ 7-2). A particularly flexible paper may be necessary to allow such narrow leaves to lie flat when the book is open.

§ 3-4 Margin dimensions

Having determined the size of the leaf, the typographer must decide how much of its area is to be given up to the text, and at the same time how much space or margin there should be between each edge of the text area and the cut or folded edge of the leaf. For design purposes, margins are usually understood to be the space round the main text on each page. The page number, if it appears alone at the foot of the page, occupies too little space to be accounted part of the main text, and for visual purposes may be considered to lie within the tail margin. The headline (§ 7-9), on the other hand, may appear on every page of text except the first of a chapter, usually consists of more than a word or two, and may be treated as though it were visually a part of the main text.

To print a page without margins would hardly be practicable. If the text were to run right up to the sewn edge of the book, the spine, part of it would disappear into the bend in the page caused by *backing* (§ 16-6). Text running up to any cut edge would be in danger of being cut away if the binder were to make even the slightest mistake in cutting.

Some books, particularly those used by libraries, are re-bound, sometimes more than once. This involves cutting the edges again, each cut reducing the margins (except at the spine). A proportion of copies in any edition either finds its way into libraries or is intended by the publisher to do so. Anything printed on the page too close to the cut edge may therefore disappear during re-binding. When illustrations bleed, no significant part of the picture should appear within a short distance of the cut edge, for the same reason.

A variety of reasons may be put forward in favour of ample margins. Margins are said to provide a frame of white round the text, separating it from the scenery beyond the open book. This frame, like that of a picture, enhances by contrast the appearance of the area it surrounds, and emphasizes the position of that area on the page. A more practical reason is that margins provide room for the fingers and thumbs of the reader. For one reason and another, ample margins are considered to be a fundamental of fine printing, and the private press movement has filled the bookshelves of collectors with large areas of costly hand-made paper, much more than half of which by area is unprinted.

Wide margins do tend to enhance the appearance of a book, but as the function of the book is to be read rather than admired, such margins are uneconomical, and the help they give to the act of reading is open to doubt. Unless the designer of industrially made books is convinced of the practical value of ample margins, he may as a rule prefer to design margins consistent with the custom of the day.

MOGUL INDIA

Already by the eighth century one hears of the Moslems having made raids on India, sailing up the Indus to Multan. Mahmud of Ghazni, or 'the Image-breaker', as he was called, must, however, be considered the first serious threat to the different Hindu dynasties among whom India was divided. It was Mahmud who initiated the series of invasions from Afghanistan and Central Asia that persisted, in some form or another, until the coming of Babur, the first of the Mogul emperors, himself one of the many invaders. 'On Friday, the first of Sofar 932,' Babur writes in his diary, 'when the sun was in Sagittarius, I set out on my march to invade Hindustan', the exact date, according to our methods of reckoning, being November 17th, 1525. Before he arrived, raiding India had become almost a yearly habit among the northern clansmen. Small but well disciplined Moslem forces formed like black clouds in the mountain fastnesses beyond the North-West Frontier. They poured 'up like goats', as one of the Hindu generals once remarked, 'and down like waterfalls' into the rich plains, demolishing both idols and shrines wherever they went. The booty they carried off with them was perhaps the richest ever seen, 'jewels and unbored pearls and rubies shining like sparks or like wine congealed with ice, emeralds like fresh sprigs of myrtle and diamonds in size and weight like pomegranates'. At Mathura, for instance, 'five idols of red gold, five yards high, with jewelled eyes', were carried off. There is also the terrible story of Mahmud's raid on the temple of Somnath, dedicated to Siva. 'It housed a massive stone lingam five cubits in height, which was regarded as being of special sanctity and attracted thousands of pilgrims. It was bathed every day in water brought all the way from the Ganges and garlanded with flowers from Kashmir. The revenue of ten thousand villages was assigned for its support and a thousand Brahmins performed the daily ritual of the temple.' From this single detail may be calculated the wealth of the entire sanctuary, with its chains of massive gold bells and jewel-incrusted columns.

Everywhere they went the invaders plundered and smashed, 'splintering the craven images'. They would take the statues out and heat them over bonfires and then pour vinegar or cold water over the heated stone until they cracked.

As in Cortes' conquest of Mexico, a handful of men overthrew a whole empire. The Moslem forces were always small, Babur's whole army, including camp-followers, amounted to only twelve thousand men; while in the battle of Talikot, in the south, we hear of the Hindus mustering three quarters of a million men, with two thousand elephants, against a Moslem army of half their number. The enormous Hindu armies were no match for the rigidly disciplined Moslem troops, and melted into headlong flight before the scimitars of the crescent moon. The Indian armies did not lack individual chiefs of great personal gallantry; but their generalship was inefficient and they were hampered by endless barriers of caste.

In vain did the squadrons of the Rajputs gather to withstand them; in vain the women of Hindustan melted down their silver and gold ornaments to fill the war chests. Never again, except in the fastness of the Rajputana hills and deserts, were the warriors of India to resist the triumphant invaders from the north.

Babur founded the Mogul Empire; but it was not finally consolidated until his grandson Akbar's reign when the Emperor had the good sense to ally himself to the powerful Rajput princes by marrying the Maharajah of Jaipur's daughter, who was to become the mother of Prince Salim, the heir to the throne.

The Moguls managed to keep their hold on India until the eighteenth century. Then, in 1739, came the sack of Delhi, which 'sounded the deathknell of the Mogul Empire, though phantom emperors continued to occupy the throne, sitting in their ruined halls under tattered canopies. Yet so powerful was the magic of the name of the Great Mogul, that he was still regarded as almost sacred throughout the country, and rival powers contended for the control over his person',[1] one of these powers being the British who, having come to India as traders, were obliged to remain as soldiers and administrators.

1. *India* by H. G. RAWLINSON.

Figure 3-4. A wide page – demy quarto – combines with comparatively narrow margins to provide ample space for illustrations when text is subsidiary to picture. In Roderick Cameron's *Shadows from India* (Heinemann, 1958) Stephenson Blake's Old

The mosque of Ajmer called the Arhai-din-ka-jhonpra, or 'the hut of two and a half days'. Originally a Jain college built in 1153, it was turned into a mosque by the Afghans of Ghor, who took Ajmer in 1192. The Moslems seem to have had no scruples about using material provided by Hindu temples. As Mr. Havell tells us in his book on Indian architecture,[1] 'they doubtless found a grim satisfaction in compelling thousands of Hindu craftsmen to wreck their own holy shrines and to rebuild according to the ritual of Islam'. Along with Qutb-ul-Islam at Delhi, it was the first mosque to be erected in India.

1. *Indian Architecture*, by E. B. HAVELL (John Murray, London, 1915).

Face Open and Monotype Bembo were printed by offset, and the author's photographs were printed by photogravure. Page numbers and chapter titles are set in the footline to keep them clear of the illustrations. Reduced to 69 per cent.

The point about fingers and thumbs is a doubtful one. A well-made book should lie open without much effort from the reader; the text area will not suffer from handling with clean hands, and few adult readers are likely to be consciously impeded by their own fingers.

A book which contains enough illustrations of suitable size may benefit from a reduced type area outside which all the illustrations are placed. A type area which is consistent from page to page, and uninterrupted by illustration, may help to maintain the reader's interest. In planning such an arrangement, the designer may do well to work out how many pages will appear without illustrations and what they will look like. If such pages will be frequent, a more flexible arrangement might be based on pages of two columns of text, with illustrations replacing the second column where they occur. A similar scheme may be employed to place sub-headings and notes beside the text area rather than within it, but unless the text is evenly as well as extensively annotated this is bound to be wasteful of space. Writing on the text pages of a book is today considered a pernicious habit, but a book is made to serve its owner, and certain works of information may be the more useful for ample margins which leave room for written notes.

Adjustment of the text area, and hence of the margins, may facilitate the fitting of the text into a marketable number of pages or an even working.

To rely on formulae when deciding margins (or at any other stage) would be unwise; the conditions of book production are too varied for set rules. Margins which are at present considered reasonably generous for the general run of books produced in traditional styles, other than paperbacks, occupy about 40 to 45 per cent of the area of the leaf; the margins of this edition, at 45 per cent, are an example. The margins of dictionaries and other reference works are usually narrower than this, and those of belles-lettres are somewhat wider. There is something a little incongruous in allotting to margins more than half the leaf area in a book made by means supposed to be economical.

A text setting in which there is plenty of space between the lines seems to need smaller margins than one in which the lines are set close together. The use of a light type-face, constructed of slender strokes, seems in the same way to suit rather narrow margins. It is as though plenty of paper showing through the text area compensates for any reduction in the amount of paper around it.

§ 3-5 Margin proportions

The proportions between margins are primarily a matter of convention and personal preference, but may also be affected by the way in which the book is to be bound. The most common arrangement is to place the rectangle of text rather high on the leaf and nearer to the spine than to the fore-edge. The back margin is therefore the smallest; the head, slightly larger; the fore-edge, wider still; and the tail, largest of all. This is because on opening a book the reader sees not one page but two — a left and a right, or, in printing terms, a *verso* and a *recto* — together described as an *opening*, and properly treated as a single item for purposes of design. The spine margins of the two pages are seen as a single channel of white; if this channel appears to be wider than each fore-edge margin to left or right, the rectangles of text will appear to lie rather too near the outer edges of the pages. The spine margin of each page should therefore appear to be half the width of the fore-edge margin, or very little more. In order to bring this appearance about, allowance may need to be made for the area of each leaf which may disappear into the spine when the book is backed.

If a rectangle of text is placed midway between the head and tail of a leaf, the eyes for some reason suggest to the mind that the text is just a little nearer the tail — in fact, that it tends to 'fall out of the page'. To avoid this, the tail margin is usually larger than the head.

The exact proportions that the four margins should have to each other is best decided by each designer for each book. One formula for a conventional ratio back/head/fore-edge/tail suggests $1\frac{1}{2}/2/3/4$. This may be useful as a starting-point, but no single formula is valid for wide as well as narrow margins, for octavos as well as quartos, or for very thick books as well as thin. A diagram of the cut leaf, or better still of the opening, showing the text area in position, is the surest resort, and if it is accurately drawn and accompanies the copy to the printer, that will be better still.

One of the purposes of conventional margins is to enhance the appearance of the opening by making a unity of it. This may be sound if the opening is in fact a unity, but there may be causes to dissociate two facing pages. Poems on facing pages, for example, should be considered separately rather than in relation to each other. If the margins are extremely wide, and if the panel of text is pleasantly shaped (tall and narrow rather than square), proportions between margins which are totally at odds with convention can be distinctly effective. The text is not in any way difficult to read, and those readers who overcome their initial astonishment may find the arrangement admirable. Margins of this kind tend to rely on dispropor-tionate width for emphasis, and are not always economically possible in the making of ordinary books.

Without excess, however, an unconventional proportion between margins may be an essential part of an unconventional design, though there is always a risk in book design that something inconspicuously unconventional may appear to be a conventional idea which has been mishandled. If the printed area of each page is not backed up in position by that printed on the other side of the leaf, type or illustration showing through the paper may disfigure the margins. Any layout so unusual will stand a better chance of successful production if each page provides printer and binder with some point of reference, as a guide to position. If for example all page numbers, whether recto or verso, are placed at a fixed distance from spine and head, there should be no mistake in the position of other elements on the page, however bizarre.

Margins of conventional width and proportion are not invariably the best even for books designed in a conventional style. The doctrine of drawing the two pages together and upward in the opening has been asserted often enough to accumulate the influence of a law. Reliance on laws or even conventions in typography may result in the abandonment of common sense. When planning a thick book which will be stiffly bound, for example, the designer can do his reader a service by moving the text area outwards from the spine, so that the text will not curve downwards into the deep channel of the back when the book lies open. *Unsewn binding* (§ 16-4) needs similar planning for a different reason. The folded spine of the section is cut off, and to allow for this reduction in the width of the leaf, the designer should allow not less than 3 millimetres or $\frac{1}{8}$ of an inch, the usual allowance for cutting, and not more than twice as much, by way of addition to the spine margin.

The specification for margins, to be transmitted as an instruction to the printer, calls for flexibility of a kind advisable in every allocation of space. Within the area of the leaf, any planned group of vertical or lateral dimensions should include one which is approximate, and which contains whatever space remains when the rest has been occupied in accordance with specification. Once the printed area of a page has been defined, it will be enough to state the head and spine margins. The tail and fore-edge margins will occupy the balance of vertical and lateral space.

CHAPTER 4

Type and composition

ALPHABETS, by which all the sounds of speech can be represented in combinations of fewer than thirty letters, can be traced back to the Sinai desert some four thousand years ago. Paper — more durable, portable, plentiful and inexpensive than other writing materials — was first made in China, a century or so before the birth of Christ. In the mid-fifteenth century Johann Gutenberg, a goldsmith of Mainz, began to make separate metal casts or *types* of each letter of the alphabet, to arrange these types into words, lines, and pages, and to print ink from them on to paper. That was the beginning of the age of the alphabetically printed book.

For more than four centuries, various alphabets, paper, and images printed from composed type have been looking much as they do today. The principal conventions of literate communication are little changed. With these ancient components and customs, the twentieth-century typographer translates a modern text into a modern edition for modern readers.

Type-founding and type-setting were successfully mechanized by the end of the nineteenth century. Machines for casting metal type in composed order, invented at that time, together with type-designs introduced in the next fifty years, enabled publishers and printers to raise the typographic quality of industrially produced books to a high level. In the composition of single types, only the initial type-setting of the text was mechanized in this way; large type in most books, corrections, page make-up, and all the rest of composition was left to the minds and hands of compositors, as it has been for several centuries.

Now type-setting can be wholly mechanized. Machines, attended to and adjusted but not guided by printers, can read typescript and translate it into typographic pages ready for printing. The pages of type images they compose are designed to look like those still printed from metal type. The type designs and composition styles of electronic typography are for the most part derived from the long history of the single, separate, movable type. Such type has now reached the end of its development. Those book designers who are familiar with the best features of that development will be well placed to foster the development of new type designs and of new systems and styles of composition which are still emerging.

§ 4-1 Type-founding

Type-founding is now understood to mean not only the foundry work of metal casting but those typographic preparations which determine the shape of the cast. The origin of any letter or other *character* to be used repeatedly in composition is a *type drawing*. Any variation on the design, even the addition of an accent to a letter, requires a different drawing. The drawing is a large one, whatever the size of the eventual type; the profile of the character is carefully adapted to the optical effects of composition, and to the mechanical effects of presswork (chapter 12). The letters o and e, for instance, may look well enough on their own, but if they do not extend slightly above and below the horizontal strokes of x, they will look slightly too small when composed into such a word as oxen. The angles contained by the intersecting strokes of x will tend to fill with ink if they are not kept artificially open in the design. Such optical and technical modifications of letter-form are better left to specialists than attempted by printers except when there is scope for experiment and informed criticism.

Hkpx Hkpx Hkpx **Hkpx**

Figure 4-1. Four sizes of Monotype Baskerville; from left to right, 72, 36, 18, and 9-point, scaled to 33, 66, 132, and 264 per cent respectively, to show similarities and differences in proportions in metal type.

The optical effect of size variation in type is conspicuous, even to an unobservant eye. Until the late nineteenth century, separate drawings were necessary for each size of any type design, for technical reasons, and the type designer naturally took the opportunity of adapting each drawing to the size in which it was to be reproduced. In large sizes, the letters tended to be slender, closely fitted together, and extended into tall verticals in such letters as l and p; small letters were constructed of somewhat shorter and thicker strokes, and stood a little farther apart from each other. When separate drawings were no longer necessary for different sizes of type, a single set of drawings for all sizes produced small types which looked thin and crowded, and large type which fitted loosely into words and lacked grace. Good type drawing of the twentieth century made allowance for size variation, though one set of drawings might still serve for two or three of the sizes designed for continuous reading. In most photo-matrix systems (§ 5-4) type-size is determined by lens distance from a single matrix for each character in all sizes. Very few type designs, so produced from a single drawing for all sizes, are satisfactory in small, medium, and large sizes, but the economy of such a procedure is irresistible.

In metal type-founding, the outline of the type-drawing is reversed from left to right, and traced by a pantograph which duplicates it in a scale reduced according to the size of the intended type. The pantograph guides a cutting tool which engraves the character in relief on the end of a steel rod; this is a *punch*. Until Linn Boyd Benton's pantograph was introduced into type-founding during the nineteenth century, each punch was engraved by hand, with the modifications for type-size already described. The punch, as a rule, is part of the type-founder's equipment, and remains his property. *Photo-matrices* and *digital matrices* (§ 5-6) are produced from type-drawings by photographic or digital means, without the intervention of the punch.

The punch is struck into a small slab of bronze alloy, a softer metal, so reproducing the engraved letter, recessed, and facing the right way round. After further preparation, this becomes the *die* or *matrix*; one punch can be driven into a considerable number of them. Matrices for mechanical type-setting are now sold to the printer by the manufacturer of the type-setting machine or system which is to set the type. Matrices for hand composition, usually in larger sizes, may be hired out by the machinery manufacturer, rather than sold, for casting separately from type-setting. A few surviving type-founders continue to supply types cast from matrices which are not offered for sale, and for a little while longer some of the best of all display type-designs may still be available in this form.

Continual casting from composition matrices tends to induce wear. In *Monotype* work (§ 5-2) this usually appears as unsightly spurs, blobs, and bends on terminal strokes of the characters. In *Linotype* composition (§ 5-1) type metal intrudes between matrices, and shows up as hair-lines between letters. Photo-matrices are not worn by use, but may be damaged in handling, causing a defect which afflicts every size of the same character on every occasion on which it appears. The printer should always replace such imperfect matrices as soon as they show signs of any of these defects, and should avoid printing any characters produced from them.

Type metal, softer than that of the matrix — an alloy of lead, tempered with antimony and tin to give the casting sharpness and endurance — is melted and poured into the rectangular orifice of the *mould*, to which the matrix is fitted for casting by hand. In mechanical casting, matrix after matrix is presented to the orifice, in the order in which the characters are to be composed, and at each presentation type metal is injected into the mould. The fixed dimension of the orifice determines the size of the type; its adjustable dimension allows for variations in the width of the letters. In most forms of metal type casting, the standard width of the casting will normally be that selected for each character by its designer, but the whole set of characters can be cast with a little more space or a little less between adjacent characters. The same is true of photo-composition, but photo-matrix systems in which the matrices are moved into position tend to suffer from irregularities in the spaces between adjacent characters.

When the metal cools and solidifies, it reproduces from the matrix the character (now reversed again from left to right) cut on the punch, and from the mould the rectangular *shank* which supports the character. This metal cast is a piece of type or a *sort*. The printing surface or *face* at the top may be a letter or some other character such as a punctuation mark, a figure, or a sign. The shank is rectangular in section and in plan. When all the components of the page are held fast against each other by inwards pressure from all four sides, this rectangularity locks every character and space into an exactly correct position within the line of type, and every line into an exact parallel with every other. When any component in a page has not been cast in correct position on a rectangular shank — as for example when display lines are pasted singly into position for the camera — there is a possibility that some part of the image will be printed out of its true position. Perhaps the displacement will be vertical or lateral or both; perhaps an image will be tilted, or a line diverge from the parallel with its neighbours; but whatever the nature and extent of the displacement, it is likely to distract and disappoint the observant reader.

In Britain, type is nearly always 0.918 inch or just over 23 millimetres high; this is the *type-height* or *height-to-paper*. The exception is at the Printing Division of Oxford University Press, where the old type-height of 0.9395 inch is retained. When in use, the sort stands on its plain end (the *feet*), with the face upwards. The dimensions of metal type mentioned elsewhere in this book refer to the *body* (§ 4-2) or to the face as seen from above and the right way up.

§ 4-2 Parts and dimensions of type

The *face* of the type is the relief surface which is to be inked and pressed against the paper. It is raised to type-height by the shank which supports it. A sort which has a shank but no face, and which does not rise to type-height, is a *space* for use between words or characters, and prints nothing.

The width of the shank across the letter is the *set*, so called because it is regulated by the setting of the adjustable dimension of the mould. The set of a *fount* (§ 4-4) is usually understood to mean the set of one of its widest characters such as M. The *fit* of a letter is the space between the outer edges of the *main strokes* (§ 4-3) and the edges of the shank. Fit, which governs the width of the space between adjacent letters, becomes significant when the letters are composed into words; if the fit is loose, narrow spaces between words contrast inadequately with spaces between letters, and the letters fail to combine into a word as a visually effective unit. The *a-z length* of an alphabet is the width of the twenty-six letters of the alphabet set close up to each other; when alphabets are shown in this way in a type-specimen, comparison shows which of them, when used for composition, is likely to occupy less space or more than others.

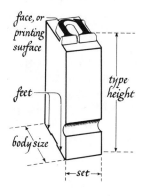

Figure 4-2a. Parts and dimensions of metal type.

The *base-line* is an imaginary line through the base of the face of roman or italic letters other than g j p q y J Q, and this line is also known as *the line* and the *main line*. The position of the base-line on the body of metal type determines whether all the letters in a word or line will appear to the reader to rest in alignment with each other. This position depends on the design of the type rather than on its size. Two different designs, both on 12-point body, are unlikely to align exactly with each other at the base-line, when set side by side; the eye is quick to observe any misalignment in this line, and is more tolerant of discrepancy elsewhere. Because of this problem of alignment in metal typography, the setting of types of different design in the same line has long been considered doubtful practice, partly because it gives rise to expensive handwork in rectifying the alignment.

This convention need not apply to photo-composition systems in which any two or more type designs and type sizes align themselves automatically on the same base-line (figure 4-2b).

Hkpx Hkpx Hkpx Hkpx

Figure 4-2b. Four sizes of Lasercomp Times — 36, 24, 18, and 12-point — set in the same line and on the same base-line.

In metal type, the distance between the base-line and the front of the shank is the *beard*. Part of the beard may be cut off the shank of a large capital when it is used as an initial, to prevent the beard from occupying part of the line immediately below the initial, since the initial will appear to fit better into the text when this space is occupied by a text line (figure 4-2c). This cutting is unnecessary in a *titling* fount, which has no lower-case, and in which the capitals occupy almost the whole depth of the body (figure 4-2c). The base-line is much closer to the front of the shank than it could be if space had to be left for lower-case *descenders* (§ 4-3). In photo-composition there is no beard, and any capital can be used as an initial with type immediately below it.

The shank of large initials in metal can be cut into or *mortised* at the side, to fit smaller letters close up against the initial's face, within the rectangle otherwise occupied by the shank (figure 4-2c). No such cutting is needed in photo-composition, but initials now seem a little out of fashion, perhaps because they belong to the long history of metal types. But the work should be easy enough for most systems, even with diagonal indents as beside the initial A in figure 4-2c.

T HIS SHOWS the full size of the body of a capital T.

T HIS SHOWS HOW the beard of the same letter can be trimmed to fit more closely into the text.

A S THIS EXAMPLE SHOWS, an initial may fit better if mortised in such a way that the adjacent text fits closely.

Figure 4-2c. Initials and text in metal, from *Monotype* (§ 5-2). Above, in Centaur. Below. Albertus Titling initial with Plantin Light.

H ERE A 48-POINT TITLING initial, some 45 points high in face, fits closely above the first line of text beneath it, although the beard has not been trimmed at all.

The *mean-line* is an imaginary line through the top of those small roman and italic letters which have no *ascenders* (§ 4-3) — a c e g m n o p q r s u v w x y z. As the top of a curved letter such as o should be very slightly higher than that of a flat letter such as z, the position of the mean-line is approximate. The same is true of the base-line and indeed of the *x-height*. The x-height of an alphabet is the distance between the base-line and the mean-line, and it is equivalent to the face height of the letter x. This letter is convenient for measurement and comparison because it has horizontal *serifs* (§ 4-3) at both vertical extremes. The x-height represents the *apparent size* of the letters, whatever the body or the length of ascending or descending strokes; the apparent size of an 11-point face may be larger than that of a 12-point.

The depth of the shank, from its *front* (below the base of the letter) to its *back* (beyond its apex) is the *body*, the extent of which identifies the size of the type. The size of a fount is that of the body for which it was designed, and that is usually (but not always) the smallest on which it can be cast. One mathematical series, for example — Monotype series 569 — consists of 10-point characters cast on a 6-point body, with the overlapping face resting on spaces placed in a second casting of the same line. In photocomposition there is neither shank nor body, and lines of typographic images can be assembled so closely together that the ascending and descending strokes of the letters touch each other or overlap. The different sizes of such images, however, are usually described in terms of body.

The bodies of metal types cast to modern specifications are described in *points.* In the United States and in Britain, this unit of typographic measurement may be abbreviated to 0.01383 inch; in some other countries the Didot point, 0.0148 inch, is used. These two versions of the point may be expressed as 0.351 and 0.376 millimetres respectively.

Before the general standardization of point systems, bodies were known by traditional names deriving from their use. In Britain and the United States these names included *pica*, and in France and Germany *cicero*. Pica and cicero have both survived to become exact units of measurement, pica as 0.166 inches or 4.2 millimetres, cicero as 4.5 millimetres, and each 12 of the appropriate points. Pica and cicero are used to express typographic measurements across and down the page.

In *Printing types*, published in 1922, Daniel Berkeley Updike, the American printer, wrote that —

At the time that the Didot system was introduced the French metric system had not been adopted, and it is only fair to say that until a type system is formulated which is in full and regular accordance with the metric system, perfection will not be attained.

A metric system of typographic measurement, capable of identifying the different sizes of any photocomposition type-face, has been drafted by the International Standardization Organization (1975), but at the time of writing is not in general use. This system would however provide alternative and new terms for typographic dimensions more likely to be observed by the typographic designer in photocomposition than in metal types.

Alternative terms, replacing those already defined here, are *character width* for *set*, and *side bearing* for *fit*. The nearest equivalent to *body* is *recommended minimum row distance*, the minimum vertical space between the base-lines of adjacent rows (or lines) of type by which all characters including accented capitals can be separated from all others. The *H height* is that of H; the *accented capital height* is that of an accented H. The *Hp height* is that from the top of H to the bottom of p; the *kp height* is from the top of k to the bottom of p; and in some type designs Hp and kp heights are the same, in others Hp is lower.

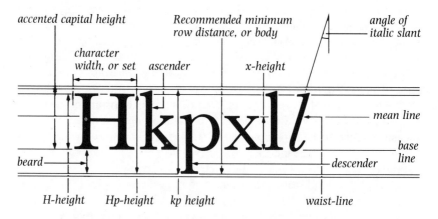

Figure 4-2*d*. Parts and dimensions of typographic letter-forms generally.

The various parts and dimensions defined in this section are illustrated at figure 4-2*d*. They refer to concepts which from time to time make a useful appearance in the assessment of type designs, whether to set them beside each other for comparison, or to examine the suitability of one for a specific task. Any comprehensive assessment of this kind would also require other measurements, not yet standardized. Two examples might be termed *waist-line* and *slant*. By *waist-line* I mean the thickness of the main strokes, as typified by the waist-measurement of the relevant forms of L — roman capital, roman lower-case, italic capital or lower-case, small capital, bold roman or italic capital or lower-case, and even perhaps figure 1 on occasion — and by *slant* I mean the angle of inclination from the vertical of italic lower-case 1 as representing the average slant of the other letters in the alphabet. Some of these measurements can be worked out approximately with the help of a magnifying glass and a sharp-edged pica ruler marked with points. The exact figures may be available from some manufacturers, but have not been publicized.

§ 4-3 Letter form

The forms of roman and italic printing types are derived, however indirectly, from letters evolved in writing. The writing-tool of the Middle Ages, when typography was invented in Europe, was the square-ended pen, which imparted to the written letter characteristics which still survive in type design. One of these is *stress* or *shading.* The natural angle of a pen held in the right hand for writing is towards the diagonal across the page. When the pen was square-ended, the thickest part of the stroke of each letter inclined diagonally across the page, towards the right elbow; the thinnest part, across the width of the pen, was at right-angles to the thickest; and at intermediate angles the stroke increased gradually from thin to thick and back again. The angle of stress, then, is the direction of the thickest part of the curved strokes. In roman types it varies between the diagonal (oblique stress, from left down to right) and the vertical (figure 4-3*a*).

abhmnp cdequ

Figure 4-3*a. Above,* oblique stress in Monotype Perpetua Bold; *below,* vertical stress in Monotype Bodoni Bold.

abhmnp cdequ

When the stress is oblique it serves to emphasize the difference between certain letters. Curves which move to the right from the mean-line, as in a b h m n p, are thicker near the top; those which move to the left, as in c d e q u, are thicker below. Vertical stress, on the other hand, thickens the curves in a manner uniform in many letters, and so emphasizes their similarity to each other. This may not be serious except where the contrast between main-stroke and hair-line is extreme; pronounced vertical stress may reduce the page to a pattern of vertical strokes, tenuously connected (figure 4-3*a*). Another characteristic present in the fifteenth-century origins of roman type design is the *serif,* the small terminal stroke at one end or both ends of a main stroke, which defines its length. Serifs may be divided into two classes; the *top serif,* at the top of a stroke, may be either horizontal or diagonal; the *foot serif,* at the foot of a stroke, is usually horizontal except in d and u. *Bracketed* serifs are joined to the main-stroke by a curve or a wedge, which makes them easier to see. *Hair-line* serifs consist of a very thin line; *slab* serifs are thicker and square-ended (figure 4-3*b*).

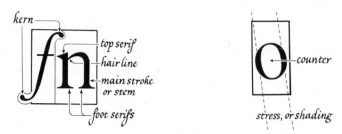

Figure 4-3*b*. Some of the parts of the letter; the letters above would be reversed from left to right in type. Above right, oblique stress (though in some obliquely stressed faces, o is drawn with vertical stress). Below, three common kinds of serif.

The serif helps to indicate the difference between such characters as ı (one), l (small el), and I (capital i). But its general function is to reach out a little way across the gap between letter and letter within a word, and by combining the strokes and letters of the word into an entity, to emphasize the separation of words even when the word-spacing is close. If a serif is to have any use at all, it needs to be clearly visible, and hair-line serifs supported by no bracket are of doubtful effect unless thickened in printing. Types without serifs are known as *sans-serif*; whether or not they are as legible as those with serifs, they are not generally believed to be, and are not popular for continuous reading in books.

The *hair-line* of a letter is its thinnest part, other than the serif, whether or not it is particularly thin. The *main-stroke* is the thicker stroke of a letter, or its main part when there are no thinner strokes. The *kern* is any part of the letter which overhangs the shank and rests on the shoulder of an adjacent sort or space. No shoulder is available when f is followed in the same word by k, and any reference to Franz Kafka is likely to lead to a typographic collision. Few roman letters other than f need kerns, but several italic letters have them, and *f* usually has two. Without these kerns, the traditional letter forms are distorted, and italic loses some of its cursive grace.

The *counter* is the space wholly or mainly enclosed by the strokes of the letter. This space establishes the position and attitude of the surrounding strokes, and assists recognition of the letter. If it is either too wide or too narrow, the identity of the letter is obscured.

Ascenders are those strokes in b d f h k l which rise beyond the x-height. *Descenders* fall below the x-height in g j p q y. Short letters are contained within the x-height, as in a c e m n o r s u v w x z; though i has its dot and t its terminal outside the x-height, they may be considered to be short letters. When ascenders and descenders (collectively, *extenders*) are short in proportion to the x-height, the x-height, and hence the apparent size, is large in proportion to the body. Extremely short extenders, except in small type, are conspicuous to the eye accustomed to the traditional forms of type. Something of the legibility and grace of well-printed books is sacrificed for the dubious convenience of those who wish to pack too many letters into too little space. Type designers chop the extenders short to allow the lines to be crowded together, and typographic designers space the lines out again to compensate for this mutilation. Extenders are rarely too long; but when they are, they disfigure the page with a pattern of conspicuous vertical strokes. Special long or short ascenders or descenders are available for use with some founts; an example is mentioned in the words of Bruce Rogers early in chapter 3.

§ 4-4 Variety of characters

The letters in which these words are composed, and which are used for nearly all text setting in English-speaking and many other countries, are known as *roman*, from their origin. The slanting letter which is used for emphasis and other special purposes is a medieval development of the roman letter and is known as *italic*. Another kind of letter, which sometimes replaces italic, has only its slant in common with it; in other respects it is a roman or mainly roman design, and is known as a *sloped roman*. Roman and italic letters collectively may be termed *latin*.

Roman and italic both have capital letters, known to the printer as *caps* or *upper-case*, and small letters, known as *lower-case*. The reference to cases derives from an obsolete arrangement of the cases in which type was kept for composition. Setting from case is now restricted almost entirely to correction and display composition.

To the printer, an alphabet is a set of twenty-six letters of a certain design and body, together with a few additional combinations of letters. A *fount* is usually made up of a set of alphabets of one size and based on one design. It may consist of one alphabet only, if no more alphabets exist in that design and size. Usually however a text fount will comprise five alphabets — roman and italic upper and lower-case, and small capitals. A seven-alphabet fount is likely to include bold or semi-bold roman upper and lower-case. In addition to their alphabets, most text founts have ligatures and diphthongs, arabic figures, ampersands, punctuation marks, and signs. Almost any work of information requires a wider variety of characters designed for compatibility with the rest of the fount: accents and foreign characters; different kinds of figure, including fractions; foreign punctuation; and some arithmetical signs.

A compositor setting metal type by hand selects each sort as he needs it from the compartments in a case, which lies ready on the canted *frame* before him. The contents of the case, the types he has placed ready to hand for immediate use, may be defined as the fount he is going to use. In the next chapter, a fount of matrices, also ready for immediate use but in mechanical type-setting, will be described. The constituents of the fount required for one book are quite likely to differ from those of the fount required for the next. In addition to the compatibly designed or *mated* characters in any fount, there may be some general-purpose characters differing in design but equal in

ABCDEFGHIJKLMNOPQRSTUVWXYZ ÆŒ ÅÅÇÐĘŁØ
abcdefghijklmnopqrstuvwxyz ffffiffflfifl æœ åȧçdęłøß
ABCDEFGHIJKLMNOPQRSTUVWXYZ
& ÆŒ ÅÅÇÐĘŁØ 1234567890 1234567890
..;;"!?-()[]–——£$$*†‡§||¶¡¿«» ÉÈÊÉÈÊÉËÈÊ éèêéèêéëèê
abcdefghijklmnopqrstuvwxyz ffffiffflfifl æœ åȧçdęłøß & ÆŒ
ABCDEFGHIJKLMNOPQRSTUVWXYZ ÅÅÇÐĘŁØ
1234567890 1234567890 .,:;"!?()[]£$$ ÉÈÊÉÈÊÉËÈÊ éèêéèêéëèê
abcdefghijklmnopqrstuvwxyz ffffiffflfifl æœ åȧçdęłøß
ABCDEFGHIJKLMNOPQRSTUVWXYZ & ÆŒ
1234567890 1234567890 .,:;"!?()[]¡¿«»-—£$$
ÅÅÇÐĘŁØ ÉÈÊÉÈÊÉËÈÊ éèêéèêéëèê
abcdefghijklmnopqrstuvwxyz ffffiffflfifl åçøß
ABCDEFGHIJKLMNOPQRSTUVWXYZ & ÅÇØ
1234567890 .,:;"!?( ) ÉÈÊ éèê
+ − × = < > ± ·/|'%°®©||

Figure 4-4*a*. A fount, from a printer's type-specimen book — 11-point Monophoto Photina series 747 with semi-bold series 748. Since there are two separate series, these fifteen lines might be termed a family, but in photo-composition all these characters would be mounted in the typesetter for use together. The lateral positions of accents over italic letters were a peculiarity of the type-specimen only; and floating accents are shown over E and e only.

body, and even characters borrowed from another type-face. The arts, sciences, and languages of the world are expressed in print by means of many thousands of different characters, and a versatile printer cannot tell which of them he is going to need next, or even whether they all exist in a suitable typographic form. For his work on one book, he will normally be able to limit his selection to a fount of a few hundred sorts at most; for his work on the next, he may have to make a very different selection. Even then he may still have to resort to characters from outside his fount; whatever case may be on his frame, for instance, the compositor may still need from elsewhere a phrase in Greek, some mathematical signs, or a word in phonetics. For the purposes of this book, then, the fount may be defined as the selection of types of matrices made ready for use on a specific book, but not as the entire variety of sorts needed even for that book, and certainly not as the whole great array of different characters needed for every book. This concept of the fount, always different in scope and constituent parts, is more significant in mechanical type-setting than in hand composition.

A *series* is a set of founts closely related to each other in design, and usually very similar to each other, but graded in size. If only one alphabet has been made in a certain design, that alphabet alone may be a series. A *family* is a group of series compatible for composition, but loosely related in design (figure 4-4b). A family may include excerpts from more than one series.

A printer's alphabet usually includes the diphthongs æ and œ, though they are not now fashionable in English spelling. A diphthong, comprising two or more letters cast in metal on a single body, may be defined as a *logotype*, since the term was invented for that meaning, although it is now more often used to indicate a trade-mark of almost any kind, as a 'logo'. The logotypes traditional in lower-case letters are fi ff fl ffi and ffl; they add grace and variety to composition, and particularly to italic. In all too many photo-composition systems, however, they have been omitted as unnecessary. They have been replaced by distorted versions of f and ƒ which stop short of the kern. Such a letter is necessary in German, to separate f and f, i, or l in certain instances, as indicated by Hart (1978), but in English composition of good quality the logotypes are preferable. A *ligature* is a joining stroke or loop, or two or more characters cast in metal on a single body and joined by such a stroke. Much of the liveliness of sixteenth-century roman typography sprang from repetitions of the ct ligature and the additional logotypes necessitated by use of the long s, and early italic types were distinctively cursive owing to a variety of ligatures. Figure 4-4a, extracted from a type-specimen book (Williamson, 1981), shows that some systems of photo-composition include diphthongs and other logotypes.

The roman lower-case alphabet provides the majority of characters in English composition; its effectiveness for each purpose deserves the closest scrutiny, and its proportions offer a criterion for all the other characters in the fount. Roman capitals, for instance, are to be judged by their compatibility with the mated lower-case; if the capitals are too tall or too heavily drawn or both, they will mar the page as though with a rash of spots.

Italic letters differ in various ways from roman, and these differences combine to provide contrast for the italic minority of letters. The greater the slant, the greater the difference in body-line (italics tending to be thinner than romans in this aspect), and the greater the difference in a-z length (italics being characteristically narrower), the greater the contrast. But the

abcdefghijklmnopqrstuvwxyz æœ åąçdęłøß
ABCDEFGHIJKLMNOPQRSTUVWXYZ
&ɞ ÆŒ ÅĄÇÐĘŁØ 1234567890 ₁₂₃₄₅₆₇₈₉₀.
..;:"!?-()[] --£$$*†‡§||¶¡¿«» ÉÈÊËĒĔĚĖÈ éèêëēĕěëè

abcdefghijklmnopqrstuvwxyz æœ åąçdęłøß &ɞ ÆŒ
ABCDEFGHIJKLMNOPQRSTUVWXYZ ÅĄÇÐĘŁØ
1234567890 ..;:"!?()[]£$$ ÉÈÊËĒĔĚĖÈ éèêëēĕěëè

abcdefghijklmnopqrstuvwxyz æœ åąçdęłøß
ABCDEFGHIJKLMNOPQRSTUVWXYZ
&ɞ ÆŒ ÅĄÇÐĘŁØ 1234567890
..;:"!?-()[]--£$$ ÉÈÊËĒĔĚĖÈ éèêëēĕěëè

abcdefghijklmnopqrstuvwxyz æœ åąçdęłøß &ɞ ÆŒ
ABCDEFGHIJKLMNOPQRSTUVWXYZ ÅĄÇÐĘŁØ
1234567890 ..;:"!?()[]¡¿«»£$$ ÉÈÊËĒĔĚĖÈ éèêëēĕěëè

abcdefghijklmnopqrstuvwxyz æœ åąçdęłøß
ABCDEFGHIJKLMNOPQRSTUVWXYZ
&ɞÆŒ 1234567890 ..;:"!?()[]¡¿«»-—£$$
ÅĄÇÐĘŁØ ÉÈÊËĒĔĚĖÈ éèêëēĕěëè

abcdefghijklmnopqrstuvwxyz åçø
ABCDEFGHIJKLMNOPQRSTUVWXYZ &ɞ ÅÇØ
1234567890 ,.;!?()ÈÈÊ éèê

Figure 4-4*b*. A family — in the sense that even in photocomposition not all machines could accommodate so many characters for use together, although for some typesetters a 12-alphabet fount is normal. This is 11-point Monophoto Univers series 685 (Light), 689 (Medium), and 693 (Bold). As usual with sans-serif type-faces, there are no small capitals or hanging figures, and italic is replaced by sloped roman.

letters also tend to be more cursive in style, and letters a f k are normally different in construction. Some italic alphabets can be accompanied by *swash* letters and ligatures (figure 4-4c), decorative in form, and best reserved for display lines — and even then used only with care. Sloped roman is not often seen in metal type; examples were introduced in the 1930s, but they lacked contrast with roman, and gained no ground from true italic. An increasing number of photo-composition systems, on the other hand, have a facility by which roman type can be photographically or digitally slanted into the form of a sloped roman. In some of these systems, the resulting alphabets have replaced italic, and in others they have been used as a kind of additional italic, on the same page as the real thing. Fortunately these confusing tendencies do not seem to have become popular.

Figure 4-4c. Swash capitals and the long s in Monotype Garamond. Three of the capitals are followed by lower-case letters to show how they fit together. Long s requires more logotypes than f.

Small capitals — SMALL CAPITALS — in metal founts are peculiar to those designed for text composition in books, and are not usually included in sans-serif series or in larger founts of any type. Small capitals are similar in x-height to the related roman lower-case letters. In proportion to their height, they tend to be a little heavier and wider than the capitals of the same fount; some are thinner in body-line than the mated lower-case roman, but most are a little heavier. Lacking ascenders, they do not look well next to aligning figures in the same line, or when followed by ? or ! In photocomposition, the ability of most systems to change reduction scales during setting has been utilized on occasion to substitute reduced roman capitals for small capitals; the resulting alphabets have proved even less acceptable than sloped versions of roman in place of italic. On the whole, the best small capitals are distinctly heavier than the accompanying roman lower-case. When this is so, and when in a photo-composition system small capitals are provided for founts above 14-point, the designer is likely to find such alphabets as 24-point small capitals offer a more emphatic substitute for 14-point roman capitals, particularly when the latter are little heavier in body-line than the mated roman (figure 4-4d).

ABCDEF **KLMNO VWXYZ**

Figure 4-4d. From left to right, 16-point roman capitals, 24-point small capitals, and 16-point semi-bold capitals. Monophoto Ehrhardt series 453 and 573, from the Monophoto type-specimen book, enlarged.

In metal type, an accented letter is usually separate, as a sort, from the same letter without an accent. Some fifty such sorts, used in the principal European and Scandinavian languages, should as a rule be readily available from printer or manufacturer as standard equipment (figure 4-4e). Other sorts are likely to be needed for other languages such as Czech, Hungarian, Polish, Rumanian, Serbo-Croatian, Slovak, and Turkish. An accent is generally understood to be visually separate from the letter; marks attached to or part of the letter are known as diacritical. In photocomposition, accents are usually *floating* or separate from the letter; in figure 4-4a, for example, ten accents are available for setting over appropriate capital and lower-case letters, except that there are only three in bold italic. There are also seven diacritical letters, and the *eszett* (ß) for German in lower-case.

Á À Â Ä Å É È Ê Ë Í Ì Î Ï Ó Ò Ô Ö Ø Ú Ù Û Ü Ç Ñ
á à â ä å é è ê ë í ì î ï ó ò ô ö ø ú ù û ü ç ß

Figure 4-4e. A standard range of special sorts, which should be readily available from printer or machinery manufacturer. Compare the accented groups of E and e in figures 4-4a and b.

Special letters, or groups of letters, or alphabets are needed at times. In English, for example, the long s (ſ) may have to be used for extracts from documents and printed works of the eighteenth century and earlier, and when it is it may need the company of seven or more logotypes (figure 4-4c). The International Phonetic Association has standardized a numerous set of phonetic symbols and accents in roman and italic. Any comprehensive mathematical fount – and there are not many of them about – includes thousands of characters, many of them of no use for any other purpose.

Of the many non-latin alphabets used in the world today, the most familiar are Hebrew, Arabic, Greek, and Cyrillic (Russian). On the whole these are separate founts; when they appear as quotations within a roman text, the roman fount may have to be selected for its compatibility with its foreign neighbour. There are few examples of mated roman and non-roman alphabets, and only specialist printers are equipped with non-roman type or accustomed to setting it.

A popular text fount is likely to include a mated *bold* or *semi-bold* roman (figure 4-4*a*). This is a redrawn version of the basic roman letters, with conspicuously thicker strokes. At its best, bold type for text use is useful rather than elegant; at its worst, it can be horrible. The shapes of the counters are usually incompatible with those in the traditional roman. Such types are best reserved for headings; within a paragraph of text, they interrupt the even texture of the page. Bold italic is less common than bold roman, but is sometimes useful and even necessary. In metal, most printers offer mated bold series in sizes up to 12 or 14-point at most, though sometimes with one or two in larger sizes. In photo-composition, most series which have mated bold at all tend to be mated further up the scale of size, as well as down to the smallest founts of all.

Figures, 0123456789, are often called *arabic figures* or *arabic*, as though to distinguish them from *roman numerals*, I II III IV V VI VII VIII IX X, and perhaps even from *text figures* (diagrams and other informative illustrations in the text). Arabic figures are either *hanging*, 0123456789, or *aligning*, 0123456789; one of these two kinds of figure is standard equipment in every series, and is normally used in the absence of other instructions, but in text founts the alternative is often available. Hart (1978) forbids mixing hanging and aligning figures in the same book without special instructions; those instructions may be provided by the typographer, who is likely at times to find both kinds useful in different parts of the same page. Aligning figures are much the same height as capitals, and look well in the same line; hanging figures look too small among words set in capitals, but are less conspicuous among small capitals and upper and lower-case. Both kinds of figure may be either *upright* or *italic*, 0123456789; the word *roman* is better not applied to arabic figures, and although inclined figures have no italic origin the use of any other term would be pedantic. Metal type sometimes has no italic figures, and the use of upright figures among italic words is traditionally acceptable. *Superior* and *inferior* figures, 0123456789 and 0123456789, and fractions, $\frac{1}{4}\frac{1}{2}\frac{3}{4}$, are not usually mated to any specific fount, but should always be reasonably compatible with the rest of the text. Common fractions — the various quarters and eighths — are usually quite easy in metal, but in appearance they tend to be incompatible with the adjacent text type, and in height they are all too often a little too tall for it. In photocomposition, such fractions as $\frac{123}{456}$ are usually quite easy.

Some punctuation marks are used with both roman and italic, others are slanted for use with italic thus , ; : ' ' [ ] ( ) ? ! The hyphen - is thicker than the en and em rules – —; the unspaced em rule—as here—seems to link words rather than to separate them, and a rarely available $\frac{3}{4}$-em rule — as here and elsewhere in this book — is useful with space on each side. Most text founts include a small number of mated reference and currency signs such as £ $ $ * † ‡ § || ¶; other signs, including + − × = < > ± · / | ' % ° ® ©, hardly need to be mated but may be part of the fount.

The design of such ancillary characters as arabic figures and punctuation marks deserves critical attention no less than the design of roman and italic letters. A page in a work of history, scattered with four-figure numbers indicating years, may be spoilt by figures more conspicuous than the lower-case letters; and double superior figures indicating footnotes can generate in a reader a crescendo of irritation if the superior figures stand at the wrong height in relation to the letters, or are set too far apart, or are incompatible in design with the lower-case letters. Unsatisfactory punctuation and similar marks may be quite enough to discourage a designer from using an otherwise satisfactory type design. Bruce Rogers, designing Monotype Centaur roman to accompany Frederic Warde's Arrighi italic, designed punctuation which differed from that of the italic not only in inclination but in weight and shape, with a different full-point and hyphen. Rogers also introduced a pretty and inconspicuous asterisk in the form of a hollow star (figure 4-4*f*).

., : ; - ' ' ? !   ☆   ., : ; ~ ' ' ? !

Figure 4-4*f*. Variations in the design of punctuation for roman and italic. Left, 18-point Monotype Centaur, series 252; right, the mated italic, Arrighi, series 352. Even the full-points differ. In the centre, the unusual asterisk.

Composing-room equipment includes a variety of other printing material. Some of this is discussed in § 9-1, and some of it may be useful in text composition. Various kinds of typographic line, circle, triangle, square, and star appear in small size, to guide the reader to different points on the page, or to different co-ordinates within a graph. Conventional signs which do not already exist can be invented, as when in a guide to hotels a crossed knife and fork commends the restaurant. Rules of various kinds can be cut to the appropriate length if they have not been set to that length in the course of text composition. Braces { } to connect from two lines to about six are always likely to be needed in any work of information, and are more difficult to obtain in their naturally graceful form in photo-composition than in metal.

§ 4-5 Spaces

In metal composition, shanks which print no character but insert unprinted space within the line of type are known as *spaces*. They are cast below type-height, so that they are touched by neither the inking rollers nor the paper. When they are cast to shoulder-height, they support adjacent kerns.

Mechanical type-setting is based on a combination of *variable spaces* with *fixed spaces* (§ 5-2). In hand composition every space, once cast in whatever body and set, is fixed. To make exact adjustments of width, the compositor sets different spaces side by side, or exchanges spaces of different widths. The set of fixed spaces in metal is related to that of the em of the fount. A *hair-space* is $\frac{1}{6}$ em or a fraction over 16 per cent; a *thin*, $\frac{1}{5}$ or 20 per cent; a middle or *mid*, $\frac{1}{4}$ or 25 per cent; a *thick*, $\frac{1}{3}$ or just over 33 per cent; and an en or *nut*, $\frac{1}{2}$ or 50 per cent. The em itself, 100 per cent, is often called a *mutton* in conversation, to distinguish it from the en. In most composition, the space between words should not be narrower than thin or wider than nut, and the thick space should be wide enough for the majority of lines. These simple fractions of the em are now being replaced by the various unit systems of photo-composition, but remain a useful basis of mental arithmetic when typographic plans for lines and pages are being visualized and compared.

Roman lower-case letters, derived from capitals incised and written in the Roman Empire, began to appear as manuscript predecessors of their present forms in the time of Charlemagne. Compact and legible, the letters combine best into words when close together. By comparison, words and lines set in capitals or small capitals are less easy to read; but the different shapes disentangle themselves before the reader's eye when there is space round each letter. Some of the letters appear to have extra space on one side or both; in some combinations such as WA, the letters appear to be farther apart than in MN, for example. In order to give words or lines in capitals or small capitals an appearance of clarity and regularity, spaces are needed between the letters within the word.

When text is being composed in metal, this *letter-spacing* is usually equivalent to $\frac{1}{9}$ em. In the larger type of *display* lines, usually understood to mean 14-point and above, this proportion — about 11 per cent of the em — looks a little wide to some eyes, which for these sizes would prefer about 6 per cent of the em. If the same width of letter-spacing is to be applied to text as to display, this 6 per cent is likely to seem narrow but adequate for text. When display lines are set by hand in separate metal types, letter-spacing can be proportionate to the shapes of the adjacent capitals and small capitals. VAT, for instance, needs less letter-space than HIM. In photocomposition this refinement would be difficult and time-consuming on a keyboard, but a method of proportionate letter-spacing by computer has been explained by David Kindersley (1976). Meanwhile, letter-spacing in photo-composition is mechanically even.

Lower-case letters are on the whole best placed in relation to each other in accordance with the fitting assigned to them by the type-designer. The letter-spacing of lower-case is generally regarded as bad practice. Letter-spacing seems to suit aligning figures better than hanging figures.

When adjacent lines of type are set on the body for which they were designed, and are separated by no added spacing material or space, there is in a sense no space between the lines. In another sense, there is some space, partly occupied by the descenders of the upper line and ascenders of the lower. Most type-faces gain legibility when this space is slightly increased, and any such increase may be known as *interlinear space*. In metal composition, interlinear space may be obtained by inserting strips of lead alloy known as *leads*. The usual thicknesses are 1 point (*hair lead*), 1½ point (*eight-lead*, because there are eight to the pica), 2 points (*thin lead*), and 3 points (*thick lead*). Thicker strips are known as *reglets*, and may be wooden. Sometimes a typographic item in a line has to extend beyond the front and back of the shanks of the rest of the line, into whatever interlinear space may have to be provided for it. Examples include underlining, and the *vinculum* or horizontal line over the characters which in mathematics represents enclosure by parentheses, particularly to indicate the extent of reference of a root sign. The compositor has to cut leads of several exact lengths, to occupy the interlinear space not already occupied by the extruding item. Underlining in manuscript or typescript is conventionally represented by italic; the convention is so well established that most designers prefer not to use underlines even when this disproportionately expensive procedure is unnecessary, as in photo-composition. Printers have achieved some modifications in mathematical notation, in order to keep the cost of composition within bounds, and accordingly the vinculum may be replaced by parentheses.

In metal composition, all blank space has to be filled with spacing material. Outside the area of the typographic page, margins and other major spaces are filled by hand with rectangular *furniture* of various sizes and materials. Any printing surface such as an illustration *block* (§ 11-3) has to be mounted on a rectangular *mount* lower than type-high. In all other forms of composition, such labour and material are unnecessary; unprinted space is simply an area into which nothing has been placed.

§ 4-6 The line of type

The width of the setting is the *measure*. In metal, every line of type has to be spaced out or *justified* to fill the measure exactly. The sorts will be held in place for printing by pressure applied to both ends of every line on the page. Any line narrower than the others, or part of it, and with it the rest of the page or part of that, can fall out of the *forme* (§ 4-9) when it is lifted, or can

rise during printing, hideously to squash the image into the paper. When
the type ends part way along the line, the rest of the line has to be filled up
with spaces. In metal type, this is so when the wording is intentionally
uneven in width, as in verse; such composition cannot strictly be described
as *unjustified*, but this term is convenient and in common use. Measure is
usually but not essentially specified in picas, to the nearest half-pica. Very
wide measures may be necessary for tables turned sideways in tall books.

Setting type by hand, from the case on the *frame* before him, the
compositor takes letter after letter in due order to form a word. He places the
sorts in his *composing-stick*, a small oblong tray held in one hand; one longer
side is open, so that lines of type can be moved across the edge without
lifting, and the distance between the shorter ends is adjustable to set the
measure. At the end of each word he places a narrow space, probably a thin,
before beginning the next word. When the line of words and spaces
approaches the end of the measure, he compares the probable width of the
next word with that of the remaining space. If the word will fit into the
measure, he sets it there. If it will not, he assesses the effect of its absence on
the word-spaces in the line, as they will have to be increased to justify the
line. He may increase them up to about a nut, but he will prefer not to do so
if he has already set similarly wide word-spaces in the preceding line, or
expects to set them in the next line. Loose word-spacing in adjacent lines
forms *rivers* of white, meandering down the page across the path of the
reader's eye. If he intends neither to omit that next word from his line
(*turning it over*) nor to fit it in complete, he will have to divide it at an
appropriate point, insert a hyphen, and include the first part only in the line
he is working on. The second part will appear at the beginning of the next
line. If he has ended the two preceding lines in this way, he will probably try
to avoid a third adjacent hyphen, in accordance with the custom of sound
typography. When he has made his decision, he increases each word-space
in the line by the amount needed to justify it.

This may seem a laborious description of a simple process. But the
compositor's type-setting line by line in this way is itself laborious, even
when it is done on a *keyboard* (§ 5-1); it is never simple, and the competence
applied to it determines the visual quality of the entire text. The style of
setting applied to every line is influenced by that of preceding and
succeeding lines; the penultimate line of a paragraph, for instance, may
have to be particularly close-spaced if the last line would otherwise consist
of fewer than five characters, and when that is not possible the compositor
may have to return to an earlier line in order to *take back* a word or part of a
word and so make room. If adjacent lines happen to begin or end with the
same conspicuous word or phrase, type already set may have to be adjusted
to prevent a distracting coincidence. No rule could be made about such
problems, as there may be times when the coincidence is intentional.

The customs of word-division derive partly from etymology, partly from meaning, partly from pronunciation, and partly from tradition. Effective communication depends upon conventions, in word-division as elsewhere, and the best conventions are those the reader is likely to expect. The first part of a divided word should not mislead the reader about the pronunciation or meaning of the second part. Now that much word-division is determined by computers, such oddities as decent-ralization, propheth-ood, and even pro-ud are to be seen in print. Words already hyphenated are provided with a second and unwelcome hyphen, as in photo-compo-sition, and unhyphenated compound words are divided in the wrong place, as in thermonuc-lear. When most lines contain ten or fewer word-spaces, close and even word-spacing may be possible only at the expense of an excess of word-division at line-ends.

§ 4-7 Galley, proof, and correction

Continuous wording, in the type-size of the main text or smaller, is normally set by machine. In metal type-setting, each sort or line, as soon as it is cast, emerges into a tray called a *galley*, which is long enough to contain the equivalent of about three octavo pages. A *galley headline* precedes the rest of the type, identifying at least the title and galley number, in the absence at this stage of *headlines* (§ 7-9) and page numbers. These galleys of mechanically set and uncorrected type are known as *caster galleys*. The type in them contains every human and mechanical mistake that has happened while keyboard and caster were at work. There may not be many mistakes, but most are conspicuous enough to unnerve the author if he were to see them. They include the most frequent mistake in composition, the *literal* — the use of a character different from that indicated in the copy.

In photo-composition, there is no galley; instead, latent image of the typographic characters is projected on to a light-sensitive *slip* which may be a great deal longer than the three-page slips used with metal type. The word 'galley' is still used for a related form of proof, and indeed many other terms survive from the long history of metal typography for use in the age of photo-composition.

An inked impression of the metal type in the galley is proofed on a long *slip* of paper similar in dimensions to the galley. This is a *slip* or *galley proof*. Proofs differ from prints in not having been *made ready* (§ 12-2), and therefore do not demonstrate the quality of impression to be expected of letterpress *presswork* (chapter 12). Proofs are said to be *pulled*, not printed; the term refers to the action of a man working a hand-press. Irregularities of level in the floor of the galley, in the proofing press, and in the unprepared type combine to cause wear in the metal after repeated pulling, with some

[48]

consequent deterioration in the printing surface. The number of proofs pulled from any galley should accordingly be kept to a minimum. The caster galleys are read by the printer's proof-reader, and *house* corrections marked; these are to be effected in the printing-house, without additional charge to the publisher, before *clean* or corrected proofs are *sent out* (of the house) to publisher or author.

Photo-composition proofs take various forms, but all are photographic. One of the most useful forms is that of photo-copying, which can without difficulty reproduce a number of copies of a marked proof including the readers' marks. Small characters may be blurred in proof by this method, and there have been occasions when a comma in text size was indistinguishable in proof from a full-point. Other forms of photographic proof may be sharper and more exact than the printed image, but more expensive than a photocopy. By these methods any number of proofs can be pulled without affecting the printing surface, but the cost per set of proofs is substantially higher than that of proofs pulled from metal type.

Like *make-up* into page (§ 4-8), which can be carried out at the same time, the correction of metal type is still carried out by hand in much the same way as it has been for the last five centuries. Correction improves the caster galleys to a point at which they are a presentable translation of the copy into typographic form. Subsequent alterations by the author may be defined as variations from the copy, for which the publisher is asked to pay, and these may include designer's alterations.

Working from a marked proof, the compositor picks out each wrong sort and replaces it with the right one. Unless the two have the same set, he then has to re-justify the line. If by way of correction he has to insert more than one or two additional sorts, omitted from the caster galley, he may have to carry over or *over-run* a word or more from the end of the line to the beginning of the next. Then he may have to continue over-running until the space occupied by the insertion has been taken up, and to re-justify each line after doing so. This time-consuming handwork is expensive; it necessitates disturbing type already set, and may introduce errors into lines previously correct. Over-run passages are sometimes re-set mechanically, and are themselves corrected and taken into the rest of the type. Apart from such extensive correction, sorts correctly set need no further attention at this stage. *Pie* sorts, not available in the fount used for type-setting, are inserted by hand, replacing spaces set to make room for them. Display lines are usually set by hand, and placed in position in the type.

A photo-composition fount has to include all the required characters, since the hand insertion of pie characters is uneconomic except on a very small scale. Display lines in larger sizes of the text fount are set on the keyboard like everything else, and emerge on the same proof. Corrections and alterations are also set on the keyboard, and in some systems corrected lines are re-set and stripped into the camera copy in replacement of defective lines. As photo-composition systems become faster, however, more and more systems rely on an entire or partial second pass through the phototypesetter, embodying the keyboarded corrections. Since the number of passes has still to be kept to a minimum, printers are beginning to return to the galley proof at the first stage, and to send out the equivalent of caster galleys, in order to correct house and editorial mistakes in a single pass.

When metal type is to be made up into page before being *pulled clean* (proofed after correction), make-up and correction are part of the same process. If proofs are to be sent out in galley, the compositor may still have to assemble and space out disparate elements of the composition. Display lines are one of these; extracts in smaller type, tables, and footnotes are other examples. Anything set in type differing in body from the main text is separately set, since it is a different fount, and has to be assimilated by hand. The compositor also inserts spaces between such items and the adjacent text.

The value of making up uncorrected matter is open to question. Even when the keyboard operator's composition appears before him as he sets it, as it does in advanced photo-composition systems, lines and even groups of lines are likely to be omitted or repeated from time to time in the course of a long day's work. Even if they are not, house errors combined with author's alterations are likely to affect the number of lines on any uncorrected proof. Author, editor, and designer are all likely to be misled by a provisional make-up. Computer programs for automatic make-up so far available seem to be well enough suited to works of entertainment which in typographic terms are comparatively simple, but when applied to more complex works of information seem unlikely to provide as well for the reader as a make-up compositor who brings the mind and eye of a reader to the pragmatic solution of each problem in turn. Good make-up pleases the eye and lends itself to the act of reading; but if book designers generally tolerate low-grade forms of automatic make-up, then bad make-up will soon be the only kind available, unless the designer makes all decisions about the make-up of all the pages for which he is responsible.

Proofs are best sent out in galley if the sizing and placing of illustrations will be influenced by the number of typographic lines in each passage, or if textual insertions and deletions are to be expected. Alterations which affect the make-up of successive pages are less costly in galley than in page. If no such probabilities attend a proof from metal type, it is now usually sent out in page, so that a second stage of proof may be unnecessary, and so that the indexer can start his work at the earliest possible moment. Authors find full-length galleys awkward to handle, and prefer galley proofs or page proofs to be presented in *half-galley* or the equivalent, on slips similar in depth to a page and a half or a single page.

§ 4-8 Make-up into page

Make-up is the process by which galleys of type are divided into pages of type. Whether or not proofs have already been sent out in galley, the designer should see proofs in page in time to make any radical alterations that may be necessary. The total number of pages, and the arrangement and placing of each, are cardinal points in the book's production and presentation. For the same reason, the author is entitled to see a proof in page, and the quality of presentation is likely to benefit from his anxious diligence. Some authors moreover may be willing to read proofs of their own work only once, particularly when they have prepared the copy carefully and intend to make no major alterations. The compiler of the index can work only from a set of proofs in page; he should do so at the earliest opportunity, and he must be advised of any subsequent changes in make-up or page numbering.

Make-up starts at the first page of the book and continues to the next in order to the end. The pages are numbered as they are made up. But as he makes up any page, the compositor working on metal type may encounter a problem which he can solve only by returning to an earlier page, already made up, and by altering that and subsequent pages. A book designer, working out the make-up of a book from uncorrected galleys, is likely to face the same kind of problem, and will do well to adopt the same solution. If for example on reaching page 100, the compositor finds that he is about to begin it with the last line of a paragraph, containing only a couple of short words, he will wish to find a way of avoiding so unsightly an arrangement. Searching back through the immediately preceding pages, already made up and still on the frame beside him, the compositor finds that on page 95 a paragraph ends with a line which is nearly full of characters. A little extra word-space in that line and the one before it enables him to *make a line* which was not there before, to *carry over* a line from page 95 to page 96 and so on until page 100 begins with the penultimate line of a paragraph instead of the short last line. This operation in support of good make-up might be termed *re-composition*; it is a matter not of textual alteration or literal correction but of rearrangement — page 95 was satisfactory until the compositor reached page 100. If make-up is planned on corrected or on marked galley proofs, re-composition can be effected when the compositor or keyboard operator reaches page 95, and he will then have no need to return to an earlier page.

When footnotes are numbered by page, footnote numbers cannot be set until the text page has been made up with its footnotes, as make-up or its planning reveals which footnotes will appear on which page. When there are *cross-references* (references to other parts of the same book) to subsequent pages, the referring page numbers can be set in the earlier pages of text only after the subsequent pages referred to have been made up and numbered. Page-numbered footnotes and cross-references by page present no problems in metal type, but footnotes are now more often numbered by chapter than by page, and cross-references kept to a minimum, for the sake of economy in photo-composition.

Footnotes, preferred to *end notes* (notes at the end of a chapter or of a book) by reviewers and authors and probably by most readers, are often difficult to present really well. A series of short footnotes, perhaps referring to page numbers only, originally set in successive lines, is better rearranged during make-up with two or more notes in each line, and that line occasionally shared with the last word or two of a longer note occupying several lines (figure 4-8).

---

[1] B.M. Add. MS. 5841, fol. 149; 5853, fol. 90.   [2] See p. xiv above.
[3] *Lit. Anecd.* v. 393 (Mores 'being intended for orders by his father').
[4] *G.M.* Oct. 1783, liii. 848.   [5] *Lit. Anecd.* v. 395.   [6] See p. lxi below.

---

[1] *Lit. Anecd.* v. 401.
[2] *Bibl. Mores.* (1779), lists 11 vols. of James's Common-places (p. 173).
[3] *Dissertation*, p. 87.   [4] *Bibl. Mores.*, p. 109; *Lit. Anecd.* v. 401, viii. 36.
[5] *Lit. Anecd.* v. 401.   [6] See p. lx below.
[7] See p. lix.   [8] *Dissertation*, p. 71.
[9] Nichols, *Anecdotes of Bowyer* (1782), p. 585.
[10] *G.M.* Dec. 1778, xlviii. 607.   [11] *Lit. Anecd.* v. 402.
[12] *Lit. Anecd.* v. 700.   [13] See p. xvii above.

Figure 4-8. Footnotes, set in Monotype Caslon, from two facing pages. After the initial type-setting, in which each footnote started a new line, the notes have been rearranged to share lines in a more compact formation.

Textual references to tables and illustrations, by relative position rather than by number (*above, opposite, on next page*), are likely to need alteration, as the item referred to may have to appear elsewhere. A typical dilemma is a reference, in the last line of the page, to an illustration intended by the author to appear on the same page. If the illustration were to be brought into that page, the reference to it would have to go over to the next, in order to make room for the illustration.

Headlines (§ 7-9) and page numbers are set before or during make-up. Any item deeper than one line of type, which cannot conveniently be divided horizontally at any selected point, can be correctly placed in the text, in the preferred position on the page, only during make-up. Verse in stanzas, for instance, may properly be divided between stanzas and even within them but not within couplets; most tables can be divided horizontally or vertically or both, though preferably across facing pages, so long as head-words and column titles are repeated on the second page; but illustration can hardly ever be divided. When type is to *run round* narrow illustrations, the type beside the illustration has to over-run into a measure narrower than that of unillustrated pages; the point in the text at which this is best begun is determined during make-up.

The compositor seeks to do more than to protect the reader against inconvenience: he seeks to attract the eye to his pages, arranging intractable material with elegance and apparent ease. Some of the traditional criteria of this kind of work are well known. Hart expresses opinions of a kind common to careful printers everywhere:

A page or column should not start with the last line of a paragraph even when the line is a full one. . . . Two successive hyphens only are allowed at the ends of lines. . . . No more than two successive lines should begin or end with the same word. . . . A minimum of five characters (excluding the full point) is allowed on the last line of a paragraph. . . . If a right-hand page is a full-page illustration or table, the facing left-hand page should not end with a hyphen. . . . A divided word should not end a right-hand page. . . . A right-hand page should not end with a colon which is introducing poetry, examples, etc. . . . When chapters start on fresh pages, the minimum number of lines left on the final page of the previous chapter should be five. . . .

Good make-up contributes to the pleasure of owning books and of reading them. If machines, or indeed book designers, are to assume responsibility for the make-up of any edition, the only standard of quality worth setting is that of a first-class compositor.

§ 4-9 Imposition and repro

Imposition, the arrangement of pages in a rectangular formation for printing, is discussed at length in § 12-1. In metal, the pages are secured with *page-cord* round the edges, and laid in position on a horizontal working-surface (the *stone*). They are surrounded by rectangular *furniture*, which holds the type square and which spaces out the margins. A rectangular frame of steel, the *chase*, is placed round them, and they are wedged against its inner edges with *quoins*. Chase, furniture, quoins, and type are collectively known as the *forme*. The compositor settles the type on its feet by *planing down*, carefully knocking areas of the page downwards with a flat wooden surface (a *plane*). When the inwards pressure of the quoins enables the compositor to lift the forme horizontally without any part of it falling out, the forme is *locked up*. The forme can then be carried to the floor or *bed* of the press, without a single letter being displaced, and with all pages arranged in position and secure against movement during printing.

It is to make this rectangular lock-up possible that the lines have to be justified, whether or not the right-hand edges of most of the lines are occupied by characters as distinct from spaces. Such items as marginal notes which are to appear beside the page of type, or wide illustrations which extend from within its area into the margins, complicate the arrangement of furniture and quoins and add to the cost.

In photo-composition, imposition is a good deal simpler than in metal. The pages exist in the form of paper or film, and are handled for the most part and placed as a single sheet of material. There is no spacing material, no planing down, no lock-up, and sometimes not enough rectangularity. Justification is unnecessary, and margins may be occupied by illustration or supplementary text or both without great expense. And the assimilation of imposition into photocomposition seems likely to be possible before long.

After imposition, metal type is ready for press except in one respect. Traditionally at this point a final proof, the *press revise*, is pulled from the forme. This is usually a house proof, which is not sent out; the printer's proof-readers should not merely check marks but should *read for press*, to make sure that the author's alterations are textually compatible with the rest of the text. They should ensure that the pages are so imposed that after folding they will appear in due order, and that they are so aligned with each other that margins will be even, and page *backs* page (§12-1), throughout. At this stage, the press reader checks everything, including the meaning of the text and the effect of the designer's instructions. He may have to raise queries about matters not observed at previous stages of composition, and any delay so shortly before printing may be serious; such queries need the immediate and careful attention of editor or designer.

Reproduction proofs or *repro proofs* are pulled from metal type with care, on carefully selected paper, to match the quality of good printing, as copy for camera. This kind of proof can be read for press in much the same way as the imposed proof already described, except that repro is not usually locked up in a folding imposition. Repro proofs are usually read for press beside the proofing press, in which the type waits for any correction and any replacement proof. Metal type has been corrected in the bed of a press since the European invention of the single movable type, and last-minute correction delays the press only a few minutes.

Press reading, and these last-minute corrections, seem to be incompatible with photographic printing processes, mainly because so much composition now dispenses with the single movable type. By the time photo-type-set pages are all but ready for press, it is too late for corrections which could be carried out only by starting the composition process again from its beginning. As a result, the composition of books is already less accurate than it was before photo-composition began to dominate book production in the place of the single movable type, and a period of world recession is unlikely to see a reversion to the best standards of the past or the increased expense and time that would be the certain result.

Book proofs are usually sent out at the first stage of proof, in place of proof in galley or slip page, as they are needed at the earliest possible point in production. The pages are made up and imposed for folding, either in chase or in the bed of the proofing press, pulled on both sides of the sheet, and folded and bound in paperback style. This is a costly operation, like printing an edition of a few dozen copies, but book proofs have been a valued part of publishing procedure. Metal type provides the least expensive form of book proof at the right time.

The metal from which type is cast is provided by the printer and remains his property. The composed order of the type is the publisher's property, since he has paid for the composition, and it is not to be disturbed without his permission. When type is kept in store, it is usually *dropped* (removed from the forme) and wrapped for safety; re-imposition will be necessary before a reprint. When it is no longer needed, type is *distributed*; display type is actually distributed back into case, ready for further composition, but the same term is used for text type which is re-melted for further casting in new forms.

In Britain the ownership of intermediate material such as film for photographic printing processes seems not to have been decided by a court case, but in law such material is probably the property of the printer unless his customer formally ordered him to produce the intermediate material as well as the finished work, and unless the printer formally accepted that order as the basis of his estimate and thereby established a contract to provide intermediate material as well as finished work. The Publishers Association has agreed with the British Printing Industries Federation that printers in membership of the Federation will treat intermediates as though they were the property of the publisher. This legal situation may have its effect on the design of an edition in Britain; other countries may have other laws. But probably in all countries the right to use intermediate material of a publisher's order for any printing rests with that publisher.

In almost all books, the main text and items in smaller type are mechanically set. The composition of single movable type is completed by compositors. After decades of development, photo-composition, in which typographic images are assembled entirely by machine, has yet to match at all points the quality of composition regulated throughout by the human mind, hand, and eye. If this standard of quality is no longer in demand, it will cease to be available.

'Not to inconvenience the reader must always be one of the main considerations.' Hart applies this maxim to the division of words in particular; a good compositor applies it to composition in general, and a typographic designer must do the same. The compositor is guided by the style of the house, but he also has to plan and assess his operations with the mind and eye of a reader, avoiding any clumsy incident of composition whether there is a rule about it or not. Some of the customs of good composition are mentioned here, but while the mechanization of composition continues the designer will need to know many more. A careful study of Hart's *Rules* is a beginning; and he will need to continue from that with a critical examination of every page of every proof set in accordance with his instructions.

Metal type is slow in composition and not very clean to handle. Its casting is hot, noisy, and dangerous. Once composed, the pages are intricate, fragile, heavy, and expensive to store. But for more than five centuries, type metal has been the foundation of book production. Authors, publishers, and typographic designers have evolved their methods, and in some ways the presentation of their texts, to make the best use of its limitations and possibilities. Most of the type-faces we read were designed for casting in metal or have been adapted from such designs. The era of metal type extended the information available to the world into unprecedented variety, continuity, and expansion. The way in which this was done has set the highest standards of quality in book production. As new techniques develop — faster, quieter, cleaner, more convenient, and more economical — the quality of reading matter they provide is to be critically compared with that of the best that could be expected of metal type — the best in letter design, in composition, and in the appearance of the printed image.

CHAPTER 5

Composition systems

From the mid-nineteenth century in various countries type-setting machines were invented and brought into use. Some were in production for decades before they were superseded. Most were based on the assembly of types already cast, and while they were substantially faster than the hand compositor, none was entirely satisfactory. In 1885 Linn Boyd Benton of Milwaukee invented a punch-cutting machine, and made possible an unlimited supply of punches and hence of matrices. The first wholly successful type-setting machine, the Linotype (§ 5-1), invented in the United States by Ottmar Mergenthaler and first used in 1890, in the form of the Simplex Linotype Model 1 on which later models were based, depended on the supply of matrices to the printer, and on their use in the machine for setting type and in the same operation casting all the types of the line as a single piece. The Monotype (§ 5-2), invented by Tolbert Lanston and developed by John Sellers Bancroft, was in action as the Limited Font Monotype in 1898, casting single types from matrices in the course of composition. Apart from the Intertype, which is very similar to the Linotype, no other machines were developed to share the twentieth-century market for mechanical text composition in metal.

By 1900 photo-typesetting machines had been invented in the United States, in Hungary, and in Great Britain. All were based on the principle of a light-generated image of a character formed by a transparent matrix. In 1946 the Government Printing Office in Washington installed an Intertype Fotosetter. Photo-composition development was accelerated in the 1950s by new applications of the computer.

After more than eighty years, photo-composition has replaced most metal type in book production, but is still in a phase of research and development. New possibilities and new systems continue to emerge; solved problems, together with systems which worked well enough for a few years, recede into history. Mechanized type-setting in metal in the twentieth century can be studied in terms of two specific machines; photo-composition machines are many and varied, and accordingly in this chapter I examine the state of the process at the time of writing, rather than its proliferation of *hardware* (as photo-composition mechanisms are now called).

As explained in chapter 4, composition comprises that part of the

[57]

printer's work which translates typescript into typographic text, made up into pages, imposed, and ready for the press. This includes some processes, such as copy preparation and proof reading, which do not seem to be capable of mechanization. The term *photo-composition* is however used here to mean a mechanical and photographic method of setting typographic images in composition order, and usually of completing some other stages of composition such as correction.

A composition *system* is the establishment of people and machinery deployed by a printer for typographic composition by any method. A Monotype system, for instance, relies on machines and men for the initial type-setting of the text, and on men for the rest. A photo-composition system may include make-up compositors. Whatever the system, a book designer who works with it will do well to know where mechanical operation gives way to human work.

Except in a Linotype system, the precision of mechanical composition depends on counting and calculation. Linotype matrices are justified in such a way that their width needs no numerical measurement. The set of every Monotype sort or space, on the other hand, is a multiple of the typographic *unit*, which is usually one-eighteenth of the set of one of the widest letters in the fount, such as M or W. The unit in other systems is likely to be another fraction, perhaps of the square em of the fount, the set then being the same as the body. In mechanical type-setting by unit-based systems, the adjustment of *variable* word-spaces (§ 5-1) to the exact set required for justification is effected by counting the units already occupied by characters other than variable spaces: comparing the total with the number of units in the measure: and dividing the difference by the number of variable spaces in the line.

In a Monotype system, a numbered revolving drum on the keyboard performs this calculation for the operator, who observes the result and taps the indicated justifying keys. In photo-composition systems, calculation is electronic, and may be capable of activating and controlling other composition processes, apart from sub-editorial copy preparation, proof-pulling, and proof-reading. The extent to which these processes are controlled by people or by electronic devices differs from system to system.

Any calculating machine — though an electronic calculator is a network rather than a machine — could be termed a *computer*. The drum on the Monotype keyboard is a simple analogue computer, using distance to represent numbers. But computers are generally understood to be electronic and digital. They count in the simplest way, as though on fingers up for one and down for nil, with positive and negative charges instead of fingers, and manipulate numbers in various forms of calculation. They have memories, to record numerical information, and they work at unimaginable speeds.

[58]

A computer of one kind or another is part of every photo-setting machine, just as a motor is part of every printing press in industrial production. We have to recognise the presence in printing of motors and computers, to understand how presses and photo-setters work as they do. But a designer who wishes to understand computer operations will have to begin with the binary system of numbering, and continue with various kinds of science and logic; this book and its author lack capacity to provide enough information about these matters. This chapter is concentrated upon those aspects of the computer which immediately concern the designer.

§ 5-1 Line-casting

Mechanical composition in metal is controlled by a keyboard operator, who selects matrices in the order of composition by tapping the appropriate keys, and who controls word-division and justification in much the same way as the hand compositor. In the Linotype system, the keyboard and the line-casting element are part of a single machine, and the speed of casting is limited to that of keyboard operation. When keyboard operation and type production are linked in this way, the system is known as one of *direct entry*. The same machine can be adapted for indirect entry by the attachment of a tape-reading device. The tape is punched on a separate keyboard, and actuates the machine's own keyboard and its caster much more rapidly than the operator could.

Whatever the body, the Linotype casts a whole line of type at a time, as a solid *slug* with characters and spaces along one edge (figure 5-1a), and so dispenses with separate types, cast character by character, which had provided a foundation for all preceding typography.

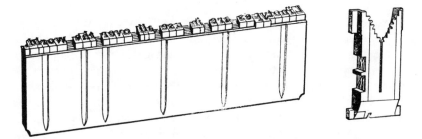

Figure 5-1a. Left, a Linotype slug. Right, a Linotype matrix. The serrations at the top form a code or key to release the matrix from the distributor bar above its own channel in the magazine. The projections at top and bottom hold the matrix steady in the channel slides.

Linotype matrices (figure 5-1*a*) are contained in separate channels within magazines or a magazine mounted on the slant over the keyboard. Each key on the board offers access to one of the channels and hence to one kind of matrix. Released by the appropriate key, a matrix slides down its channel into the *assembler box*, which functions as the machine's composing-stick. At the end of each word, the operator selects a wedge-shaped adjustable *space-band* which takes its place next to the preceding matrix. When the measure, to which the assembler box has been set, is nearly full, the operator initiates an automatic sequence of justification, casting, trimming, and matrix distribution which continues while he sets the next line.

When the matrices stand side by side in the assembler box, ready to be filled with molten metal, each recessed letter must be in effect an island, within its own boundaries; one letter cannot overhang the next character or space, or intrude into its rectangle. In short, there can be no kerns. A close fit is difficult, as the side-walls of each letter need some thickness for strength to withstand repeated casting. Most roman letters present a tolerably conventional appearance without kerns, but once the 'button-hook' shape of a line-cast f has been noticed, it seems to intrude whenever it appears. In traditional italics, kerns are obligatory; *f* and *j* are examples. For these and other reasons, sloped romans are better suited to line-casting than italics.

Different models of Linotype machine have different matrix capacities in the magazines mounted at one time. The standard magazine of 90 channels, for founts from $4\frac{3}{4}$-point to 18 and sometimes 24-point, is controlled by the main keyboard of 90 keys; some models have an additional side keyboard of 34 keys, controlling an auxiliary magazine of 34 channels which is normally used for display alphabets. Other makes of Linotype, usually found only in newspaper printing, can carry additional main and auxiliary magazines, but cannot mix type from all of them in the same line.

Text matrices for line-casting are usually *duplex*; each has two characters punched into it, one above the other, and either of these can be used for casting, by means of a *shift* or *function* key which in this instance adjusts the position of the matrix in the assembler box but provides no access to the magazine. A common arrangement is to have each of the roman letters paired with the same letter in the italic alphabet, and small capitals paired with ligatures, figures, and punctuation. Line-cast italics, matching the set of roman, tend accordingly to be loosely fitted or widely drawn, losing some of the contrast natural to italic, and the small capitals do not always appear well-suited by their set. The standard magazine of 90 channels therefore offers 180 characters, and would not for example have room for the characters in the fount shown in figure 4-4*a*. Single matrices can however

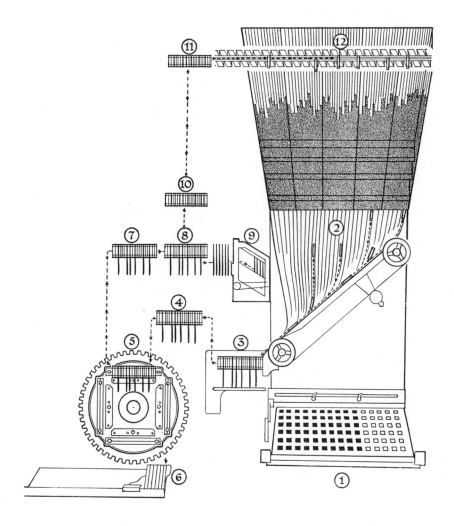

Figure 5-1*b*. A diagram of the Linotype composing and casting machine. 1: The keyboard; each key, when pressed, releases a matrix from one of the channels in the magazine above. 2: Matrices leaving the magazine. 3: The matrices are assembled in order, with space-bands from 9 above. 4: The composed line of matrices and space-bands is automatically transferred to the mould. 5: The line is presented to one of the four moulds in the mould wheel. The space-bands are pressed upwards until the line exactly fills the measure, and metal is injected into the matrices. 6: The cast slug is trimmed and ejected into the galley. 7 and 8: The matrix line is removed upwards from the mould, and the space-bands are separated from the matrices. 9: The space-bands are collected ready for further use. 10, 11, and 12: Each matrix passes along the distributor bar until released by its combination of teeth at the appropriate point, where it falls into its channel in the magazine.

be placed in the assembler box by hand, at the expense of keyboard speed. Three of the channels provide fixed spaces — thin, nut, and mutton — and fixed spaces of $\frac{1}{2}$, 1, and 2 points are available for hand insertion.

When the operator actuates the casting sequence, the line is first justified by upwards pressure on the lower ends of the space-bands, which extend below the line of matrices. Under this pressure, each band widens until the line is justified. The extra thin space-band preferable in bookwork expands from a set of 2 points to $5\frac{1}{2}$ points; a 2-point space is the equivalent of a thin in a 12-point fount but a thick in 6-point.

After justification, the line is automatically transported to the oblong orifice of the mould. Each line-casting mould is fixed in dimensions, as wide as the measure and as deep as the body. The widest mould on a standard Linotype offers a measure of 30 picas, which is not always enough for quarto books, but special models extend the measure to 36 picas. Molten type-metal is injected through the mould into the matrices. For casting a whole line at a time in this way, the alloy is rather softer than that in which single types are cast, and is more vulnerable to local damage during make-up and to wear during make-ready and printing.

The line is automatically trimmed, and is ejected into the galley. At the same time, the matrices are raised to the distributor bar above the magazine; they run along it, and drop off each into its own channel. The space-bands are returned to their separate container (figure 5-1b).

---

4. Interpretation of the Parable of the Tares (13:36-43)

36. Then he left the crowds: The interpretation is addressed to his disciples; so are the parables which follow.

37. Son of man in the apocalyptic sense.

38. For Matthew and his church the missionary **field** is the entire **world** (as in 28: 19). A few Jewish parables compare the world to a garden. **Sons of the evil one** is the harsh judgment passed on the Jews in John 8:41, 44. Does Matthew perhaps believe that some people are inherently evil?

39. Συντέλεια αἰῶνος (or τοῦ αἰῶνος) is a favorite phrase of Matthew's; in fact these are the last words in the Gospel (28:20). The Hebrew word 'ôlām (עולם) can mean **world** or **age.** Its **end** or consummation is the beginning of the age or world to come. Similar expressions are found in Dan. 12:4, 13; Test. Levi 10:2; II Baruch 13:3; 27:15; etc.

41. A distinction is drawn between the **kingdom** of the **Son of man,** i.e., the church, and the Father's kingdom (vs. 43). Here the **angels** accompany the Son of man, as they accompany God when he comes to judgment in Enoch 1:3-9. Jesus therefore is given the prerogatives of the Father. Men are spoken of as "scandals" or **causes of sin,** as in 16:23, where RSV translates the word as "hindrance" (cf. on 11:6). **Evil-doers,** literally

---

Figure 5-1c. Linotype setting in a ten-alphabet fount, which includes roman and italic upper and lower-case, small capitals, bold upper and lower-case, Greek upper and lower-case, Hebrew, two sets of figures, and accented sorts in italic. Part of a page designed by George Salter and printed in the United States. Reduced to 68 per cent.

The characters on the slug cannot be altered in any way; the line has to be completely re-set if there is a single correction. The operator is in danger of another mistake when he does this, but he can read the matrices in the assembler box before justification.

Linotype systems were well suited to the newspaper market. Mixer models provided a variety of alphabets for text variations and display in the newspaper page, and a measure of 30 picas was usually more than wide enough for such composition. The slugs were convenient to handle during make-up. The comparatively soft alloy was not subjected to the wear of printing, as newspapers were rarely printed from the composed lines. Weaknesses of type design, inherent in line-casting, are insignificant in ephemeral printing. Book printers in the United States, where books were commonly printed from *plates* (§ 11-4), demonstrated their ability to undertake tabular composition and other difficult work on slugs, but line-casting seems to have contributed to a minority of British editions.

§ 5-2 Type-casting

Until the 1970s, most book printers in Britain relied on the Monotype system, making good use of the advantages of separate and movable types by combining mechanical type-setting with the traditional skills of the composing-room. In this system, the keyboard is separate from the type-caster, and controls it only indirectly, by means of punched holes in a comparatively wide spool of paper. Introduced in the nineteenth century, this was an early example of punched tape in industrial use, though the 31-channel width of the spool is very different from the compact 6 or 8-channel tape of today. The *channels* are imaginary lines along the length of the spool, and every punch is struck into one of them. As the spool is punched, it is automatically wound into a detachable roll. When the finished spool is removed from the keyboard, the last-made of the perforations is on the outside of the roll. In due course the roll is fixed on the caster and drawn through it.

The keyboard is big enough to contain one key for every character and fixed space in the fount. The operator cannot read the composition as he sets it, but characters can be identified from the holes punched in the spool.

The matrices are contained in a rectangular frame known as a *matrix-case* (figure 5-2a). Matrices can be removed from the matrix-case and rearranged within the limits imposed by the capacity of the case for characters of differing set. The illustration shows a matrix-case with 15 horizontal unit rows and 17 vertical rows of matrix positions, and a total of 255 positions; other sizes of matrix-case have 225 or 272 positions. Each unit row corresponds to a single setting of the mould, so the characters in it

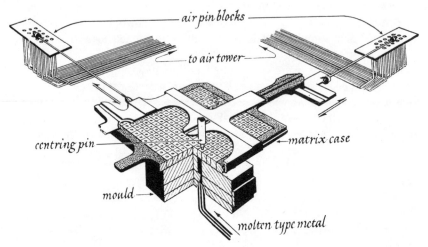

Figure 5-2a. Part of the Monotype caster. Each key on the keyboard which represents a letter causes two holes to be punched in a spool of paper. When the spool is run through the caster, compressed air passing through one of these holes causes a stop-pin in an air pin block to rise.

This pin controls the length of travel of the draw-rod, and so determines which row in the matrix-case is to be aligned under the centring pin and over the mould orifice.

The set of the mould orifice is adjusted according to which unit-row of the matrix-case is selected for casting. In the matrix-case (below), all the matrices in any one row have the same set.

Air passing through the other hole in the spool controls the other air pin block and draw-rod, and so determines which matrix in the row is to be presented to the mould.

Below, the matrix-case, with the narrowest letters in the top row and the widest at the bottom.

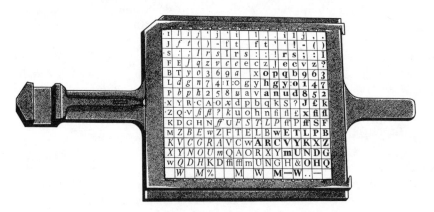

are all cast on a shank of the same set. A typical pattern of units of set for the fifteen rows is 5, 6, 7, 8, 9, 9, 9, 10, 10, 11, 12, 13, 14, 15, and 18. The *set of the fount* is the set in points of the widest character, which is usually 18 units wide. The unit values of fixed spaces are 5, 6, 9, and 18 units; blank matrix positions for these can be seen in the appropriate rows in the illustration.

The variable space has a minimum value of 4 units, halfway between a thin space and a mid. At the end of the line, the operator taps the indicated justification keys, which allocate to each of the variable spaces that widened set which will justify the line. The resulting justification punches adjust the spacing wedges in the type-caster before the rest of the line is cast, as the end of the spool is fed first into the caster, and the whole book is cast in reverse order of composition. The wedges in turn adjust the set of the mould for each variable space.

Letter-spacing is carried out by means of the *unit-adding attachment*. This takes time to adjust, and letter-spacing therefore tends to be even throughout each book. Application of the unit-adder increases the unit value of every character and space. Letter-spacing of one unit is usually considered narrow, two-unit medium, and three-unit wide. Among letter-spaced words, word-spacing needs care; in figure 5-2b the word-spaces are approximately equal to the set of N in the capital lines and N in the small capitals, in each instance combined with the letter-space itself. Unit spacing, whether fixed or variable, is proportionate to the body of the fount, in much the same way as fixed spaces in hand composition.

SMALL CAPITALS LETTER-SPACED ONE UNIT

CAPITALS LETTER-SPACED ONE UNIT

SMALL CAPITALS LETTER-SPACED TWO UNITS

CAPITALS LETTER-SPACED TWO UNITS

SMALL CAPITALS LETTER-SPACED THREE UNITS

CAPITALS LETTER-SPACED THREE UNITS

Figure 5-2b. Close, medium, and wide letter-spacing in metal (Monotype Ehrhardt). Compare photo-set letter-spacing in figure 5-7a.

A five-alphabet fount, with upper and lower-case roman and italic, and with small capitals, can be fitted into a matrix-case of 225 positions. The matrix-case in figure 5-2a contains a seven-alphabet fount, with bold upper and lower-case in addition to the usual five alphabets. But even seven alphabets are enough for a variety of texts only because anything else that may be needed can be dropped in by hand. Figure 4-4a, for example, shows a nine-alphabet general-purpose fount with aligning and hanging figures, seven or eight letters with diacritical marks and ten accents, and other sorts not included in figure 5-2a.

Monotype characters and spaces are cast one at a time. The shanks are shaped by the mould. The body dimension of the mould is fixed, so that each mould casts type of one body only. The set in units of each sort is determined by the automatic set adjustment of the mould to each unit-row in the matrix-case. Molten type-metal is injected through the mould into the matrix; controlled by the perforations in the spool, the matrix-case moves to and fro, to present for casting first one matrix and then the next. The letters can be closely fitted, and the set of italic alphabets usually contrasts with that of the mated roman. Kerned letters overhang and rest on adjacent shoulders as in founder's type. The limited number of unit values available for all characters gives rise to no distortion of letter-form. Since only one small piece of type is cast at a time, the metal used in the Monotype caster is hard enough to withstand the preparatory stages of printing and the presswork itself; if a long run from type is expected, extra hard metal can be used.

The composition caster can produce composed type in sizes from $4\frac{1}{4}$ to 14-point. When fitted with a special low-speed gear, it can cast composed type up to 24-point, but only with about 90 matrices in the case. Mould sizes of the smaller bodies are graded to the nearest $\frac{1}{4}$-point, and those of the normal text bodies to the nearest $\frac{1}{2}$-point, but moulds are expensive and not every printer possesses every size. As in Linotype casting, interlinear space can be added by casting on a larger body than that for which the fount was designed, so that instead of placing 2-point leads between lines of 11-point the printer casts 11-point type on a 13-point body. The maximum measure is 60 picas, wide enough for a folio book and for almost any tabular composition which is to be turned sideways on the page.

Display lines are usually set by hand from types cast on the supercaster. This casts type, decorative material, spaces, leads, and so on in sizes up to 72-point, but in sets from one matrix at a time, not in composition order.

By means of repro proofs, metal type can provide camera copy for *offset* reproduction (§ 11-5), and photo-composition can also be converted into a relief image. But metal was chiefly preferred when the book was to be printed from type or *plates* (§ 11-4). The *flat-bed* press (§ 12-4) was ponderous and slow, but at its best it was capable of printing characters which were admirable in profile (though not necessarily faithful to the type drawing), densely black, and consistent in colour and impression throughout the book.

Mechanical typesetting in metal deserves study for three reasons. The first is that at its best the single movable type sets a standard of quality for other processes to match in the future. Secondly, style and technique in

[66]

typographic composition are derived from centuries of metal typography. And the third reason is that typesetting in metal can be seen and heard and, indeed, smelt, and is more easily understood through the senses than are the silent and hygienic mysteries of electronics.

§ 5-3 Typewriters and composers

A *typewriter* may be defined, for purposes of this chapter, as a machine which strikes images of letters and other characters on to paper one at a time from a relief surface through a pigmented ribbon. The term *composer* is used here to indicate a machine or system which satisfies this definition of a typewriter but is capable of producing images and arrangements more similar to those of typography than to those of the simpler kind of typewriter.

UNSHIFT

1234567890 $\frac{1}{2}\frac{2}{3}\frac{3}{4}$ =-;,.

abcdefghijklmnopqrstuvwxyz

SHIFT

*"/@£ &'() $\frac{1}{4}\frac{1}{3}$% +?:,.

ABCDEFGHIJKLMNOPQRSTUVWXYZ

Figure 5-3a. A typical typewriter fount, equivalent to 12-point in body, and to 6 points in set, including 88 characters of which full-point and comma are duplicates. Typed through a wax carbon ribbon, on an electric typewriter.

The product of most typewriters is unlikely ever to become popular with readers. All characters share the same set; accordingly, they are unevenly fitted, and their counters are uneven in width. Lacking stress, for which there is no room in such crowded letters as m, they look like diagrams of letters, but unlike the letters to which readers of books are accustomed. Lines cannot be justified to even length. Interlinear space is either too little or too much; when there is any interlinear space at all, it is rarely less than six points or about half a line. Letter-spacing is not always possible. The set of word-spaces is the same as that of the characters, so that all lines are rather loosely spaced in comparison with typographic lines. Uneconomical interlinear space is necessary to legibility, as the words in any line should appear to be closer to each other than to words in adjacent lines. The fount of some 88 characters is limited to two alphabets (upper and lower-case), eight, nine, or ten arabic figures (since lower-case letter l and upper-case O are sometimes used to represent figures), and at most eighteen other characters including punctuation marks. These are not always those most likely to be useful in a book: some are ambiguous, including the single quotation mark which stands for both opening and closing quotes: and some are duplicated for use with either upper or lower-case. Some typewriters have a limited facility for extraneous sorts. Only one size of character is fitted to each machine. The weight of impression tends to vary from character to character in manually powered typewriters.

Some electric typewriters exhibit fewer deformities. Set may vary from character to character, opening up enough width for stress and for counters of even width, and fitting the letters more evenly together. Justification may be possible, but a preliminary typing of each line is necessary to establish the set of the required word-spaces, and justified matter has to be typed twice. The set of word-spaces and the depth of interlinear spaces are likely to be variable and closely graded. On some machines the operator can detach the whole fount and replace it with another, whether for a single extraneous sort or for a change of alphabet, for instance to italic. On electrically powered machines, impression is comparatively even from character to character.

Even with these advantages, typewriters of this kind cannot be expected to achieve typographic quality, and they cannot approach the versatility of typographic composition. There may however be economic or other reasons for using typescript as copy to be photographed for printing; for example, typewriter characters with a single set may be apt company for a text reproduced from computer printout which is similar in appearance. When typescript is to be reproduced for printing, the designer may have an opportunity to mitigate the asperities of its appearance. A wax-coated ribbon and a smooth hard paper (exactly the same throughout the typescript) can combine to produce a sharp and even impression. Some photographic reduction of the text is often an improvement, and display can be separately typed and enlarged. The arrangement of detail — paragraph indents, spacing between sentences and lines, headings, page numbers, and so on — can be derived from typographic style. Justification is widely preferred in typographic composition but is inessential in most pages, and some evenness in the length of the lines can be gained by conventional word-division.

Optical character recognition (OCR) is being used by some newspapers but is not one of the principal techniques of book production. Few typewriter faces are compatible with the system; the electronic scanning-head identifies either specially designed characters or an individual sign automatically typed with each character. Typographic instructions are

[68]

typed into the copy in the form of codes. Copy can be corrected before scanning, and the printer is likely to dispense with house corrections and proof-reading. Authors generally are unlikely to possess the requisite make of typewriter, to master the typographic conventions and codes, or to spell and punctuate with the printer's accuracy and consistency, so they cannot be expected to provide machine-readable copy, which will usually be re-typed by the printer.

The advantage of all typewriters over all other composition systems is the immediate visibility to the operator of the image which will be reproduced for printing. This enables him to see and correct his work as he types it. When there are variations in the lateral and vertical placing of characters, as in fractional and exponential expressions in mathematics, he can place the characters by eye.

The IBM Electronic Selectric Composer is an example of a type of machine which starts with the additional advantages of electric typewriters, summarized above. The operator corrects his work as he does it; the first typescript will not be used for reproduction, so a wrong character or word need not be deleted from the typescript before being replaced by correct matter typed over it. The corrected text is electronically recorded, and is then automatically *played out* (re-typed by the machine, not by the operator) several pages at a time, at high speed and with automatic justification. Interlinear space is adjustable to the nearest point; the IBM point is one-twelfth of the IBM pica, which is exactly one-sixth of an inch, or a little under 4 per cent more than the typographic pica. There are seven type sizes, from 6 to 12-point, and a variety of type series which include separate founts of italic, mated bold roman and italic, and occasionally other variations, but not small capitals. The set of each bold and italic character is the same as that of the same character in the roman alphabet, so italic letters are rather wider and bold letters rather narrower than in Monotype. For the seven sizes of type, there are only three sets of the fount. In IBM Century, for instance, 6-point has the narrow set: 8 and 9-point have the same medium set as each other: and 10 and 11-point are wide-set (figure 5-3b). These limits on variety of set have their effect on the type designs, and different founts and alphabets in one series may differ from each other in letter form and fit, or may be very similar in apparent size. Founts, each of 88 characters, are changed manually when necessary during operation, and when founts of different sizes are used side by side — as when the capitals of a smaller fount are used to replace small capitals — they align with each other on the base line.

Well-produced composer work, set in carefully selected series and founts, can be all but indistinguishable from typographic composition, at least to the layman, but it cannot be as good as the best composition. The effect of the difference on the reader's preference as a buyer and his experience in the

[69]

(C-11-M) ABCDEFGHIJKLMNOPQRSTUVWXYZabcdefghijklmnopq
rstuvwxyz1234567890£.,-'':;!?*½¼¾—()[ ]=†/+%&@   (C-11-I) *ABC*
DEFGHIJKLMNOPQRSTUVWXYZabcdefghijklmnopqrstuvwxyz123456
*7890£.,-'':;!?*½¼¾—()[]=†/+%&@*   (C-11-B) **ABCDEFGHIJKLMNOP**
QRSTUVWXYZabcdefghijklmnopqrstuvwxyz1234567890£.,-'':;!?*½¼¾

6-POINT

The short extending strokes of this design provide it with a large apparent
size; the 8-point, for example, looks just about as big as 9-point Baskerville.
There are five sizes of Century: 6, 8, 9, 10, and 11-point, *each with italic
and bold.* The 9 and 11-point founts are not as well-rounded as the others,

8-POINT

The short extending strokes of this design provide it with a large apparent
size; the 8-point, for example, looks just about as big as 9-point Baskerville.
There are five sizes of Century: 6, 8, 9, 10, and 11-point, *each with italic
and bold.* The 9 and 11-point founts are not as well-rounded as the others,

9-POINT

The short extending strokes of this design provide it with a large apparent
size; the 8-point, for example, looks just about as big as 9-point Baskerville.
There are five sizes of Century: 6, 8, 9, 10, and 11-point, *each with italic
and bold.* The 9 and 11-point founts are not as well-rounded as the others,

10-POINT

The short extending strokes of this design provide it with a large apparent
size; the 8-point, for example, looks just about as big as 9-point Baskerville. **There are five sizes of Century:** 6, 8, 9, 10, and 11-point, *each with
italic and bold.* The 9 and 11-point founts are not as well-rounded as the

11-POINT

The short extending strokes of this design provide it with a large apparent
size; the 8-point, for example, looks just about as big as 9-point Baskerville. **There are five sizes of Century:** 6, 8, 9, 10, and 11-point, *each with
italic and bold.* The 9 and 11-point founts are not as well-rounded as the

Figure 5-3*b*. The IBM Century series of five sizes, each including italic and bold. In
the 11-point synopsis at the top, the short titles in parentheses indicate Century
11-point Medium (C-11-M) and so on. The specimen settings are all leaded 2 points,
and are set by IBM Composer in three measures for the three escapements. Reduced
to 90 per cent.

[70]

act of reading cannot easily be assessed. The designer's task is always to seek the best, and composer composition is better described as adequate for most purposes.

Since he can see and measure his work as he sets it, the operator of any typewriter or composer can type certain kinds of text in page; verse is an example — make-up can be planned on the copy, since the lines to appear on each page of the edition can be counted. Most work however is typed in galley and cut between lines for pasting up in page form. For this reason typewriter and composer work should usually have not less than two points of interlinear space, so that no extending strokes are abbreviated by the knife when adjacent lines are separated. Cutting of this kind is also necessitated by corrections and alterations in proof, when a whole line is usually re-typed and pasted on.

§ 5-4 Photo-composition matrices

The design of letters and other characters for printing from metal type has always allowed for the effect of *squash* (§ 11-1), which thickens all parts of the image to a degree determined by the pressure between paper and metal, the nature and quantity of ink, and the characteristics of the paper. Squash is peculiar to printing from a relief surface; it renders the fine details of a metal type-face stronger and more visible, even when printed on a hard smooth paper. Photo-composition founts are rarely printed from a relief surface, and some of them are deprived of the advantage of squash. They seem to have been adapted from type drawings (§ 4-1) designed for metal type, rather than from the image printed from such type. They tend accordingly to appear too thin and sharp for comfort in reading.

Other kinds of adaptation are necessary in the design of characters for photo-composition, which is normally printed from a plane surface. The necessity for these adaptations can be identified at various points in the rest of this chapter. The most conspicuous of such adaptations is caused by the difference in function between the metal matrix and the photo-matrix. In metal, a change of size in any character necessitates a change of matrix: but any single photo-matrix may be used to generate an image in any size. Variations in design to match gradations of size are preferable (§ 4-1), but in composition from photo-matrices they are normally dispensed with for economic reasons. The product of the photo-matrix is therefore likely to be different from that of the metal matrix; even if in 12-point the two images are similar, there will be differences in 8-point and 18-point.

A *photo-matrix* consists of a transparent image on an opaque background, usually developed on film. When any system offers a facility for the insertion of matrices for *pie* sorts (those extraneous to the usual fount), and most bookwork systems do, the printer may be able to make his own matrices when necessary. His doing so may however offer, to the typographically perceptive eye, evidence that type drawing is a specialist skill, and that its requirements have not so far been part of the stock-in-trade of the general run of printers.

Figure 5-4a. Monophoto 400/8 photo-matrix case with 400 separate and exchangeable matrices.

Photo-matrices are small and light on the whole, and most photo-composition systems can be prepared for setting an edition with a substantially larger fount than the largest Monotype case of 275 matrix positions. This is just as well, since fount changes by hand during setting at photo-composition speeds are an expensive interruption of progress. Nearly all newspapers and magazines and a majority of books could probably be set from a standard fount of some 500 characters, but a minority of books and learned journals would require differently constituted founts of the same size. Some publications require a larger fount; figure 5-1c shows a ten-alphabet fount in action, and any post-graduate work of pure mathematics may require as many sorts again, and even more. A Monotype system, including the handwork of compositors on the frames, is capable of any kind of composition ever undertaken in metal type; a photo-composition system which is economic for novels may be incapable of mathematical expressions, phonetics, *exotic* alphabets (such as Greek) interspersed with text in latin characters, and even of chemical expressions with horizontal bonds and inferior figures.

No fount of photo-matrices for book composition seems so far to be large enough or variable enough to match the visual quality of metal typography at all points. Such a fount would have to be based on at least three kinds of matrix for most of the characters — one for display, one for the main text, and one for small type — in order to compensate for the optical effect of variation in typographic size. Such variant matrices have been designed, but much of the economic advantage of the photo-matrix over metal lies in the use of a single set of matrices which has not enough matrix positions for variants. The problem is one of space and movement; within existing systems, a fount of several thousand photo-matrices cannot economically be accommodated, nor could each be moved into position quickly enough when required.

The same problem imposes a kind of poverty on many photo-composition founts. Evidence of this can be seen in various kinds of typographic eccentricity. A shortage of logotypes is usually an early symptom of matrix-poverty. Electronically slanted roman capitals instead of italic capitals indicate serious deprivation; traditional italic lower-case letters are from about 3 to 30 per cent thinner than roman in *waist-line* (the lateral thickness of the main vertical stroke). If they are to be mated with slanted roman capitals, they either lose this contrasting lightness of design, or they marry incompatible partners. A fount which relies on electronically diminished roman capitals, in place of small capitals, is hardly suited to the exacting requirements of a well-produced book.

Photo-composition founts are designed to cope with composition methods developed over centuries for the single movable type. A fount of photo-matrices is in competition not with the limited capacity of the

Monotype matrix-case but with the ample resources of the composing-room and its compositors. A system such as the early Pacesetters, in which the entire fount of matrices was part of a glass disc, could be stultified by the repetitive need for a single character of an unexpected kind. The variety of books, and even of book-printers, is such that every system needs its founts to include a section of initial choice by the printer and another section of subsequent choice of special book-by-book requirements. A system in which a single alternative character causes the exchange of a whole segment of the fount containing a number of characters may prove inadequately flexible for certain kinds of bookwork.

Digital matrices, the outcome of development in computers rather than in photo-composition, are programs recorded in electronic memory. The program may be compared with an image expressed in co-ordinates on a graph. Marked co-ordinates combine to form the image, as a pattern of dots blending into each other (figure 5-4b), or as outlines to be filled in by progressive linear scanning switched on at entry into the image and switched off at departure.

Figure 5-4b. Diagrammatic reconstruction of a letter in digital form. In practice the diameter and spacing of the dots would combine to produce a solid image. Reproduced from a digitizing screen.

The digital matrix occupies part of a capacious memory, but takes up no other kind of space, and makes no mechanical movement. Like computers, some photo-setters are networks rather than machines; the Monotype Lasercomp, for example, has only one moving part other than the input tape reader and the output film roller. When each size of every character is generated from a different matrix, the matrix capacity of such a photo-setter may be described as a series rather than as a fount. A standard fount of 500 characters, in a series of ten sizes of type image from 6 to 24-point, is likely to be enough for many editions, and a digital series is likely to be very much larger than this. There should be no matrix-poverty here. Every character could be designed for the size in which it is to be reproduced, if machinery manufacturers did not seem to have lost the will or perhaps the skill to do so. Access to every matrix is instant; in some photo-matrix systems, repeated access to certain matrices may drastically reduce the speed of composition.

Floating accents are not so much a symptom of inadequacy in photo-composition as an advantage. In figure 4-4*e*, accent and character are cast together from a single matrix, although floating accents are sometimes used in metal. A floating accent is set separately from the letter to which it is applied, and can be set with any other equally accommodating letter or even by itself. The letter i is less accommodating than other vowels because of its dot, and a dotless ı usually accompanies floating accents. Since the position of any such accent, in relation to the letter above or below it, is allocated during composition rather than in the type-drawing office, that position should be checked by the designer in any specimen setting and at random points in any proof. Unless otherwise indicated, the lower point of a diagonal accent and the centre of any other accent should appear to be centred on the letter to which it refers. Figures 4-4*a* and *b* provide examples of floating accents and also of their misplacement.

§ 5-5 Photo-composition by flash

The first generation of photo-composition systems is based on principles derived from photography. Within the photo-setter, a photo-matrix is moved into position in front of a light-source, and a lens for enlargement or reduction is moved into position in front of the matrix. A flash of light passes through the transparent image on the matrix and through the lens to reproduce the matrix image, enlarged or reduced, on a sensitized *output* surface which may be transparent film or opaque paper. A lateral adjustment of relative position, whether of lens or prism, graded according to the set of each character, directs the next image to its position beside its predecessor (figure 5-4*a*).

A photo-setter of this kind is likely to be at least four times as fast as a

Monotype caster. Its machinery however is by comparison delicate, complex, and precise, rather than simple and hard-wearing, and such a speed ratio is only just competitive. At most, flash systems seem to be capable of about ten times the speed of metal casting, or something over 35 *cps* (characters per second). This may prove to be near the maximum speed of reciprocal movement and of photographic exposure. Some degradation of photographic image may begin to occur when the whole image at once is exposed for less than 0.03 of a second. If this is so, flash may still compete effectively with metal-casting, but not with succeeding generations of photo-setters.

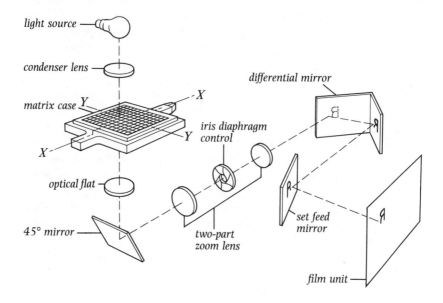

Figure 5-5. Working method of the Monophoto 400/8. The matrix-case moves X to X and Y to Y to bring the selected matrix into line with the stationary light-source. The optical flat adjusts the character's vertical alignment. The zoom lens governs type size. The set-feed mirror places the characters in the line. Working speed about 11 characters per second.

In terms of quality, however, the principle of a single flash through each matrix has shown itself to be in some ways the best so far. One of the criteria is the precision of the image profile or outline, not only as seen by the reader but as examined by the typographer, with a magnifying glass and in comparison with the best available print from a similar type-face in metal. A flash through a stationary matrix on to *positive* film (§ 11-6) is capable of an exact and faithful reproduction of any matrix image, and tends to maintain this standard of image quality evenly throughout any book.

§ 5-6 Cathode-ray tube and digital systems

Speeds several times greater than that of any flash system are achieved by photo-setters of the second generation, based on principles utilized in television receivers. The matrix image is not reproduced as a whole, but is analysed into a pattern of parallel lines by scanning. This circumvents the length of exposure needed by a larger image area. In a *cathode-ray tube* photo-setter (CRT), an electron beam thinner than a needle vibrates up and down the matrix as it advances across it, at electro-magnetically induced speed. The scanned lines, which pass through the matrix image and combine to correspond to its shape, are registered in a *photo-multiplier*. This device can modify the image in various ways, before projecting a corresponding though possibly modified pattern of parallel lines which compile the latent image of a character on the output.

In comparison with that in any flash system, the incidence of matrix movement is reduced. The electron beam, unlike a beam of light, can be and is magnetically deflected horizontally and vertically, far enough to scan a set of matrices without any mechanical movement. This provides, for example, enough primary matrix positions for roman and italic upper and lower-case with ligatures, roman punctuation, and a set of arabic figures. All other characters are contained in secondary sets of matrices, which have to be moved into position by mechanical means. This comparatively slow movement disproportionately interrupts the rapidity of composition, and output speeds normal in newspapers and fiction are not to be expected in more varied texts. Any symptoms of matrix poverty in a CRT system are likely to be due to the limited number of primary positions.

Having compiled an image on the output, the electron beam continues to the space to be occupied by the next, and cannot return to add a kern under the completed image. In the same way, if a matrix were to be designed with an overhanging kern on the right of the letter, the beam cannot return from the outer edge of the kern to insert a character under it. The issue has been confused by electronic engineers who use the term *kerning* to mean what printers would recognize as adjustment to typographic fit. The image modifications of the photo-multiplier, however, include the application of slant, and italic characters in CRT systems may show a vestigial kern.

In CRT and digital systems the image profile is imperfect. The imperfection may not be noticeable, or even easy to find, but the type-designer would have known it could appear, and may have distorted the letter-forms slightly to minimize its effect. Even so, the strokes of the image may all be slightly thinner or thicker than those in the original design. Other evidence of the scanning effect, known as *indexing*, shows itself in jagged edges on diagonals and curves, and in uneven thickness in straight strokes. Verging

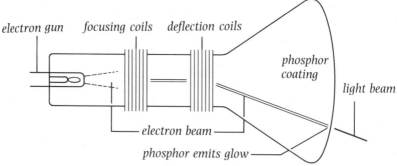

Figure 5-6*a*. Above, the Linotron 303 matrix-grid with 144 characters; the machine can be fitted with up to 24 such grids, offering a maximum of 3,456 matrices. Below, cathode-ray tube in a photo-typesetter.

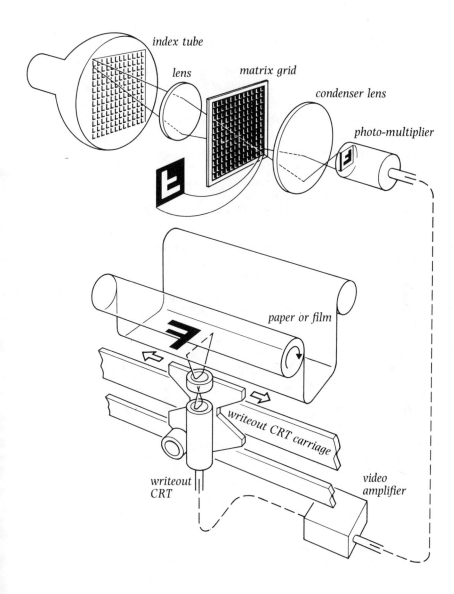

Figure 5-6*b*. The working method of the Linotron 303 photo-typesetter. The scan-generating cathode-ray tube (CRT) or index tube generates scan lines in any one of 144 positions corresponding to individual characters on the grid. Passing into the photo-multiplier, the scan lines generate a video signal which is amplified by the video amplifier. The signals activate the write-out CRT which moves along its carriage to place the characters correctly in the line. Working speed is equivalent to about 100 cps.

[79]

on the imperceptible in the best conditions, indexing may appear as little more than a fractional irregularity or softness of edge.

The frequency of output scan lines, or *resolution*, is usually between 1000 and 1500 lines to the inch. The quality of profile, however, is determined by the number of lines to the typographic body of the character. Holland (1967) has analysed the resolution needed for reasonable clarity —

Somewhere between 100 and 150 lines per character body seem to resolve typography reasonably well. The ultimate threshold of mere identification of alphabet characters should not be deduced from a's and b's. The important and essential conditions are those where differentiation must be made between an inch mark and a double quote mark, between an apostrophe and an acute accent, between an italic and roman comma or full point. . . . One thousand lines per inch is about 150 lines per character in 12 point — it is 75 in 6 point.

A resolution of 150 lines in 12-point is approximately 900 to the inch. The body of the matrix image, and the scale of optical enlargement or reduction if any, must therefore be known before the probable effect of indexing can be assessed. The designer who means to make sure the profile will be good enough for a specific edition will have to undertake a close and thorough examination of printed specimens which include a variety of type-sizes.

Figure 5-6c. A scanning experiment (Holland, 1967). In each of the three groups ÝjÝj, the first two characters are scanned vertically as in a CRT system, and the third and fourth are scanned horizontally as by the Lasercomp. The scanning resolution applied to the groups from left to right is 72, 144, and 288 lines per character body. The profiles in the three groups represent differential enlargements of 7, 14, and 28-point type experimentally scanned at 1,000 lines to the inch.

As Holland points out, however, 1000 lines per inch is 75 lines per 6-point body, and indexing at this resolution is likely to be visible in small type on smooth paper. The finest resolution available at the time of writing is the 5300 lines per inch of the Alphatype CRS Photosetter, in whose output indexing is invisible. Probably 2000 lines per inch would be fine enough, and is not too ambitious a hope for the future.

The grain of the *offset plate* (§ 11-6) and the fibrous surface of most book papers tend to impart a slight irregularity to the printed image. The imperfections of a profile scanned with a reasonably fine resolution such as 150 lines to the character body tend to be hidden among the irregularities natural to such printing.

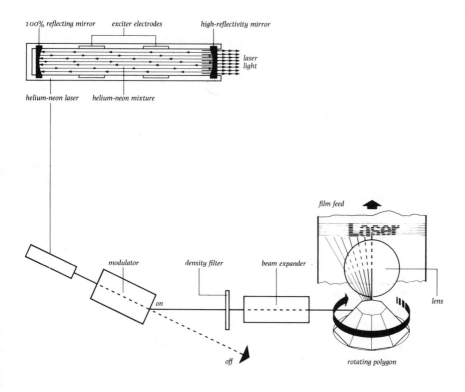

Figure 5-6d. Working method of the Monotype Lasercomp. Above, the helium-neon laser (light amplification by stimulated emission of radiation). The light is intensified by electrodes during reflection to and fro between mirrors, until it escapes through one of the mirrors which is designed to permit such emission.

Below, the path of the laser beam, magnetically deflected into the system by the modulator to generate an image, or passing straight through to leave its target area blank. The beam expander increases the precision of reduction focus. Reaching one facet of the revolving polygon, the beam is aimed at the beginning of the line of typographic images, and is focused by the lens to a diameter of 0.001 inch. The next segment of deflected beam reaches the same facet after enough rotation to place its image dot to the right of and touching the first. This process continues, interrupted when the undeflected beam leaves white spaces between images, across the full width-capacity of the machine. The next scan begins at the left from the next facet, after the film or paper has advanced upwards by 0.001 inch.

In figure 5-6c the horizontal scan demonstrates an inadequate reproduction of the serifs of Y, even at 288 lines to the character body, but this could probably be improved by editing the digital matrix. The effect of scanning in either direction at 72 lines is probably more dire in the experiment than it would be in practice, and there is not much to choose between the two directions.

The advantage of the horizontal scan lies not in the profile but in the kern. A CRT system can imitate kerns by relying on a numerous array of logotypes, or can generate kerned letters with a duplicate scan which reduces the speed of output. The Lasercomp has no more difficulty with kerns than its predecessors in the Monophoto range or its ancestor the Monotype metal system, and all Lasercomp type-faces are originally designed accordingly.

§ 5-7 Units and spaces

A unit value is assigned to the typographic set of every character in all photo-composition as in a Monotype system. In a Monotype matrix-case (§ 5-2), unit rows provide about twelve unit values to be shared by all spaces and characters. Photo-composition lends itself to a wider variety of unit values, and benefits from a much smaller unit. The Monotype Corporation, for example, which introduced the 18-unit em for metal composition, and Mergenthaler-Linotype which followed suit, now rely on 96 and 54-unit ems respectively for photo-composition.

Adjustments to the fit of metal characters are usually possible before Monotype casting. For example, when a matrix is transferred from the 5-unit row to the 7-unit row, the extra space of 2 units appears on the character's left in the printed result. Figure 7-1 shows that this facility can be useful, in separating punctuation a little way from adjacent characters. On the whole, however, the best set for Monotype characters is that for which the characters were designed; each set is appropriate for the size in which the character is to appear. Type drawings for photo-matrices are usually designed for reproduction in a range of sizes, and if the smaller sizes are not to look cramped, the set of the fount may be a little too wide for the 10 to 12-point group, and a good deal too wide for display. In some photo-composition systems, set can be reduced in selected sizes, and the loose fit of the larger sizes closed up a little. When the unit is as narrow as 96 to the em, such adjustments can be minute, but can still improve the appearance of the page without causing any letters to touch each other.

The vertical position of any photo-composed character is even more reliable than in metal type, but lateral positions in some systems of photo-composition are more likely to depart from the visual centre between adjacent characters, even after correct placing earlier in the text. The equivalent of a mortise is practicable in photo-composition, and text can be arranged round initials as in figure 4-2c, but the necessary planning can consume time and is best reserved for such occasional items of display.

Most photo-composition systems offer access to fixed spaces of any unit value down to a single unit, though only a few variations can be provided by single key-strokes. This is an advantage over the Monotype system, in which fixed spaces of less than 5 units out of 18 would be useful, for example in the space between groups of figures in a large number such as 1 000 000 set in the continental style. For such purposes one-sixth of the em would be enough, but separate spaces of one-ninth and one-eighteenth, instead of an even number of units, would improve the apparent evenness of unit-added letter-spacing. Letter-spacing by unit-adding, fixed in set before the letter-spaced words are composed, is common to Monotype and to photo-composition, but the 96-unit em offers the typographer a more exact choice of space. In figure 5-2b close letter-spaces are $\frac{1}{18}$ of the em or about 5.6 per cent of the em; medium letter-spaces are $\frac{2}{18}$ or about 11.1 per cent; and wide letter-spaces are $\frac{3}{18}$ or about 16.7 per cent. Medium letter-spaces of about 11.1 per cent of the em are familiar enough in text, but to some eyes are a little wide for display. In figure 5-7a the close letter-spaces are $\frac{6}{96}$ of the em or about 6.3 per cent. Between words as closely letter-spaced as this, a thin space, augmented by the letter-space, may prove enough. In figure 5-7a, the word-spaces are $\frac{24}{96}$ units augmented by the 6-unit letter-spaces to 30 units or about 31.3 per cent of the em.

SMALL CAPITALS CLOSE LETTER-SPACED

CAPITALS CLOSE LETTER-SPACED

Figure 5-7a. Close letter-spacing in photo-composition (Lasercomp Ehrhardt). Compare metal letter-spacing in figure 5-2b.

In metal type, display-sizes of capitals are letter-spaced by hand, to an average width governed either by the designer's instructions or by the rule of the house, but preferably with spacing differentials according to the profiles of adjacent characters. Such combinations as MN, for instance, in which the adjacent strokes of the two letters are close to each other along the whole length, need more letter-space than AV in which the adjacent strokes are separated by space already. Good letter-spacing in metal display appears to be even and is in fact uneven; in photo-composition, when the operator cannot see the actual letters as he spaces them, even letter-spacing conversely appears to be uneven.

[83]

Variable spaces can be regulated in photo-composition as in metal. The usual minimum variable space in Monotype is four units, but by means of a special attachment this can be reduced to three. Use of this attachment demonstrates that word-spaces of three units (or one-sixth of the em) are identifiable by the reader, and provide a page of agreeably close texture, but bring some of the words perilously close to collision on each side of a space approximating to 16.7 per cent of the em. Probably a minimum variable space of about one-fifth of the em is preferable, and this book has been set with a minimum variable of 20/96 of the em or about 20.8 per cent. This is not too narrow for so closely-fitted a fount as the Photina used here, but a more loosely-fitted fount may need more space, if the narrowest of word-spaces is to be easily distinguished from the fitting of the letters within the words.

In an emergency, a Monotype keyboard operator or compositor may well use word-spaces wider than a nut, to avoid breaking a difficult word, or to make an extra line of text where one is needed for make-up purposes. He is most likely to do this when adjacent lines are closely spaced; if three loosely-spaced lines coincide, the text may be interrupted by rivers. When justification decisions are taken electronically, in the absence of human imagination and judgement, avoidable irregularities are likely to occur, to a degree at which even an imperceptive reader may be distracted. The options of the machine should therefore be more limited than those of the man. Too narrow a maximum will cause too many words to be divided at line-ends. The best maximum variable space for any book is proportionate to the average number of word-spaces in its lines; for most purposes, the equivalent of a nut will serve well enough, but in double-column work something wider is more likely to prevent a rash of hyphens breaking out down the right-hand edge of the text.

Some photo-composition systems are capable of reducing excessive word-spaces by a more or less discreet application of letter-spacing within words in lower-case. In some languages, lower-case letters are spaced for emphasis, and Bernard Shaw introduced this mode into his printing in order to keep italic out of it. But lower-case letters look their best, and are most easily read, when they fit quite closely and very evenly together. As long as this tradition remains the popular preference, uneven fit in lower-case letters is likely to be considered to be a typographic blemish.

In metal type-setting, interlinear space can be cast with the type, as when 11-point type is cast on a $12\frac{1}{2}$-point body. But this facility is limited by the sizes of moulds held by the printer; if he has no $12\frac{1}{2}$-point mould, and many printers do not, he will have to cast on 11-point body and insert $1\frac{1}{2}$-point leads, whether by means of an automatic leading attachment or by hand. Almost any photo-setter can add $\frac{1}{2}$-point spaces between lines, or multiples of $\frac{1}{2}$-point, and some can add $\frac{1}{4}$-point or less. The $\frac{1}{4}$-point of interlinear space

EXAMPLES OF TYPOGRAPHIC SPACING

The outlook for typography is as good as ever it was—and much the same. Its future depends largely on the knowledge and taste of educated men. For a printer there are two camps, and only two, to be in: one, the camp of things as they are; the other, that of things as they should be. The first camp is on a level and extensive plain, and many eminently respectable persons lead lives of comfort therein; the sport is, however, inferior! The other camp is more interesting. Though on an inconvenient hill, it commands a wide view of typography, and in it are the class that help on sound taste in printing, because they are willing to make sacrifices for it. This group is small, accomplishes little comparatively, but has the one saving grace of honest endeavour—*it tries.*   DANIEL BERKELEY UPDIKE

The outlook for typography is as good as ever it was—and much the same. Its future depends largely on the knowledge and taste of educated men. For a printer there are two camps, and only two, to be in: one, the camp of things as they are; the other, that of things as they should be. The first camp is on a level and extensive plain, and many eminently respectable persons lead lives of comfort therein; the sport is, however, inferior! The other camp is more interesting. Though on an inconvenient hill, it commands a wide view of typography, and in it are the class that help on sound taste in printing, because they are willing to make sacrifices for it. This group is small, accomplishes little comparatively, but has the one saving grace of honest endeavour—*it tries.* DANIEL BERKELEY UPDIKE

The outlook for typography is as good as ever it was—and much the same. Its future depends largely on the knowledge and taste of educated men. For a printer there are two camps, and only two, to be in: one, the camp of things as they are; the other, that of things as they should be. The first camp is on a level and extensive plain, and many eminently respectable persons lead lives of comfort therein; the sport is, however, inferior! The other camp is more interesting. Though on an inconvenient hill, it commands a wide view of typography, and in it are the class that help on sound taste in printing, because they are willing to make sacrifices for it. This group is small, accomplishes little comparatively, but has the one saving grace of honest endeavour—*it tries.*

DANIEL BERKELEY UPDIKE

Figure 5-7b. Three specimens of composition in 10-point Lasercomp Photina. From the top, the minimum variable word-space is 16/96, 20/96, and 24/96 units. The leading is 2, 2½, and 3 points. The letterspacing is 6/96, 12/96, and 18/96 units.

is more useful than conspicuous; in a page of about 40 lines, a reduction of $\frac{1}{4}$-point in the interlinear space provides room for one additional line in much the same type area. A more visible amount of interlinear space, from 3 to 6 points, is often useful between printed items of several lines; in metal, the printer can then only insert leads by hand, and any quantity of such work will be less expensive if photo-composed.

§ 5-8 Body and measure

The body of a metal sort is the distance from front to back of the shank. The body of a metal fount is that on which all the characters and spaces are cast, but the body for which those characters were designed may also be known as the body of the fount, and is normally used to identify the fount within its series. In photo-composition there is neither type nor shank, and the fount may be said to consist of matrices only. But for purposes of identification and measurement, typographic images of different sizes are usually graded in *points*, identified by *body*, and known as *founts*, as though they were metal types. Each size may in fact represent a certain scale of reduction or enlargement from a single set of photo-matrices; a 12-point image, for example, may be enlarged to 150 per cent from an 8-point matrix.

Among printers who have worked with metal type, interlinear space in photo-composition tends to be described as though it were metal. These words are set in 10 on 12-point Lasercomp Photina, or 10-point *line-spaced* (spaced between the lines) 2 points. Other terms such as *line-feed* (for *body*) and *line increment* (for interlinear space) are also in use but may not be clear to some printers.

The body for which characters are designed is usually one of the smallest on which it can be set. The *gauge* of the characters is the vertical distance from the top of the longest ascender to the bottom of the longest descender. If a fount with a gauge equal to its body is set without added interlinear space, ascenders will occasionally touch descenders in the line above; if the body is less than the gauge, the extenders collide or merge. Some founts, however, can be set on bodies slightly smaller than those for which they were designed; this is advisable only when the pages will be made up mechanically, as extenders would be vulnerable when the lines are manually separated by cutting.

The range and gradation of photo-composition founts in any system tend to be wider and finer than in metal-casters. When different sizes are derived from different lenses, fewer sizes may be available owing to the cost of lenses and the space they occupy. More often, adjustments of lens position determine size, and the range of a bookwork system may begin with half-point gradations from 5 to 8-point, continue with one-point gradations to 14-point, and end with two-point gradations to 24-point or more. In some systems, quarter-point gradations are available throughout the range. When a single set of photo-matrices is used for all sizes from 5-point upwards, the characters begin to lose their elegance from about 18-point. Some systems however, designed for newspaper composition rather than for that in books, can produce images far too big for book pages.

Changing body during composition is easy; photo-composition is the natural process for a text in which two or more sizes of type frequently interrupt each other, as for instance when the main text in one size connects extracts in another. Usually with equal ease, founts of two or more bodies can be set together in the same line, automatically aligning with each other on the base-line. A pleasant variety of superior figures can be set in a high alignment from the ordinary figures of the fount, appropriately reduced; and when small capitals are to be used in text paragraphs, they may gain in emphasis from an extra enlargement of about a point. In mathematical setting, for instance in 10 on 12-point, such larger characters as the integral and summation signs can be set in 24-point.

The vertical position of characters is adjustable within the depth of the body. Most systems offer at least high or low alignment, used for superior or inferior characters, and some provide several alignments. Characters are usually reduced from the text size when their vertical position is to be high or low.

The maximum measure in most systems is quite wide enough for almost any bookwork text composition. The only restraint is likely to apply to turned tables; on a crown quarto page, such a table may fill a measure of 57 picas. In metal, each line of such tabular material can be divided into two or more lines for type-setting and reassembled into a single line by the compositor; this reassembly might be difficult to do well line by line in film, but the right-hand half of a very wide table could sometimes be set and manually reassembled as a separate but adjacent item. Newspaper systems are often capable of a maximum measure of 100 picas, equal to the over-all text width of a broadside page. Such systems, and others, are capable of reversing the *feed* (the measured onward movement of the output material line by line), so that at the end of one column the type-setting recommences to the right of that column's first line, as the first line of the next column. Measures can be changed during composition, but usually in bookwork the designer will find it simpler to specify indents at left or right or both.

§ 5-9 Modified letter forms

In some systems, the shapes of characters can be modified between matrix and output, as for example by the photo-multiplier in a CRT system. This facility is commonly used to relieve a shortage of matrix positions; for example, by changing the attitude of a character from the vertical to a slant of about 12 degrees, figures and some punctuation and signs can be converted for use with italic type. The horizontal aspects of the character remain unchanged, like those of a man leaning forward against a high wind, with his shoulders level and both feet on the ground. Roman letters can be converted in this way into sloped roman, but this has not become a popular substitute for italic. In form, sloped roman capitals are an adequate replacement for italic capitals, but may appear too heavily constructed to accompany a slender italic lower-case. Sloped small capitals are likely to be useful from time to time, particularly in headlines and headings which include a word intended to be in italic.

The proportions of the characters, as well as their attitudes, can be modified. To save space, a whole fount can be *condensed* or made narrower; or to occupy more space they can be laterally *expanded*. The curved strokes of the letters are altered in shape but not in thickness; the counters are conspicuously affected. A fount modified in this way may be capable of being read but will hardly encourage admiration or continuous reading, and is best reserved for works of reference. Readers tend to prefer the proportions to which they are accustomed.

§ 5-10 Input and correction

The *input* of a composition system is the coded data into which the text is translated for reading and action by the machine. In many systems, input takes the form of either punched or magnetic tape; in Linotype, the matrices are the input; and in direct-entry photo-composition systems, input is electrical current.

Input is either justified or unjustified. In justified input, word-spacing and word-division at line-ends are controlled by the keyboard operator. In unjustified input, word-spacing and word-division are under the control of a computer in or linked to the photo-typesetter.

In order to make any correction or alteration to justified input, the operator may have to repeat the whole line on his keyboard, together with any subsequent lines which have to be *over-run*, their content having been affected by a change in the word-content of a previous corrected line. The compositor, using a scalpel, then cuts out (*strips out*) the corresponding lines in the first output and substitutes (*strips in*) the correction lines.

Quality may be jeopardized by this procedure. The correction lines, which are not set at the same time or in exactly the same conditions as the original output, may differ, however slightly, in image profile. The stripped-in lines may on occasion be fractionally out of parallel with the first output, or fractionally out of lateral or vertical position. The cut edges of film may be recorded as hair-lines on the printing plate. The reader's eye is capable of observing extremely small irregularities of this kind, and of recording an indefinable dissatisfaction with the quality of the text.

A keyboard operator setting unjustified input relinquishes control of word-spacing and word-division. The coded data may be recorded without assuming its typographic character, into which it will be fitted only when it is translated into output. Its typographic specification may indeed be altered between input and output.

Within pre-determined limits, the number of units required for justification is automatically allocated to each variable space. If this allocation, combined with the units occupied by the maximum number of complete words, is not enough to fill the measure of a justifiable line, part of the next word is added at the end. To identify the point at which this word is to be divided, the system's computer may be able to refer to a list of several thousand words, recorded in an electronically accessible dictionary, and each articulated with discretionary hyphens. When the position of one of these hyphens is appropriate for the line, it is inserted, and the others are omitted. But there are many more than a few thousand divisible words, so the dictionary is unlikely to determine every word-division; and in any case not all systems have such dictionaries. The computer may then refer its problem to a program of hyphenation rules which separate certain prefixes and suffixes from the main body of the word, insert a hyphen between consonants or after certain consonants, and so on. Word-division for the benefit of the reader, however, is best determined by a reader's perceptions; different customs apply to different words, and a few simple rules are not enough to find the right place. Accordingly, word-division from unjustified input is likely to mar the quality of text composition, and therefore calls for the attention of the designer. Conspicuously misplaced word-divisions should be amended in proof.

In any form of composition, the frequency of divided words at line-ends is inversely proportional to the average number of word-spaces per line. Narrow-measure setting, for instance in two-column pages, tends to result in either loose word-spacing or an excess of word-division, and in such work the printer may do well to widen the maximum word-space. Word-division usually presents few problems when there is an average of twelve or more variable spaces in the lines.

The advantage of unjustified input is that a system capable of mechanical word-division and justification can employ this capability not only on the first input but on subsequent correction input. The keyboard operator does not have to set each corrected or over-run line complete; only the incorrect letter or word has to be set. When a correction alters the content of a line, over-running and re-justification is automatic. Since over-run passages have not been altered except in terms of line-content, they need no special attention in proof; but the amendment of word-division at one line-end, or indeed over-running for any other reason, may give rise to another unacceptable word-division further down the paragraph.

Correction procedures differ from system to system, and from stage to stage. All corrections pass through the system to appear as output, whether separately from the first output or embodied in it as a second and corrected output, and both methods may be applied at different stages to any text. The speed of many photo-composition systems provides time for such a second output, whether of the whole text or of corrected passages. The method of correction may have its effect on typographic quality when any stripping-in is necessary.

There should be no need for the designer to examine all proofs item by item or page by page; but a single visible irregularity in correction is evidence of the probability of others, and the designer's understanding of the system and its procedures will guide him to those parts of the proof among which he needs to be wary. On the other hand, the absence of such irregularities from a few corrected pages, as when the whole text is made up from a single corrected output, may indicate that a system can be relied on for an acceptable standard throughout, so that the designer need not spend much time on precautions.

§ 5-11 Make-up

Good composition, and good make-up in particular, conduct the reader smoothly through the text, for all its division into successive openings, visually separate but textually continuous. At every point the reader can see where he is going, and deduce what is before him; if a surprise lurks on the other side of the page, it springs from the meaning and not from its presentation. Each opening contains, as far as possible, whatever tables, illustrations, footnotes, or other disparate items illuminate the adjacent passages; if they are not in sight, the reader will be confident they are not far away, and he will know how to find them. The pages are economically arranged; spaces are used purposefully, as signals or emphases or even as decoration — they do not occur at random, and above all they do not mislead. One or two lines of the main text are not separated from the rest by an illustration or heading, since they might look like a footnote or a caption; nor are they left alone on the chapter's last page, which might then appear to be neither occupied nor empty.

Each page is similar to its neighbours in area and in position on the leaf, a uniform part of a larger unit; and while no two pages may be alike in the structure of paragraphs, or of extracts, headings, and so on, typographic detail is consistent throughout, so that no irregularity in printing distracts the reader's attention from the author's communication. Elegance may be the designer's contribution; the compositor's skill can be seen in the neat, purposeful arrangement of the page as a whole and of every detail in it. Through any book so made up, the reader can advance from page to page with confidence and ease, and need never recognize the craftsmanship that smoothed his path.

Some of the conventions of make-up have been compiled and published, as in Hart's *Rules* (1978). These provide guidance in commonplace occurrences of make-up, but the infinite diversity of books and their pages cannot be enclosed by any number of customs or rules. Conventions only indicate in outline the governing principles of make-up, and show how certain problems are traditionally solved. The standard of make-up characteristic of well-made books has always depended less on specific conventions than on the purpose they serve, and on the taste, literacy, and conscience of trained compositors.

The compositor's best work is applied to his own language. He observes the reader's requirements not as a mechanic but as a reader himself, equipped with common sense, technical knowledge, and experience. Given the same intellectual equipment, a book designer can make up well enough, marking proofs or cutting and pasting, when for instance the sizing and placing of illustrations needs to be co-ordinated with the division of text into pages and parts of pages. De Vinne (1904), writing of book production in the United States at the beginning of this century, recommended that difficult make-up might be guided by the author, who could be expected to add or delete words to solve make-up problems in galley proof. Most make-up, however, like word-division, has generally had little to do with authorship or book design, and has been left to good hands in the composing-room.

A division into pages which is approximate or provisional has a limited value only. One false step in such a tentative division may have to be rectified by making up preceding or subsequent pages a second time. Page proofs marked by the printer to indicate such rectification are awkward and unconvincing. Author, indexer, and designer are entitled to see proofs in which everything is shown correctly placed in a make-up intended to be final, in order to assess the quality of make-up no less critically than literal accuracy. For many editions, and for many readers of proof outside the printing-office, an earlier stage of proof in galley should be an unnecessary expense of time, and a later stage little more than an opportunity to verify corrections and alterations.

The procedure for making up a text already set begins with proof-reading and the marking of corrections on a house proof of the first output (in Monotype terms, the caster galleys). A line or more may have been omitted or repeated, and the insertion of a missing word may alter the number of lines in a paragraph. Until these corrections have been marked, the number of typographic lines in the proof cannot be known with certainty. The make-up of uncorrected output is therefore bound to be provisional only.

Type-setting into galleys — and most composition systems produce their first output in the equivalent of type in galley — is the only accurate method of *casting off* (§ 18-1), or calculating the typographic area of a specific text. When the first proof has been corrected, it can be cast off page by page, and provisionally divided into page lengths. These lengths include spaces for such items as tables and illustrations which do not always appear on the same galley as the main text, and which cannot be correctly placed until make-up.

This first marking of divisions between pages reveals whatever make-up problems will have to be solved. For example, if the last line of a page so marked refers to a footnote of two lines, the division will have to be placed elsewhere; that line must become either the third line from the end of the page (to allow room for the footnote under it), or the first line of the next page. This is managed by marking selective local re-composition in preceding lines, if necessary several pages back. Spaces between items may be altered by a line or more; a paragraph ending in a full line may be word-spaced more widely towards the end, in order to make an extra line; or a short break-line may be taken back into the rest of the paragraph by closer word-spacing in preceding lines, in order to save a line. A compositor working on Monotype galleys has six or more octavo pages in galley on his frame at one time, and can easily enough return to earlier pages for manual re-composition. In photo-composition, all corrections and alterations involve keyboard operation, and make-up re-composition might as well be marked on proof at the same stage.

In the course of this division into pages, items of composition which have been set separately from the main text have to be marked to indicate the position on the page to which each is to be transferred as a whole. Tables, for example, are transferred in this way after they have been set. Like the spaces to be left for illustration, and unlike the text which can be divided between lines, tables may be difficult to divide and cannot be fitted into position until the page has been planned to accommodate them. Footnotes may also need rearrangement into a compact formation, particularly when some or all are short; a towering edifice of short notes, stacked up one above the other beside a yawning space on the right, is unacceptably clumsy. Headlines and page numbers can be placed only when the text has been divided into pages; *page headlines* (§ 7-9) may have to be written by the author on the page proofs.

When the divisions between pages are planned, footnotes which are to be numbered by the page, to avoid double or even treble superior figures in a long chapter, can be linked by the correct numbers, marked in text and note as a correction. Cross-references to subsequent pages can be inserted only when those pages too have been planned.

All this re-composition, the assembly into position of items set separately from the main text, and the literal alterations of which cross-references are one example, can be marked as alterations to a proof of the first output, to be effected at the same time as corrections of literal and typographic errors. When this has been done, some systems will automatically transfer disparate items and divide the text between pages at appropriate points in a second, revised output, in accordance with human solutions to make-up problems. If the second output is in galley, page division and assembly will be manual. In either event, the planning quality of photo-composition make-up is capable of matching that of metal, subject only to the possibility in manual work that stripped-in lines may be out of position or variant in reproduction.

Automatic make-up at first output, and indeed any form of make-up applied to uncorrected setting, must be suspect. The planning of make-up by a computer, programmed with however many conventions, seems likely to be questionable; a few logical rules may be enough for conversational fiction, but for more complex pages the range of problems and solutions would be immense.

Whether the established standards of composition, and of make-up in particular, are to be maintained in photo-composition on behalf of author and reader, will be determined by book designers collectively. This maintenance will be made probable only if publisher and printer equally assert by their actions that no lower standards will be offered or accepted in the long term. At the time of writing, personal decision, page by page, appears to be essential to the make-up of most editions of respectable quality.

For centuries, word-division and make-up have been controlled by printers; in no other area of composition have master-printer and designer left such fundamental initiative to the compositor. If that initiative is now to be delegated to electronic machinery, the improvements of five hundred years, devised for manual typography and hardly affected by mechanical type-setting, may be found economically inconvenient. Inferior make-up may then be offered in proof, to be corrected only if the publisher insists (and will pay for doing so); alternative standards — mechanically simpler, and therefore cheaper — will be widely tolerated; and conventions of book design and typographic quality will be forgotten, at the expense of clarity and elegance.

§ 5-12 Output and store

The speed of any photo-setter is a multiple of that of any metal-casting machine. This enables some systems to provide a second pass, which embodies house corrections and make-up. Monotype metal-casting is of course a single-pass method, with the advantage that type lends itself to manual correction in detail, without recourse to the keyboard or composition caster, and without disturbance or repetition of material already correctly set. In photo-composition, all correction is a keyboard operation, and in a fast and fully developed system corrections marked on a proof of the first pass are assimilated into the second. When output is to be used as proof but not for reproduction, a coarser resolution may be used in scanning systems. The speed of a scanning photo-setter is inversely proportionate to the number of scanned lines per inch. A resolution of 1200 lines to the inch may be fine enough for a sharp profile of the image, but 600 lines to the inch may do for proof purposes and will be twice as quick. In the same way, paper may be used for a first proof pass, and film for a second reproduction pass.

Even with a fast photo-setter, a second pass is expensive in machine time and in sensitized material, but handwork would be the only effective substitute. A computer printout, typewriter proof, or other fast record of literal data is typographically inadequate. Proof-reading, whether by printer, editor, designer, or author, is a matter of verifying all the visible characteristics of the printed page, not merely its literal accuracy. A proof is intended to prove that the whole edition is likely in every way to be properly printed, and typographic appearance is equal in importance to literal accuracy. If the appearance of the text is unsatisfactory, its accuracy may be wasted.

Output is either on sensitized opaque paper or on sensitized transparent film. Paper is less expensive than film, and for manual correction and make-up it is easier to handle. Paper seems to be excessively sensitive to light; the slightest variation of intensity or exposure time or development chemistry affects the image, so that unevenness of profile from character to character, from galley to galley, and between passes is likely to occur. An image on opaque paper, moreover, has to be exposed to a process camera, and further variations of profile from exposure to exposure tend to show themselves at this intermediate stage. As in other printing operations, the most direct method is usually the most effective. When paper output has been exposed, a negative is developed, and is utilized in the most economical form of *printing down* (§ 11-6). Light passing through a transparent image spreads fractionally, and characters reproduced from negative print-down are sometimes slightly thickened and uneven in outline.

Film is less light-sensitive than paper, perhaps because light is transmitted through its emulsion instead of being reflected back through it. Evenness of image, and its stability in storage over a period, is more easily achieved in film. The image passes directly from the photo-setter, and the processer in which the latent image is developed, to the offset plate without camera intervention. The spread of light through the transparent background of the opaque positive image tends to sharpen it without inducing profile variation. Negative film can be converted to positive by contact exposure, so that paper output can gain one advantage inherent in film, but only at the expense of additional work and material, and of another intermediate process.

Unjustified data stored in an electronic memory or recorded on magnetic or punched tape can usually be retrieved in different typographic form. An expensive stage of composition — keyboard operation — is thus separated from commitment to a specific design or even a specific edition. Data compiled for first publication of a quarto edition, for example, can subsequently be brought up to date and retrieved in a different typographic form for a popular version as a small octavo paperback. Photo-composition systems, however, tend to be incompatible with each other, and storage and retrieval will as a rule prove to be essentially functions of a single system.

At the time of writing, photo-composition has not yet emerged from its long period of development into its final form. For certain kinds of book, some systems already offer the best available form of composition, and some offer and seem sure to offer facilities never expected of metal type. But for all other editions, photo-composition in general is neither as versatile nor as effective in quality terms as metal composition. Nor is it significantly less expensive in reality, though commercial enterprise in pricing may cause it to appear so.

Until photo-composition development is complete, it cannot be expected to match at all points the quality standards established during the era of metal typography. Those standards will remain in existence as long as books printed from metal type survive, if only as examples of a historic craft. The book designer who forgets them is in danger of lapsing into mediocrity through an excess of tolerance. For books of continuing value, only the best composition is good enough. Nobody is ever likely to write more eloquently about printing than Updike, whose words of 1924 about the selection of type deserve to be applied to the whole task of book design:

How is a man to arrive at a right selection of types? The answer is, by a mixture of knowledge and taste. This knowledge must come from a trained mind and experience. Where is the taste to come from? It may as well be admitted that some persons have no taste at all, but such persons would not be likely to try to produce a well-made book, or to know one when they saw it. Most men who go into printing have some sort of taste and a few an almost impeccable taste — which is a gift of the gods. It seems to me that a right taste is cultivated in printing, as in other forms of endeavour, by knowing what has been done in the past and what has been so esteemed that it has lived. If a man examines masterpieces of printing closely, he will begin to see why they were thought masterpieces and in what the mastery lay. He will perceive that all great printing possesses certain qualities in common, that these qualities may be transferable in some slight degree to his own problems, and then he will find himself braced and stimulated into clearer, simpler views of what he can make out of his task. When he sees the books that have delighted all generations and begins to comprehend why they were great pieces of typography, he is beginning to train his taste. It is a process which once begun is fed from a thousand sources and need never end.

CHAPTER 6

Text types

$\mathbb{F}$IVE CENTURIES of printing history have seen many hundreds of different type-faces come into use and fall out of favour. In a sense, not one of them survives from previous centuries for mechanical composition, except the very few cut for composing machines in the last years of the 1890s. Founders' matrices cannot be used in typesetters, and all commercially produced text is now set almost entirely by machines. Old designs have been copied and adapted for mechanical composition; some of the revivals can hardly be distinguished from the original, others have been changed almost beyond recognition, for better or for worse. The period of revival seems now to be over. Most new arrivals are new designs, based on no single historic fount, or they are adaptations for photo-composition of a type already established in metal. Occasionally some classic oddity is resuscitated for a few spectral appearances before being quietly reinterred. Perhaps there are now in use for bookwork text more type designs and adaptations than a typographer can come to know intimately through frequent use; there are many text series unlikely to be seen in books, though available for them, and useful for other purposes; and there are display designs which may include no text founts at all.

While he can hardly be expected to master such a variety, the typographer benefits from some idea of the history of type design; letter forms are based on convention, and when he intends to weigh the value of a convention he will do well to examine its origin. He will need to know enough about type-faces in general to assess the qualities of each in particular. He will have to find out which designs can be used with which machine, and to learn the constituents of each fount and series. Through practice and a habit of visual awareness, he will come to recognize at least the types he sees most often, in the books he reads and examines as well as in those he designs. Knowledge of this kind can be gained from the various specimen sheets and booklets published by firms which produce matrices for text composition, and from the type specimen books in which printers catalogue and display their repertory.

Types now in general use are capable of being read without difficulty, or they would not have remained in general use for long. But the legibility of any series depends to some extent not only on the letter forms but on the way in which they are arranged and printed, and factors other than

[97]

legibility have to be taken into account in the choice of text types. A reasonably good choice may be easy; it will be the best possible choice at which the typographer will wish to aim.

§ 6-1 Availability

If the typographer intends a particular make of machine to be used for the composition of the text, his choice of type-face is limited to the range of matrices available for use with that machine. Some type-faces are common to more than one kind of machine, usually with differences in design of varying importance.

If he feels that one type-face only would be the right one for the text of the book, he will have the composition carried out by a printer who has the necessary founts of that series. If on the other hand the work is to be placed with a certain printer — and since the choice of a printer does not as a rule depend primarily on his list of type-faces, this is perhaps the more usual approach — the series and fount will have to be chosen from among those with which that printer is equipped. To install a new series is a costly venture, not to be undertaken for a single book of the ordinary kind; but a printer may be willing to buy a new fount for a series of which he already holds some founts, and should certainly be ready to add some matrices from the manufacturer's stock if he is using a system which allows them to be introduced. The availability of different sizes of text type is hardly a problem in photo-composition, but not all metal series include a full range of founts, and not all printers hold all the founts of any series.

If alphabets other than roman are to be used to any extent, a series may have to be chosen not only for its own qualities, but for the number and nature of its relatives. Not all text series have mated variants such as bold roman and italic, bold condensed, and so on. If the text requires unusual sorts, the designer's choice of series may have to start among those which include these sorts. For this reason, much of the setting of the higher mathematics was for many years carried out in Monotype Modern Extended (§ 6-11) or Monotype Times (§ 6-13). If the text requires the use of two or more unrelated series, such as roman with Greek or Hebrew or both, the various series will have to be chosen first for their ability to be mechanically set together with each other, and second for their affinity in terms of apparent size and weight.

The quality of the related alphabets deserves only slightly less attention than that of the roman; this is particularly true of italic, which may be used for phrases, sentences, or passages of some length. The design of figures may well influence type-face selection for a work containing a high proportion of tabular setting. Not everyone can feel much enthusiasm for a bold type of any kind, but some are distinctly better than others; the clearer tend to have rather open counters and loose fitting, to compensate for the extra weight of stroke, and semi-bold is usually preferable in this way.

§ 6-2 Letter-form

Good letter-form does not admit of easy definition, if indeed it can be defined at all. Other arts, under the scrutiny of generations of critics, have accumulated a technique and terminology of appraisal: the art of letter design has not. Certain plain virtues may, however, be described as essential to good letters. Some of these have already been touched on in chapters 4 and 5.

The first of these virtues is certainly the quality of familiarity. In over-all proportions, in thickness of stroke, and in shape of outline, every letter of a good fount is similar, within reasonable limits, to the form of that letter to which readers are accustomed. The prime function of the letter is to communicate instantly, and any novelty of basic form may be an obstruction to smooth reading and even to recognition. Each letter should be distinct from all the others, and capable of being recognized at a glance; but no letter should be so distinctive as to attract more than a glance.

Precise criteria of form and proportion do not apply, but some of the qualities of good letter form can be defined. In practice, the decision as to what is and what is not good rests with the taste of the individual typographer. Most printing designers agree to a large extent about this, and agreement has its influence on the purposes of type design.

Stress has already been defined as the difference between thicker and thinner in the different strokes or the different parts of a stroke of a letter. One effect of stress is that two lines never meet or cross at their thickest points; where two lines meet, one is thick and the other thin, so that the counters are more open than if they were enclosed by strokes of full width along their whole length. By providing contrast at points of intersection, and by enlarging the counter, stress imparts clarity as well as grace to the letter. These qualities are not conspicuous in particularly light founts, which have not enough contrast between the thickest and thinnest strokes. On the other hand, too marked a contrast between thick and thin may render a fount severe and dazzling in appearance; the printer's problem may then be to strengthen the attenuated hair-line without distending the already overblown main-stroke.

Neither shading nor serifs are essential parts of the roman letter, which can be read without either; and unstressed or slightly stressed sans-serif letters are now quite often seen in use as text types. Serif and stress are incidental results of the methods of engraver and scribe; but they have become useful parts of the letter's structure, contributing to the comfort of reading and the elegance of letter design, which should not be discarded without good reason.

The length of the ascending and descending strokes is less important but is worth examination.

If the fit of the letters is close enough, they will appear to combine clearly into words, even when there is — as there should usually be — little space between words. A loosely fitted fount seems to call for extra space between words and between lines, if the separate existence of words and lines is to be evident, and so tends to occupy substantially more space.

These are among the design characteristics which may indicate the best use of a type-face; its legibility is primarily a matter of its main proportions, discussed in § 6-7.

§ 6-3 Legibility

Legibility is the most valuable quality in a typographer's work. After a good deal of research, some of it vitiated by ignorance of science among typographers and ignorance of typography among scientists, legibility remains a problem too subtle for measurement and analysis, and too abstruse for influential guidance.

Legibility may be defined as ability to be read continuously, by the kind of reader for whom the text is intended and in the kind of circumstances in which he may be expected to read, with the greatest possible speed, accuracy, and pleasure, and with the least possible effort and distraction. In various stages of book design, legibility appears and reappears as different kinds of problem, but always as one of the main purposes of the operation. This section concerns assessment of the legibility of a text fount.

A thick line is more easily seen than a thin line, but a bold fount is not necessarily more legible than a light one. A letter is a pattern of lines, and the shape of a legible letter, as well as the lines themselves, is clear to the reader. Clarity is determined not only by the thickness of the lines, but by their distance from each other and from the lines of adjacent letters in the same word. If the lines are too close together or too far apart, their relative positions, and therefore the shape of the letter, are not instantly recognizable; this effect may be observed in black-letter founts. The white spaces within the letter (the counter) and between the letters (the fit) are not less important than the thickness of the lines. In an ideally legible type design, the thickness of the lines which make up the letter would be exactly appropriate to the letter's lower-case x-height and set.

There is no known method of calculating the proportions appropriate to all book-production circumstances, or the pleasant variety of type designs now in use might have shrunk to half a dozen or less. The typographer has to choose between existing founts rather than to design new ones; his task is one of comparison rather than of precise definition. One of his decisions must be whether a fount is too bold or too light to be entirely satisfactory for a specific edition. Perhaps the consensus of opinion among successive type designers, demonstrated by their designs, may be a safe guide, and the clearest founts for most purposes are those which are neither particularly bold nor particularly light in relation to the majority of text types. But for those who are learning to read, for elderly, short-sighted people, and for reading by dim light (in some churches, for example), strongly drawn letters are surely best. In such circumstances, speed of recognition matters less than ability to discern every part of each letter.

Type size probably has more influence on legibility than the proportions of the letter, and it is the apparent size, not the body, which the reader observes. The smallest of the old type-bodies was *minikin* or 3-point, which very few people could read without discomfort for more than a line or two at a time. An apparently large 6-point will do for reference, but for sustained reading by practised readers 10-point or an apparently large 9-point is now generally the minimum. For learners and for the short-sighted a larger type is essential, but anything too big may prove disconcerting, and teachers and other specialists may be the best guides. The matter is one of many in book design on which the opinions of laymen — publishing colleagues, and in particular the author — may usefully be consulted even though they may not be decisive.

§ 6-4 Space

Among the most carefully observed conventions of book production is that which governs the over-all shape of the book, in section and in plan. This shape is usually related to the classification of the text; the squat, stout shape which may be right for a dictionary might prejudice the sales of a novel. Booksellers are likely to report that the public is balking at a book of unusual shape. Part of the designer's aim, then, will be to fit every text, whatever its length, into a printed book with dimensions approximate to those usual for books of the same kind.

The problem is sometimes acute in the production of novels, which are more limited in style of production than most classes of text. Modern novelists tend to write to a length which can be fitted into something between 160 and 320 octavo pages, not so large as to look unlike a novel, nor so thick as to be intimidating, nor so thin as to seem meagre. This tendency probably results from the preferences of the book trade rather than from the impetus of creative effort. Occasionally, however, some book of enormous length has to be presented in saleable form. Novel readers are intolerant of small type; two or more volumes and two or more columns seem now to be unacceptable in fiction, though these may be the only solutions consistent with good typography. The reader has only to accustom himself to buying novels produced in two columns and a quarto format — he will have no difficulty in reading them. But the designer may not always, or even often, be permitted to devise what he believes to be typographically best; the pressure of book-trade preference may indicate the smallest acceptable type in the widest practicable measure without any leading at all.

Usually a less drastic solution will do. Slight variations in the size of the page may be possible, and particularly in its width which is less conspicuous on the shelf than its height; margins may be expanded or diminished; particularly thin or thick paper may compensate for a particularly large or small number of pages; interlinear space, the arrangement of chapter openings, and other variables can be adjusted. A careful choice of fount is only one of the means of making the text appear to expand or contract, but the width of the fount in a–z terms is one of its main factors.

When a choice is to be made between founts from two or more series, the founts to be considered should be as nearly as possible equal in apparent size, as for instance a large x-height 9-point, a medium 10-point, and a small 11-point. To restrict the choice to three 10-points, one apparently large, one medium, and one small, and then to choose the most economical by looking for the shortest a-z length, might result in choosing for economy a letter too small for comfortable reading. Exact adjustments of apparent size, whether for purposes of economy or not, can be made by a change of series rather than of fount within series. A medium 11-point, for example, may be intermediate in apparent size between the 11 and 12-point founts of a series which has a smaller x-height. Different series will require different amounts of interlinear space; nothing will be gained from choosing a fount for its narrow set if it requires extra lead, as the line will contain more characters but there will be fewer lines on the page.

§ 6-5 Paper and printing

A careful matching of type-face with printing process and paper surface is always necessary, particularly with metal-cast types and when printing by *letterpress* (§ 11-1). The principle of this process is that the printing surface touches the paper, and the non-printing surface, corresponding to the unprinted parts of the paper, does not. When the printing surface, covered with ink, is pressed against the paper, pressure forces some of the ink outwards, towards and even beyond the edges of the printing surface. The resulting extra thickness of ink at the edges of the printed letter gives it a slightly hard, sharp appearance; the edges are outlined in a deeper black than the rest of the letter. When forced outwards, beyond the area of contact betwen printing surface and paper, the ink thickens the printed image. This thickening, known as *squash*, varies in extent according to the pressure used in printing, which depends in turn upon the nature of the paper surface, and according to the resistance of the paper surface to this pressure. Strong pressure is needed to transfer ink evenly to the uneven surface of the softer and more fluffy kinds of paper, and the ink tends to spread round the resulting recess in the paper surface. This thickening of all the strokes favours those types which were designed for it, with main strokes which benefit from fattening up. A harder and smoother paper resists the impression and yields only a comparatively shallow recess, hardly thickening the image at all, and so suits types with no very thin hair-lines and with stout main strokes.

The offset process causes no squash. The ink is transferred to the paper by a flexible rubber blanket, without enough pressure to force the ink outwards; and since blanket and paper are in contact over the whole area of the sheet, the area round the printed image is enclosed between them and admits no spreading of the ink. Thickening may still be seen in offset printing, but this is likely to be caused by uneven repro proofs or erratic camera work.

A generation ago, qualified typographers were well aware that certain types were unsuited to printing without squash, whether by letterpress on glossy art paper or by offset. Today all too many revivals of old designs seem to be based on the design itself rather than on its printed reproduction at its best; all too many pages are pallid, and all too often the shapes of the letters are difficult for the short-sighted to discern and the details are all but invisible.

Photogravure (§ 11-8) is not particularly apt for type reproduction of the best quality. The ink is contained in a diagonal pattern of tiny rectangular cells of equal area, etched into the surface of the printing plate. This pattern tends to thicken the strokes and to serrate their edges, and may break up or thicken very fine lines. The best types for use with the process have no very thick or very thin strokes, and are not very sharply cut in serifs or other details.

§ 6-6 Aesthetics

Besides those already described, there are other and more subtle factors in the choice of a type-face. The letters of a really good series have a grace and rightness of form which can be identified by the practised eye but which cannot easily be defined. The surest guide to perception of this quality is knowledge of the construction of letters, and this can best be gained by making them with care and precision. The practice of formal lettering with a chisel-ended nib may lead to that certainty of eye the typographer needs. Nor are letters the only criterion, since every character that is to be printed should first be assessed by the designer.

In the selection of text type, aesthetic aspects of printing have some practical effect; printed matter does seem more legible, and reading does seem to become more accurate and quick, when the material is set in a type which the reader, perhaps without realizing it, finds aesthetically pleasing.

The character, origin, or 'atmosphere' of a series may exert a subtle influence upon selection. Whether the letters are delicate or sturdy in appearance may be obvious enough; there are less obvious qualities of detail, which can be perceived but which can hardly be described without recourse to a vague extravagance of language. To a limited extent the character of a series may be derived from its origin; for a text concerned with a period or a country, a type design of that place or time may be a particularly suitable choice. Allusive preference of this kind must, however, be weighed against practical considerations.

In this choice, and indeed in book design generally, compromise cannot be avoided. Any typographer with a sharp eye and a critical faculty can identify flaws in the best of series. Any fount of type suitable for text composition can by definition be read without difficulty, though some are more popular than others. Every possible series, not merely the favourites of the moment, should be allowed to enter the field of choice, if more than a very few kinds of book are to be composed; minor imperfections must not be supposed to disqualify.

§ 6-7 Classification and specification

The rest of this chapter provides examples of some historic type-faces and of the twentieth-century revivals based on most of them, together with a commentary on history and revival. Hundreds of different type-faces are available for text composition, but there is room here to consider only a few. Those selected have been used in books for some time; most have achieved, though not all have retained, some degree of popularity — some British preferences in type design in the 1980s are listed in § 6-15. These types occupy a significant place in the development of printing types for mechanical setting. Some are obsolete by now, in the sense that they are unlikely to appear in any form in more than a very few of the books of the foreseeable future. Perhaps a few never really found a place in the main-stream of typographic development. But each represents an achievement, if only a minor one, in more than half a century of revival and innovation in type design. The examples of type-faces old and new have been set in 18-point or enlarged to an approximate equivalent of that size for purposes of comparison.

In other chapters I have examined processes, procedures, machinery, materials, and other elements of book production, using specific items as examples of my theme, in order to demonstrate a method of approach to such things which is likely to avail the designer faced by elements not described here — perhaps not even invented yet. In the same way these few examples of text types, each with its brief commentary, are intended to show how other type-faces could be assessed.

The type-faces are set out in chronological order of first appearance, and they are divided into the conventional classes. Type-face classification indicates characteristics shared by a stated group of series, and is a convenient description of fashions past and present. British typographers on the whole continue to use long-established class titles; a more elaborate system of classification, set out in British Standard 2961 (1967), has not yet become popular. When the names of a class differ from system to system, both are given in §§ 6-8 to 6-12, with the British Standard title second. Apart from the application of the term Old Style, practice in nomenclature is much the same in the United States as in the United Kingdom. Type-face classification is more likely to be useful in discussion and study than in book design practice.

The most effective selection of a type-face for the text of a specific edition will usually be based on examination of all the characteristics of the fount which are likely to be deployed in the expression of that text. These may include the quality of large and small sizes, special sorts, accents and diacritical characters, italic, small capitals, superior and other arabic figures, punctuation marks and other signs, and bold or semi-bold roman and italic. Each of these items deserves to be studied not in isolation but as part of composed text.

It is an axiom of book production, as of other human activities, that if anything is left to go wrong it is likely to do so. Much of the designer's work entails explaining his requirements to distant strangers. Industrial design, like any other form of planning, is ineffective without precise and comprehensive communication. The description of a selected type is an example. To say that a book is to be set in Baskerville is a waste of effort; there are many versions of Baskerville, and some of them look quite different from all the others.

A specific fount of matrices, or in photo-composition a specific size of reproduction from such a fount, is identified by its 'body', actual in the former, nominal in the latter. Where a series of photo-matrices includes more than one design, the use of each design must be indicated. In some

Monophoto series, for example, 'A' matrices can be used for small type, 'B' matrices for text, and 'C' for display, to avoid the unseemly effect of drastic enlargement or reduction from a single design. Next to be stated is the machine or the range of machines for use in which the type-face was designed. Type designs vary from manufacturer to manufacturer, but not always from machine to machine. The name of the type-face is essential, but there are now so many different versions of successful designs that the name of each version is meaningless without ascription to the machine. If there is a series number, it should be stated, since one name may have been assigned to more than one series; in §§ 6-11 and 6-12, for example, *Modern* and *Old Style* are classes used as names, and comprise more than one design. However similar in title, each separate series calls for separate assessment, as dissimilarities of detail, of proportion, and of dimensions may be conspicuous.

§ 6-8 Venetian, or Humanist

A Venetian or Humanist type-face might appear to be any roman type originally used in Venice for humanist works. Certainly it must be roman, since this class is limited to the fifteenth century in which there was no italic type. Certainly it must be Venetian, and preferably it will be humanist. But after 1495 there were types of this kind which have been assigned to another class. The 'Venetian' types may be distinguished from the merely Venetian ones by capitals which rise to the full height of the ascending lower-case letters, and by the oblique cross-stroke of e. But there is no evident advantage in so distinguishing one type-design from another, and the terms Venetian and Humanist are perhaps better defined as titles occasionally applied to the earliest decades of purely roman type in Venice. During this period the general appearance of roman types, derived from the formal book-handwriting used in many manuscript books of the fifteenth century and from other forms of handwriting and lettering, resolved itself into a tradition which was to last for more than five hundred years.

Monotype Centaur roman and Arrighi italic

Within a quarter of a century of the European invention of typography, roman type design shed the last of Gothic influence, and emerged with purely roman letters to suit the humanist preferences of renaissance scholars. The best of these early romans was that introduced by the printer and publisher Nicolas Jenson in 1470. Jenson was born in France, and may have learned engraving in metal at the royal mint. In 1458 he seems to have gone to Mainz to find out what he could about the recently invented craft of typographic printing. He went on to Venice, and starting with one of the earliest and best of all roman types, which he probably cut himself, he continued during the next ten years to become one of the most successful printers, publishers, and booksellers in one of the most commercially competitive cities in Europe.

The roman of 1470 excelled other romans of the time not only in letter-form but in precision of cutting. Some of the letter-forms were clumsy, including g and the long s, but the generous width of the lower-case letters and the well-proportioned spaces between them lend themselves to the reader's eye. Apart from the tiny dot of the i there was no sharp detail; and whether it was due to the cutting or to the presswork of the time, there were no hair-lines. There was not much contrast between thick and thin either; but the legibility of the type demonstrates the probability that a thick line is more easily observed than a thin one, and that the shape of characters is more easily discerned when they are ample in width. The capitals were constructed of strokes almost exactly as thick as those in the lower-case, following the example of manuscript in which the same pen would be used in both kinds of letter; and the capitals were equal in height to the ascending lower case letters.

One size of handwriting was usually enough for a manuscript book, and in the same way a single size of type was enough for the early printed edition. The 1470 roman was about 16½-point in body. It was unaccompanied by italic, which would not appear in typographic form for more than thirty years, and which would begin to work as an auxiliary to roman only in the latter half of the sixteenth century. Jenson's success as a publisher combined with the excellence of the design to start a fashion, and at least thirty imitations were used in various parts of Italy during the next fifty years. It was not the only fashion of the time; roman types of the older and cruder style were still to be seen, just as the proportions of Jenson's roman could still be found in books printed in France a hundred years after its first appearance.

In 1914 the Metropolitan Museum of Art in New York commissioned the design of a new type-face from Bruce Rogers. He produced a delicate and graceful adaptation of Jenson's 1470 roman, as a single roman fount of 16-point, first used in an English edition of Maurice de Guérin's *Le Centaure*. In 1929 Rogers offered the composing-machine rights in the design to the Monotype Corporation, and a full-scale series of roman and italic founts appeared in the same year. The chronology of this account postpones consideration of the Arrighi italic, designed to accompany Centaur roman, until after that of the Poliphilus roman.

In the original 16-point and larger sizes, Centaur and Arrighi can be seen to possess a calligraphic grace unequalled by any other design. Much of this

iuſtitiā quā non a moſaica lege(ſeptima eīm poſt H
Moyſes naſcitur)ſed naturali fuit ratione conſecutu

Figure 6-8. Above, Nicolas Jenson's roman of 1470, 16.5-point × 109%. Below, the
Lectern Bible version of Monotype Centaur, 19-point × 95%.

that purpose. (He eventually chose one printed so
ago, of which a copy was still in stock at one of the

has been derived from Jenson, but Centaur has an individuality of its own
which makes it more than a revival. The punctuation deserves study; the
asterisk, for example, combines originality, wit, and effectiveness, and the
difference in shape between roman and italic points is conspicuous (figure
4-4*f*). Below 16-point, the style of type drawing begins to become
imperceptible, and the text sizes are too thin and irregular to have achieved
popularity. The 19-point shown in figure 6-8 was specially adapted for a
lectern Bible, in which the type was pressed hard into the paper, leaving an
impression of engraved magnificence.

§ 6-9 Old Face, or Garalde

This class of type designs is defined as having begun with the 1495 roman of
Aldus Manutius and adaptations of it by Claude Garamond in the sixteenth
century, and as having come to an end (until the era of revival) with the
types of William Caslon. Characteristics of the 1495 archetype were
inherited by successive designs until the eighteenth century and Caslon,
under whose name later on the meaning of the term *Old Face* is explained.
The term Garalde combines the names of Garamond and Aldus. Identifying
features include capitals lower than the ascending lower-case letters, and a
horizontal cross-bar in e, together with the oblique stress already seen in the
1470 roman of Jenson.

Monotype Bembo

Among the younger men who learned their trade in Jenson's substantial
organization was Andrea Torresani, who started his own firm with types
bought from Jenson, and who like his master established himself as printer,
publisher, and bookseller. In 1493 he published a book by a scholar of about

[108]

his own age, Aldus Manutius; and it was probably in 1495 that Andrea admitted Aldus to partnership, junior but operationally independent, in order to exploit the growing demand for Greek texts.

For this venture into Greek publishing, Aldus commissioned a revolutionary Greek type-fount, so cursive as to be almost illegible to the modern eye, so elaborately equipped with ligatured and accented sorts that it was almost uneconomic to cut the punches or set the type, but still the pattern for Greek typography for centuries thereafter. Aldus was not equally interested in Latin or Italian publications, but he would need a roman fount as well. This was cut for him by Francesco Griffo da Bologna, and in comparison with the Greek it was a conventional design. The body of the fount was only half a point smaller than that of Jenson's 1470 roman, and the shape of the lower-case letters was much the same — as might be expected, in view of Jenson's great prestige and of Andrea's connection with him and his types. But the lower-case letters are narrower for their height, closer together, and larger on the body. The design is more accomplished, as for instance in g, there is more contrast between thick and thin, and the serifs are thinner and sharper. The capitals are unattractively designed by comparison, more roughly cut, disproportionately thick in main-stroke, significantly shorter than the ascending letter, and not always well aligned at the foot. They look as though Andrea, who was notoriously mean, had advised Aldus to combine Griffo's new lower-case with a set of redundant capitals from Andrea's shop; such things were known to happen in those days.

This Aldine roman was cut in Venice for use in humanist works, but is

meridianas horas die:ubi ea, quae locuti
sum⁹ inter nos, ferè ista sūt. Tibi uero nūc
oratione utriusq; nostrū, tanq̄ habeatur,

Figure 6-9a. Above, the 1495 roman of Aldus Manutius, 16-point × 113%. Below, the first two lines, Monotype Bembo, 18-point; the second two, Lasercomp Bembo series 270C, 18-point.

Far too often people read books and just lay the
the diversity, quality, and the exactness of the pr

the diversity, quality, and the exactness of the
Far too often people read books and just lay

excluded from the class of Venetian or Humanist types. Jenson's roman of 1470, on the other hand, is the oldest roman type to have been revived for use in our own time, but is not classified as an Old Face. There is no obvious utility in the distinction between Venetian and Old Face types.

The 1495 roman, the earliest and apparently the archetype of all Old Faces, was not particularly successful in its own day. One possible reason for this may have been that it was in advance of its time. Jenson's roman of 1470 seems to have set the fashion for Venice and the neighbouring region of Italy, and a few comparatively unimportant Latin publications — since at that time Aldus was concentrating on Greek — would hardly have been enough to displace anything so well established. Aldus was to become a famous publisher, but his roman type was not among his claims to fame; his cursive Greek, on the other hand, was to supplant the plain and practical Greek types such as those used by Jenson, whose cultural standing was lower than that of Aldus, but whose commercial success in a competitive city endowed his types with a prestige of their own.

The Monotype Corporation revived and adapted the 1495 roman in 1929, and named it after Pietro Bembo, the author of the 1495 book in which the type first appeared. The adaptation was a masterpiece of type drawing; the original roman was about 16-point in size, and from this single fount the Corporation's type drawing office produced a range from 8 to 72-point, every fount in which is excellent in the style of letters of that size. The roman capitals have been rectified rather than adapted; the large sizes are noble, the text sizes legible, and much of the individuality and grace of the ancient original can be seen in the revival. No more popular and widely useful type has ever appeared in metal.

The italic, based on types used by Giovantonio Tagliente, is examined further on, since its design was some thirty years later than that of the 1495 roman.

The roman and italic alphabets were extended into a complete twentieth-century text fount by the addition of small capitals, inclined or 'italic' arabic figures, aligning versions of the figures both upright and inclined, and a mated bold type. None of these additional features appeared in the original fount or anywhere else in the fifteenth century.

Of the various adaptations for photocomposition, a digitized version, Lasercomp Bembo, is shown in figure 6-9a. This adaptation and all others which rely on a single type drawing to reproduce all sizes of each character fall short of the excellence of the metal designs.

Monotype Poliphilus roman and Blado italic

In 1499 Aldus printed one of the most famous of fifteenth-century illustrated editions, the *Hypnerotomachia Poliphili*, 'Poliphilo's strife of love in a dream.' Martin Lowry (1979) has described the text as 'a linguistic and literary debauch, choked with recondite imagery, erudite periphrases, and exotic verbiage; a work so bizarre that many critics have felt a certain uneasiness at Aldus's agreeing to print it.' The fame of the book springs from the incomparable woodcuts and from the elegance of their arrangement with the type on the page; illustrations and text are alike erotic, and Aldus was taking a risk in publishing both at a time when official attitudes to the lubricious were severe.

The roman lower-case in the *Hypnerotomachia* is that which first appeared in the Bembo tract of 1495. The incompatible capitals of that fount were replaced in the 1499 book by a much better alphabet — a little fanciful in some details, perhaps, but a relief from the dreariness of their predecessors. The presswork though erratic was heavy on the whole, perhaps because of the blocks, and the type may have been somewhat worn and thickened before printing began.

In 1923 an English translation was planned, and the printed image of the type in the book was faithfully copied and adapted for a limited range of sizes. Since in 1499 italic type had yet to appear, an italic from a later period had once more to be found to accompany the revived roman, and this is examined in the next section. The Poliphilus revival may appear to be little more than an antiquarian exercise, but the Monotype series has two virtues which may ensure its survival. The basic design of both roman and italic is excellent, and both are copies from prints thickened by heavy impression. All too many of their competitors today have been diminished from large size to small, or have been precisely copied from a light print or from punches. As a result they would always have been unsuitable for printing from metal type on glossy coated paper, and today they are unsuitable for the offset process because they are too thin. Poliphilus is likely to survive its antiquarian origin and its eccentricities of detail, and to be found useful and

tached no religion. When old Kit had found hi
self pressed in that matter of the majority of
Nineteenth Dragoons, in which crack regiment his second

Figure 6-9*b*. Above, Monotype Poliphilus and Blado italic, 13-point × 139%. Below, Lasercomp Poliphilus and Blado, 18-point.

tached no religion. When old Kit had found him
self pressed in that matter of the majority of the
Nineteenth Dragoons, in which crack regiment his second

pleasing by designers who may know nothing of its history. This is all the more likely because of the affinity of the design for enlargement and reduction from a single matrix.

Italics: Arrighi, Bembo, and Blado

Early in the fifteenth century before the invention of typography, a scholar who lacked a book for his own use and that of his friends was likely to copy an existing manuscript rather than to buy one. For the sake of speed the formal roman lower-case of the original was copied in a more cursive style, which tended to lean to the right and to be narrower than roman; the capitals were basically unchanged. This scholars' cursive became the model for a less personal form of handwriting, that of professional scribes writing out formal documents in the papal chancery and elsewhere. The functional simplicity of the scholars' cursive was replaced in the hands of experts by exaggerated extenders and by decorative flourishes.

In fifteenth-century Venice books for study were kept at home or in a library; the only books designed to be carried about were those for use in worship, and these were the only small books which then appeared in print. The most common sheet size for printing was about 450×315 millimetres or 17.7×12.4 inches, so the folded size of a quarto book was 225×158 millimetres or 8.86×6.2 inches, and that of an octavo was 158×113 millimetres or 6.2×4.4 inches; octavo sizes were used only for religious works.

In style of type-face publishers were still finding their way forward from the manuscript tradition; religious and legal books, for example, were usually set in gothic or black letter, and the Latin classics in roman, and most books of all kinds were set in a single size of type.

The publishing programme, consisting mainly of Greek works, on which Aldus Manutius had started in 1495, had not come up to expectations. Editorially a good deal of new ground had to be broken, and publications were slow to appear. The Greek founts were enormously elaborate, combining the full range of accents with an unprecedented number of ligatures in an imitation of the Greek cursive handwriting of the day; the Aldine Greeks were expensive to install and expensive to set. The books were expensive too, and the expected demand does not seem to have materialized, for they were slow to sell. Perhaps under pressure from his partners, Aldus adjusted his editorial and marketing policies. He started a series of Latin classics, editorially already established and as publications reliably popular. He introduced the octavo format for general reading, as these classics were then considered among the limited proportion of the populace which made a habit of reading. And he set these little books in a new kind of type, cursive, informal, personal — the first of all italic types, initially a text type in its own right, with roman capitals intermediate in height between capitals and small capitals. The new type was about $11\frac{1}{2}$-point in size. The price of these portable editions was forced down by increasing the average number to print from about 1,000 to about 3,000; the number of books sold was increased by international agencies and contacts.

The Aldine italic, based on the scholars' cursive lower-case, was copied and adapted all over Europe during the next half-century. In Italy during the same period another kind of italic type began to appear, based on the handwriting of the professional scribe and the chancery clerk.

The first of these chancery types was printed by Ludovico degli Arrighi in Rome in 1524. From about 1510 Arrighi was a scribe employed in the papal chancery. In 1522 and 1523 he published writing-books, the second of which included some pages in his first chancery fount which has not been observed elsewhere. The second fount was the foundation of his achieve-ment as a printer from 1524, and of Frederic Warde's italic versions to accompany the original Centaur roman of Bruce Rogers and the Monotype series. It was the first of all chancery italics to be used for the whole text of a book. As in the Aldine italic, the capitals were upright, but they were taller in proportion to the lower-case; and in this fount swash capitals appeared for the first time. The lower-case characters were narrower and less steeply inclined than those of the Aldine italic.

Giovantonio Tagliente of Venice, one of the greatest of all writing-masters and the most influential exponent of chancery handwriting, published his writing-book in 1524, and later editions continued until 1678. To accompany Monotype Bembo roman, Tagliente's type was regularized and modernized, so that much of its chancery origin has been disguised.

By 1526 Arrighi had introduced his third chancery fount, but he seems to have lost his life when Rome was sacked by mercenary armies in 1527, and a few years later Antonio Blado began to use Arrighi's third fount. Blado was a younger relative of Aldus, had been associated with Arrighi, and was one of the best and longest-lived of sixteenth-century printers in Rome where he worked from 1515 to 1567.

When Monotype Poliphilus was cut in 1923 at the suggestion of a publisher who planned a translation of the *Hypnerotomachia*, Stanley Morison proposed as model for the accompanying italic the fount used in a book printed by Blado in 1539. Monotype Blado seems to have been the first fount introduced by Morison after his appointment as typographic adviser to the Monotype Corporation. It was a learned and imaginative idea; nothing could have been more suitable, but hardly anybody in the world knew anything about chancery handwriting or types as that time, and no new chancery type had been introduced since Blado's day. The original chancery types were more upright than successors of the Aldine italics; in the revival, the slant was increased from between 5 and 8 degrees in the original to 12 degrees.

In all the original chancery types, the capitals were upright; in all the revivals they were slanted in conformity with the lower-case.

Chancery types were cut in Italy only, but the publications of the writing-masters had their effect on punch-cutters in other countries. By the mid-sixteenth century italics in France were combining the characteristics of the Aldine and chancery schools of design, and were taking on a function subordinate to that of roman.

Linotype Granjon

The importance of the Aldine roman of 1495 was that of an ancestor who transmits dominant characteristics to the next generation and to those that follow it. Claude Garamond, one of the first independent type-founders in France, copied and adapted the fount, including variant sorts found only in the Bembo tract and introduced in order to relieve the page of the appearance of excessive regularity caused by the repetition of identical letter forms. The fifteenth-century founts already described were cut in a single size only; those of Garamond and his contemporaries flowered into series of founts, and the 1592 type-specimen of Conrad Berner (reproduced by Dreyfus, 1963) showed Garamond romans from about 10-point to about 48-point. The grace and clarity of all sizes, and the gradation of design from size to size, combine to suggest that the history of type design since the sixteenth century may have been one of decline. Garamond's copy of the Aldine fount was instantly copied by his competitors, and set a fashion and standard for roman types all over Europe for the rest of the century and longer.

the modern printer's equation is *quality* plus *sp*

Figure 6-9c. Linotype Granjon, 24-point × 75%.

In 1925 the Linotype company revived a handsome roman used by Jean Poupy of Paris in 1582 for Bouchier's *Historia Ecclesia*. The fount, on a body of about 16½-point, was later recognised as having been cut by Garamond. It was shown on the Berner specimen as his Romain Gros Text. By then however the new Linotype face had been named Granjon, to commemorate another French type-founder. It is shown in figure 6-9c because we have no other revival of one of the most influential and enduring of all type designs. The elegance of the revival was admired, particularly in the large sizes, but the text sizes were too light to become popular, unlike the parallel sizes cut by Garamond himself.

The development of sixteenth-century roman type-design was led by Claude Garamond, and that of italic by Robert Granjon. Both cut designs of more than one kind. Granjon was the more versatile and imaginative. By the mid-sixteenth century italic capitals had risen on the whole to full height, and were generally inclined in parallel with the lower-case letters. Characteristics of the Aldine innovation were combined with those of the chancery designs. Italic was becoming a junior partner of roman type instead of being a text face in its own right. Specific italics were cut to accompany specific romans, sometimes by the same hand and for the same printer. A style for Old Face italics began to emerge — light, vigorous, steeply slanted — and Granjon was in the forefront of its designers. Evidence of his versatility is provided by a different kind of italic which he cut in 1565 to the order of Christophe Plantin of Antwerp, one of the most memorable of all printers. Less steeply inclined than most italics of the period, and with a more pronounced contrast between thick and thin, this was an admirable companion for roman types. In the Berner specimen it was named

Le befoin n'avilit que les cœurs fans cour
Moi, plein du fentiment dès forces de mo

ef̌,vt Marcellus,feparatis ædibus, Honoris ac Virtu-
tis fimulacra ftatueret. Neque aut collegio pontificum

Figure 6-9*d*. Above, first two lines, Granjon's Gros Cicero (large x-height pica) roman of 1568, 12.5-point × 144%; third line, Granjon's Scholasticalis italic of 1565, 11.7-point × 154%. Below, the first two lines are Monotype Plantin, 12-point × 150%, and the second two are Lasercomp Plantin, 18-point.

his domain without other encumbrances t
he himself was then already burdened. And

Les besoin n'avilit que les cœurs sans
Moi, plein du sentiment des forces de m

Scholasticalis, and as the Fell Pica Italic it is still in use at the University Press, Oxford; but it has not been revived.

In 1568 Granjon cut a roman distinctively different from the fashionable romans of Garamond. It was well-rounded, closely fitted, and above all large on the body, with short extending strokes and capitals. Plantin used some of the sorts with one of his Garamond founts, but the entire fount was not used in his printing-office during his life-time. Plantin's use of sorts to modify a Garamond fount shows a sixteenth-century preference for types which were large on the body, and romans of similar proportions had been cut in the fifteenth century; Carter (1969) reproduces a bold and tightly-fitted Italian roman, obtained by Johann Trechsel of Germany but used at Lyons in 1492, in which as in the Granjon roman the x-height is about 45 per cent of the body. In 1905 Granjon's *gros cicero*, as it was later called, appeared in a type-specimen book of the Plantin-Moretus Museum in Antwerp, and in 1913 the Monotype Corporation revived and adapted a new version for twentieth-century printing.

The new version is slightly larger on the body even than the old, laterally more compact, and in order to print well on glossy coated papers stronger in main-stroke and hair-line. Monotype Plantin was one of the first text faces specifically designed for twentieth-century use, and its popularity seems to have been growing for seventy years after its first appearance. The display sizes lack grace, and the Monophoto version is less compact, but in one form or another type-faces known as Plantin seem still to be gaining ground.

Monotype Garamond

In 1621 the typefounder Jean Jannon of Sedan produced a specimen of his types. The specimen disappeared from the sight of typographic scholars for more than three centuries. His types survived in use to the present day, and are still to be found at the Imprimerie Nationale in Paris, where for many years they were ascribed on good authority to Garamond. In 1918 the American Typefounders Company, looking for an Old Face which could compete with that of the Caslon foundry, settled on that of the Imprimerie Nationale and entitled its revival, reasonably enough, Garamond. In 1922 the Monotype Corporation, under pressure from the British printing industry, cut a 'Garamond' of its own, apparently basing all sizes on a fount approximating to 18-point as printed in 1918 at the Imprimerie in *France-Amérique 1776, 1789, 1917: Méssage du Président Wilson.*

Garamond's text romans, based on the 1495 fount of Aldus, were his highest achievement. His larger display founts were in proportion much more lightly drawn with longer extenders; they were very good as display, but not good models for a range of sizes. Jannon's romans, cut in the following century, followed the fashionable style in the display sizes. It is a pity that they appear to have been the model for all sizes of the revivals, except that the extenders were drastically shortened for twentieth-century use. The result is a distinctly thin design, well-rounded, somewhat sharp in detail, and not well suited to offset printing.

Jannon cut italics to accompany his romans, and these too were revived as part of the new series. They were in the Old Face tradition of italic which Granjon had made popular during the sixteenth century, but Jannon lacked Granjon's skill in balancing the slightly different slants of different letters on each side of a single angle of inclination, and the result looks erratic in larger sizes. The Monotype Corporation accordingly cut an alternative italic of even slant, series 174, and this like the roman and its original italic is available in a photo-composition versions.

Stanley Morison provided examples of swash letters cut by Granjon during the sixteenth century, and these were added to the Monotype italic, together with a number of ligatures (figure 4-4c). Appropriately placed and used with reserve, swash sorts add life and variety to a purely typographic page, but are all too rarely available with other series. The Monotype series includes bold roman and italic. The 'Garamond' design became popular, and other versions have appeared since that of ATF.

que defir de leur cognoiffance. Il eft certain,

La Grâce dans sa feuille, & l'Amour se repose,

Figure 6-9e. Above, Jannon's Gros Parangon (about 22-point) roman and italic, cut by 1621, 22-point × 82%. Below, the first two lines are Monotype Garamond, 12-point × 150%, and the last two are Lasercomp Garamond (roman series 156, italic series 174), 18-point.

An unprejudiced examination of two sides of t
Their skilled craftsmen will improvise any unorthod

An unprejudiced examination of two sides
Their skilled craftsmen will improvise any uno

The names of type designs seem in some instances to have been assigned in the hope of causing confusion. Linotype Granjon is based on a design by Claude Garamond: Monotype Plantin on a design by Robert Granjon: and all the Garamonds on a design by Jean Jannon. At the time of naming, the originals had not been correctly identified in terms of punch-cutters; and now the names have been in use for more than fifty years, they hardly seem to need to be changed.

Monotype Van Dijck

Christoffel van Dyck was the leading typefounder of his period, the middle years of the seventeenth century. As Garamond's roman types over-shadowed others in Europe in the sixteenth century, so did those of Van Dyck in his own time. Cut in the Garamond style, and thus maintaining the Aldine tradition of roman type design, these romans were widely used in England where they were imported with italics, one of which seems to have formed a model for William Caslon sixty years after Van Dyck died in 1669.

The patterns for the Monotype adaptation, introduced in 1935, were the Augustijn Romeyn (roman, about $13\frac{1}{2}$-point) and Text Cursijf (italic, about $16\frac{1}{2}$-point) shown on the 1681 specimen of Daniel Elsevier's widow, reproduced by Updike (1937).

landis, Zelandiſque̱ atque Burgundis Præfe-
ctum deſignaret quando ſe hiſce prefecturis ce
Æadem, is admonenti Gubernatrici ut abiret

Figure 6-9 f. Above, an Augustijn (about 14-point) roman and italic attributable to Christoffel van Dyck, 13.6 × 132%, the italic has been described as the model for Caslon's. Below, Monotype Van Dijck, 14-point × 129%.

bodied roman and the Great Primer italic are
Augustyn Romein and *Kleine Text Curcyf No.*

In the text sizes, the letters are small on the body and narrow. The capitals and small capitals are heavy for their height. The smaller composition founts tend to look thin even when printed from metal type. Like Centaur, the series is elegant, particularly in the larger text sizes, and has not been seen as often as it deserves.

Linotype Janson

Nicholas Kis, a Hungarian punch-cutter, type-founder, and printer, cut a series of founts in Amsterdam between 1680 and 1689 when he was working there. In 1689 he left the founts in Leipzig on his way home to his native country, and they seem to have been acquired by the Janson foundry there. In about 1720 they appeared in a specimen of the Ehrhardt foundry, from which Updike reproduces some of the larger founts. In the twentieth century the matrices were still in use at the Stempel foundry of Frankfurt, at a time when manufacturers of composing machines were searching for distinguished survivals to adapt. In 1933 the Mergenthaler Linotype company in America based its revival version on types cast by Stempel from matrices of about 14-point size, and named the series after Anton Janson to whom the cutting was then attributed.

lographic *Eyn neu modelbuch* (Zwickau, 152 notaries and chancery officials had been simi

Figure 6-9*g*. Above, the Stempel Janson, 14-point × 129%. Below, Linotype Janson, 12-point × 150%, followed by 18-point Mergenthaler Linotype Janson.

pes are referred to as *Hollaendische Schriften,* wn that Anton Janson, a Dutchman, was ow descended from the foundry of Anton Jan *provides a useful and vital old style in a planne*

Linotype Janson is of medium width, and rather large on the body. Alternative long descenders were cut, and to overcome the deficiencies of the duplex matrix in casting small capitals of appropriate set there was a special fount of small capitals in which roman and italic small capitals were duplexed together. Italic small capitals are examples of a rare form of letter for which good use can be found once it is available. Some of the italic letters are a little distracting in detail but are copied from the original.

The Lanston Monotype company in America designed its own version of Janson, and Mergenthaler Linotype adapted their own design for photo-composition.

Monotype Ehrhardt

In 1936, at the request of a Leipzig printer, the Monotype Corporation started work on another Janson, but this work was stopped by Stanley Morison in favour of a different approach. Instead of starting from types cast by Stempel from the Kis matrices, the type drawing office was to work from a reproduction of the Ehrhardt specimen, on which the characters were particularly closely fitted. Such types, heavier and more compact than those of Garamond, had been appearing since the fifteenth century — Carter (1969) shows an example by Johann Trechsel used in 1492 — but were never so widely popular as the more open designs. Morison selected the Tertia size, about 17-point, and required an x-height similar to that of Monotype Imprint (later in this section) and a weight intermediate between that and Monotype Plantin. In closeness of fit, trial versions fell short of the old type-specimen, and at Morison's insistence the characters were closed up. The first sizes appeared in 1938, the largest and smallest at intervals until 1952.

DIxi tefte Deo, experientia, confcientia quæ de

Vitanda imprimis fummo ftudio Simoni a fceleriatiffim

Figure 6-9h. Above, the Tertia (about 17-point) roman and italic from the Ehrhardt foundry, probably cut before 1689, 17-point × 106%. Below, the first two lines are 18-point Monotype Ehrhardt and the next two are 18-point Lasercomp Ehrhardt.

Eighty reams of double demy antique wove bo
The invitations were printed in jet black and turq

Eighty reams of double demy antique wove b
The invitations were printed in jet black and

Monotype Ehrhardt roman and italic are both particularly narrow and closely fitted, heavy, and large on the body, suitable for printing from type on glossy coated paper and hence for printing on any paper by offset. The design is popular and economical, and has been selected for many kinds of book, including the 1956 and 1966 editions of this one. Ehrhardt, in company with Plantin, appeared more often in the British National Book League's exhibitions of book design and production between 1978 and 1980 than did the previous favourites of the 1945–1963 period, Bembo and Baskerville.

The Monophoto version of the Ehrhardt design is a little heavier still but less narrow and less closely fitted.

Monotype Caslon

Harried by restrictions imposed by successive governments and sovereigns, British typefounders produced few if any designs of lasting value until the eighteenth century. Great quantities of matrices and type were imported from Holland, and Dutch types were in general use among British printers; but English punch-cutting was almost at a standstill. Towards the end of the seventeenth century, during which much English printing had been vile, the stringency of the censorship was relaxed, but no initiative in type design was to be seen for some years.

Early in the 1720s William Caslon of London began to cut roman and italic letters in the style of some of the founts then in use in Britain. These were either Van Dyck's or copies of his designs, and Caslon followed some of them so closely that even Updike confused a copy of a Van Dyck fount with one cut by Caslon (Updike, 1937, figure 290; the italic is Van Dyck's and the roman appears in a specimen of about 1710 − Johnson, 1946).

Caslon was the first of the really able British punch-cutters; his types emancipated Britain from dependence on foreign typefounders, and initiated a great era of British typography. His punches and matrices survived him, and in the 1840s his types, by then known among printers as Old Face, began to reappear in British printing otherwise dominated by Modern Face (§ 6-11). Within fifty years the Caslon founts were generally preferred for bookwork intended to present an appearance of high quality; their antique irregularities competed effectively against the mechanical precision of Modern and Old Style types (§ 6-12), and they seem to have been more widely used than they had been in the eighteenth century.

In 1915 the Monotype Corporation introduced its machine-set version. Series 128 is the only example in these pages of a set of Old Face founts based size for size on the original designs of the typefounder; the other revivals are based on a single foundry fount, adapted in the twentieth-century type drawing office for casting in different sizes. The Monotype version does Caslon something of an injustice by fitting the letters more loosely together than those shown by William Caslon in his first broadside specimen of 1734 (Updike, 1937, figure 262).

Even in the 1840s the Caslon founts seemed a little old-fashioned in

A reprint of this specimen, but with a change of impr
edition of Chambers' *Cyclopædia* in 1738, and a not

Figure 6-9i. Above, Caslon's Great Primer (about 18-point) on 18-point body. Below, Monotype Caslon, 14-point × 129%.

At his death the remainder of these purch
gone. Family arrangements required complet

appearance, though the *english* and *great primer* founts display a generous width too often absent from romans generally. Even when most books were printed from metal type the popularity of Caslon was waning, and the survival of these historic founts in photo-composition form seems unlikely.

By the end of the seventeenth century, the Old Face tradition of roman type design, which began with the Aldine founts and continued with those of Garamond and Van Dyck, was already petering out, though roman types of this class were still used by leading printers for another hundred years. In Caslon's type the tradition made one of its last appearances before the twentieth century, and by no means at its best, as may be seen from comparison with the romans of Garamond and the italics of Granjon on the Berner specimen of 1592.

Monotype Imprint

The twentieth-century return to Old Face sprang not from the printing industry but from the Arts and Crafts Movement. Gerard Meynell of the Westminster Press, Edward Johnston the calligrapher, and J. H. Mason of the London Central School of Arts and Crafts devised a new type-face intended to be 'as useful as Old Style series 2 (§ 6-12) but more distinguished in pedigree and less anaemic in appearance' (Morison, 1973). They started with Caslon's great primer, the best of his romans, shortened the extending strokes for economy, and strengthened the main-strokes for printing with half-tones on coated paper. This was the first of such adapted revivals to earn popularity in book production; it is as unobtrusively seemly and useful a type as any version of Caslon. The periodical for which the type was designed, and in which it first appeared in 1913, was intended to promote the cause of excellence in printing, but in due course went the way of other publications of its kind, however good. But the type-face, released to the trade by its proprietors, survives in general use and esteem.

The letters are larger on the body than those of the original Old Faces, and

he himself was then already burdened. And
entail. The idea of an entail was not in acc

Figure 6-9j. Above, Monotype Imprint, 11-point × 164%. Below, Monophoto Imprint, 14-point × 129%.

a camera and that its function is to pre
The first thing to remember about a film

they are sturdy enough for any kind of twentieth-century use. An extensive range of special sorts has been cut for foreign languages and other special settings.

§ 6-10 Transitional

In 1692, some two hundred years after the first use of the first Aldine roman, Louis XIV of France ordered a new series of types to be designed for the use of the Imprimerie Royale. André Jammes (1965) has recorded the research that followed, in impressive detail. Printed books, theoretical studies of letter-forms, and handwriting manuals published since the European invention of typography, together with the most elegant manuscript books of the time, were studied by a committee of the Académie des Sciences and other scholars, and discussed with the royal punch-cutter, Philippe Grandjean. Elaborate drawings were made and engraved. The pursuit of symmetry introduced vertical shading into typography, and extended horizontal top-serifs to left and right of the upper extremities of b d h k and l. Manuscript examples added a vestigial cross-stroke to the left of l, probably from much the same origin as the cross-stroke in sterling £. Logic recommended a sloped roman in place of italic, not for the last time. Some of these innovations, combined with an undoubted excellence of proportion, appeared in Grandjean's *romain du roi*, begun in 1693 and completed by his successor in 1745.

preſt à lancer ; le fleuve de l'Eſcauld effrayé s'app
Légende, HISPANIS TRANS SCALDIM PUL
ſignifie, *les Eſpagnols défaits & pouſſez au-delà de l'*

Figure 6-10a. Grandjean's *romain du roi*, 1693–1745.

No copying of the *romain du roi*, ancestor of the Modern Face, was permitted, and perhaps nobody outside France wanted to copy it. But the influence of this innovation on French punch-cutters was evident, and the intervention of the state and of methodical scholarship in matters of type design demonstrated the possibility of changes which may not have been wholly welcome to punch-cutters and printers everywhere.

Monotype Baskerville

John Baskerville of Birmingham formed his ideas of letter design during his early career as a teacher of handwriting and engraver of inscriptions. By the eighteenth century handwriting was hardly used at all for book production; formality and display returned to penmanship in shorter inscriptions, and brought with them a rounder hand for which the pen was held pointing straight up the page. Top-serifs were still drawn obliquely, in imitation of printed letters. Type-faces which in the same way retain some character-istics of Old Face, particularly in the serifs, while using the Modern vertical stress, are known as Transitional. Baskerville's was among them; he made a fortune in japanning, retired in middle age, set up a press of his own, had punches cut to his own designs, and printed his first book in 1757.

His romans were wide and open, like the 'round hands' of the early eighteenth century. His italics were designed expressly as auxiliaries to his romans; their letters were wider, and more evenly and less steeply inclined than those of Old Face italics. Some of the italic capitals had an engraver's flourish and looked dandified when composed together. Baskerville's fame in his own time sprang not from his letter-design but from the precision of his punch-cutting, paper finishing, and presswork, which were combined in a printed image of a new kind.

Pitifully behold the forrows of our hearts.
Mercifully forgive the fins of thy people.
Favourably with mercy hear our prayers.

Figure 6-10*b*. Above, Baskerville's Great Primer, 17-point × 105%. Below, first two lines, Monotype Baskerville, 12-point × 150%. Last two lines, 18-point Lasercomp Baskerville.

worship had been carried on without fai
fire had never gone down upon the hearth, I s

worship had been carried on without fail,
fire had never gone down upon the hearth, I should

The Monotype Corporation, which may have been less than fair to Caslon, did more than justice to Baskerville. The model was the great primer fount used in Baskerville's 1772 edition of Terence's *Comedies*; the same type appears in Baskerville's broadside specimen of about 1762 (Updike, 1937, figure 270). The letters were regularized, the fit was tightened, and the series was completed in 1923.

Monotype Baskerville is rather wide for its x-height, but the letters are closely fitted and economical of space; their proportions are such that the series is one of the most unobtrusively readable of the twentieth century. Baskerville designed his letters with the intention of leading them, rather than setting them solid in the Old Face manner, and like all vertically stressed letters they benefit from this extra space between the lines, which breaks up a potentially obtrusive pattern of vertical strokes. The capitals, though not particularly tall, are substantially heavier than the lower-case.

Monotype Fournier

Between 1740 and 1770 Louis Luce, the third of the royal punch-cutters of France, designed a series of founts of Transitional style, part of the originality of which was their extreme narrowness. These types, which were sold to printers generally, had a certain amount of success in the open market and were commended by the Academie des Sciences. Among those who noticed this demand for narrow types was the Parisian typefounder Pierre Simon Fournier, also known as Fournier *le jeune*, since he was a member of a family of eminent craftsmen. Fournier derived the proportions of some of his thin narrow romans from Luce, together with the vertical or almost vertical stress of some of his letters.

others. As his *Manuel Typographique* proves, to be taken for an *érudit*. Its author was cert

Figure 6-10c. Above, Monotype Fournier, and below, Monophoto Fournier, both 14-point × 129%.

that its function is to present language in all *The first thing to remember about a filmsetter is th*

The Monotype Corporation produced a facsimile version of one of Fournier's founts in 1925; the originals were the *St Augustin ordinaire* 46 (roman) and 47 (italic) in his *Manuel typographique* of 1764–66. This revival, series 185, copied Fournier's very tall and heavy capitals. For the first Nonesuch Press *Shakespeare* shorter and lighter capitals were cut, and with these the series number is 285. The italic and the inclined arabic figures are decorative after the manner of the handwriting of the day.

Monotype Bell

Improved methods and materials in printing, including the smooth *wove* paper (§ 15-3) used by Baskerville, enabled types to be still more sharply cut

son édition de LA GERUSALEMME LIBERATA deman-
dent avec confiance aux souscripteurs de cet ou-

Figure 6-10d. Firmin Didot's roman and italic of 1784.

Et les vers de Virgile et les leçons d'Horace;
Qui, plus sublime encor, plus noble en son emploi,

in the latter half of the eighteenth century; the *romain du roi* was no longer a novelty, but its influence had not declined. In 1784 Firmin Didot, a member of an illustrious family of Parisian typefounders and printers, produced a true Modern Face, with vertical stress and horizontal serifs. He had already improved the point system invented by Fournier, and brought something of the same mathematical approach to designing his types. The contrast between main stroke and hair line was more marked than had been usual, but was not excessive; the letters were admirably clear and open, and by no means deserved to be travestied as they have been. The italic harmonized with the roman, having the same kind of top-serif, and being wider than Fournier's italics. If the 1784 fount had been revived, the popularity of 'modern' faces might have lived longer than it has.

In the year after the new roman appeared, John Bell, a London typefounder, printer, publisher, and newspaper proprietor, visited type-founders and printers in Paris. He was already known to be an innovator, having popularized the short s within words to replace the long ſ. In 1788 Bell produced the first British Modern face, cut for him by Richard Austin.

sell than to save—seeing that that which he s
his own and not the patrimony of the Dales.
death the remainder of these purchases had gone.

Figure 6-10e. Above, Monotype Bell, 14-point × 129%. Below, Monophoto Bell, 12-point × 150%.

camera and that its function is to present
beauty, variety and intricacy. Its products
The first thing to remember about a filmsetter

Although evidently influenced by Didot's design, Austin's letters retained some Old Face characteristics — the stress was not consistently vertical nor the top-serifs uniformly horizontal; the italic too was closer to the style of Baskerville's than to that of Didot. An original feature was that the figures ranged with each other instead of differing in height, and figures of this kind are known as Modern, lining, or ranging figures. The Monotype Corporation's facsimile revival, copied from Bell's original punches, appeared in 1931.

The letters are large on the body, and narrow; the capitals are heavier than the lower-case, but are commendably low. The small capitals have the same thickness of main stroke as the italic lower-case, which is substantially less than that of the roman.

§ 6-11 Modern Face, or Didone

Monotype Walbaum

By 1791 Didot's new roman and italic were being copied in Germany; I have an ugly little book printed from such types that year in Gottingen. Among German typefounders who followed this French fashion was Justus Erich Walbaum, who is known to have been at work between 1799 and 1836 at Goslar and Weimar. His types were still available from German foundries in the twentieth century, and Oliver Simon of the Curwen Press imported some founts and induced the Monotype Corporation to cut a machine version in 1933, based on one of the smaller sizes which differed from that shown in figure 6-10d.

pletion, and Christopher Dale required re

outlying farms flew away, as such new pur

before; but the old patrimony of the Da

Figure 6-11a. Monotype Walbaum, 12-point × 150%.

Monotype Bodoni series 135

Although it was the Didots who introduced the Modern Face, and although versions of their types are available for machine composition, the most widely used series of Didot-style types are adapted from the designs of Giambattista Bodoni of Parma. The British standard title of this class of type design combines his name with that of Didot. Bodoni was apprenticed to Fournier, and in 1768 took charge of the Duke of Parma's press. He started with Old Face types, and before long began to prefer Transitionals, some of them Fournier's. Under the influence of Baskerville and of the duke's taste for ostentation, Bodoni liked to arrange these types in a manner more spacious than had been typical of the Old Face era. By the last years of the eighteenth century he was moving away from Transitionals and all forms of typographic decoration towards increasingly spectacular Moderns set in a stark and ample style. The machine series which now bear his name are not direct copies of his designs but adaptations of typefounders' copies of Bodoni types. Some of the subtlety of the originals has been lost in the transition, but the general appearance is similar.

betrogen, endlich zu dem Ueberblick
eines *dauernden Uebels* (dessen Heilung

Figure 6-11b. One of Bodoni's founts, 20-point × 90%, above, Below, the first two lines, Monotype Bodoni series 135, 11-point × 164%. The last two lines, Monophoto Bodoni, also series 135, 14-point × 129%.

Of the documents judged, six will be taking
The invitation printed in jet black and tur

camera and that its function is to presen
The first thing to remember about a filmsetter

Series 135 is the most familiar of the 'Bodoni' revivals. It represents him not necessarily at his worst and by no means at his best but certainly at his utmost. More often seen than lighter adaptations, it is more typical of his later work than they, since he did much to set the fashion for increased contrast between thick and thin strokes. Copied in 1921 from types cast in Italy, series 135 was one of the first 'Bodoni' series to be cut for machine composition.

Typography seems at this point to have strayed farther than it should from the archetypes of the sixteenth century in France. New techniques in typefounding, paper-making, and printing were exploited in accordance with theories of design, a taste for show, and an indulgent esteem for the printer as artist. Precedent and practice, economy, and accurate and effective communication were neglected. Such types as this were designed for extravagant pages, and when typographic extravagance went out of fashion, the popularity of 'Bodoni' types began to go with it.

Monotype Scotch Roman series 46

Not all British typefounders were enthusiastic about the Modern Face in its developing form, but they followed the fashion. Bell's British Letter Foundry was sold in 1797, and its chief punch-cutter, Richard Austin, was free to work for other foundries. His own idea of Modern was perhaps best expressed in the founts he cut for Bell, but he went on to produce rigid versions of the Didot and Bodoni style of letter. Founts of this kind were supplied, probably by Austin, to the Wilson foundry, and to the Miller foundry in Edinburgh — Miller had once been Wilson's foreman. Both foundries later sent type to the United States, where the name 'Scotch Roman' was first applied to a Wilson shipment of 1837 and later to other types in this style.

topher Dale required ready money. The flew away, as such new purchases had *but the old patrimony of the Dales rema*

Figure 6-11c. Above, Monotype Scotch Roman, and below, Monophoto Scotch Roman, both series 46, both 12-point × 150%.

camera and that its function is to presen beauty, variety and intricacy. Its produ *The first thing to remember about a film*

Series 46, cut in 1907, was the first British revival of this kind, following the example of the Mergenthaler Linotype Company of New York which adapted a Miller face in 1902. The original Scotch faces were difficult to define as a class, and generalization about their letter forms would have little value. Monotype Scotch Roman series 137, based on Miller's Old Roman Number 3 in a specimen of 1814 and cut in 1920 for the Edinburgh printers R. & R. Clark, is quite different in general appearance from the Wilson 'Scotch' of 1837.

Monotype Modern Extended series 7

The new Miller foundry, later to become Miller & Richard, was perhaps the first in Britain to concentrate on Modern Face to the exclusion of others. Austin cut a number of the foundry's punches, and he may have been responsible for the original of this series. Early in the nineteenth century *The Times* began to use a Miller & Richard Modern for its main text; when that newspaper installed Monotype machines to replace the earlier forms of typesetter previously used by the newspaper, the text face was carefully re-cut for the Monotype and brought into use in 1902 as series 7. Not particularly distinguished in letter form, the face has become familiar to readers of scientific works; for some years this was one of the few series equipped with a full range of mathematical and other special sorts.

money. The outlying farms flew away, purchases had flown before; but the old *the Dales remained untouched, as it had*

Figure 6-11*d*. Above, Monotype Modern Extended series 7, and below, the same design in its Monophoto version, both 12-point × 150%.

camera and that its function is to presen beauty, variety and intricacy. Its produ *The first thing to remember about a film*

§ 6-12 Old Style, or Réale

Monotype Old Style series 2

Throughout the first half of the nineteenth century, Modern Faces held the typographic field against nearly all comers. The same years, however, had seen the rise of the London publisher William Pickering and of Whittingham's Chiswick Press. Pickering was an innovator in book production; he introduced cloth as a binding material during the 1820s, and was one of the first publishers to instruct his printer in details of book design. Equally original was his liking for old types at that time neglected and unfashionable. He had already made some use of Baskerville's types, and in the 1840s he and Whittingham began to make use of type cast from Caslon's matrices.

The reappearance of an Old Face, as Whittingham called the Caslon types to distinguish them from those then considered Modern, revived interest in letters of this style. The mechanical regularity of the Modern Face had its effect on the taste of the day, and printers tended to consider some of Caslon's founts too irregular in form for satisfactory use.

In 1858 Miller & Richard, from the forefront of Modern Face production, led the way to a new development by issuing specimens of a regularized Old Face which they named Old Style. The new class of Old Style types, of which this was the first, reverted to gradual shading and to oblique top-serifs, but retained vertical stress. It was a good deal thinner in stroke and larger on the body than the best of the Old Faces. Types of this kind became popular in the second half of the nineteenth century, without ousting Modern from the market. The Monotype was introduced to British printers in 1901; the first two type faces cut for it in England were Modern (series 1), which appeared in 1900, and later in the same year Old Style (series 2), adapted from a successful design more recent than that of Miller & Richard.

money. The outlying farms flew away
purchases had flown before; but the ol
the Dales remained untouched, as it had

Figure 6-12. Above, Monotype Old Style, and below, Monophoto Old Style, both series 2, both 12-point × 150%.

The first thing to remember about a film
camera and that its function is to presen
The first thing to remember about a film

As in other types of its class, the design is thin and colourless, but has shown a certain affinity for printing by photogravure, and has been skilfully modified for small sizes. The series gained ground by being early in the field and by being equipped with a large range of special sorts for foreign language composition.

§ 6-13 Innovations 1895–1971

The type designs described so far in this chapter were originally cut for hand composition. In the Monotype range of metal types, display founts are still set by hand, but this chapter is concerned with text sizes. Metal types for hand composition in text sizes were still being cut at various foundries throughout the first half of the twentieth century, but those shown here were soon adapted for use with mechanical typesetters. With an effort of the imagination I could place some of these in one of the preceding classes of type design, but none of them looks very much as though it belongs there.

The first founts of the Century series were cut by Linn Boyd Benton, inventor of the first punch-cutting machine, in collaboration with Theodore Low De Vinne of the De Vinne Press in New York. The type was intended for the *Century Magazine* which De Vinne printed, and in 1895 it replaced a less robust and legible Modern. Century Expanded and other variations followed, including Century Schoolbook, adapted for the Monotype in 1927; by then Century had been cut for the Linotype and used by a number of the larger American newspapers.

When jobs have type sizes fixed quick determining calculations are based up *When jobs have type sizes fixed quick*

Figure 6-13a. Above, Monotype Century Schoolbook series 227, 14-point × 129%; below, the Lasercomp version, 18-point.

When jobs have type sizes fixed quick determining calculations are based up *When jobs have type sizes fixed quickly*

Like other Century designs, Century Schoolbook is unusually large on the body, and has drastically abbreviated descenders. The letters are strongly drawn, with emphatic serifs; their clarity, welcome to young readers, is partly due to the large x-height and ample width — best balanced by extra interlinear space — and partly to open counters and firm strokes. The individual letters are ugly by comparison with those of the Old Faces already described, but they combine well enough to show that vertical stress can be compatible with legibility. The angle of stress seems in fact to make very little difference until there is a marked contrast in weight between the shaded part of the stroke and the hair-line, as in Monotype Bodoni series 135. In the same way the angle of top-serifs, between horizontal and oblique, seems to have less effect than the presence or absence of serif brackets.

Monotype Perpetua roman and Felicity italic

The Monotype Corporation's first commission for a new type design by a living artist was entrusted to Eric Gill by Stanley Morison, the Corporation's typographic adviser. Gill was a sculptor and engraver who cut letters but made no claim to be a type designer. In his first essay in this new medium, he made no attempt to shape the letters to meet the requirements of the machine or indeed of printing; the letter forms were those he had evolved in the 1900s for his engraving and sign-writing. Morison's historical studies had revealed to him the influence of these two crafts on letter forms in fifteenth-century typography, which were then believed to be derived from formal handwriting only. He had also observed the punch-cutter's modifications of letter forms to suit the printing process. Logical as always, he had Gill's drawn designs translated into punches by Charles Malin of Paris, and the Corporation adapted the resulting designs to fit their machinery.

topher Dale required ready money. The outlyi
away, as such new purchases had flown before
patrimony of the Dales remained untouched, as it had

Figure 6-13*b*. Monotype Perpetua roman and Felicity italic, 13-point × 138%.

The first founts of the series were completed in 1928, and the first printed specimen was a translation of *The passion of Saints Perpetua and Felicity*; the italic was accordingly named Felicity. Gill was a man of ideas and of strong personality, and the original form of some of the letters shows that the design of alphabets for printing still has achievements to offer to designers of the stature to reach them. Gill, for example, had been cutting italics in sloped-roman style for some twenty years before the appearance of Felicity; the proportions of his capitals were based not on typographic precedents but on engraved inscriptions of imperial Rome; and the length of his extending strokes in proportion to x-height was greater than in any other design examined in this chapter. Typographers of today are likely to prefer Perpetua and Felicity to Grandjean's equally innovative *roman du roi*, and the Académie des Sciences might well have done so too; but neither has matched in influence and popularity the 1495 roman of the Aldine Bembo or its twentieth-century revival, and in the 1930s the demand was for such faces as Plantin which printed well from type on hard smooth papers. The demand today is for similar types because they print well by offset (§ 11-5); as a text face, Perpetua arrived at the wrong time.

One of the characteristics of a good text face is the absence of conspicuous contrast in weight between capital and lower-case letters. The Monotype Corporation, at least in Morison's time, seemed to prefer contrast; in Morison's revived series the thinnest capitals, those of Ehrhardt, were 20 per cent thicker in main-stroke than the lower-case, and the thickest, Fournier, 40 per cent. The capitals of Perpetua were little more than 10 per cent thicker in stroke than the lower-case, and when used for display in sizes below 30-point they look distinctly thin. Accordingly Perpetua Titling was designed; the use of a single series or family of type throughout any book was becoming fashionable after the Victorian period of confused typographic mixtures, but Perpetua Titling has been used effectively with many different text designs. This shows that the general proportions of a face can be of greater value than details of profile, as the titling capitals are less well drawn than those of the 30-point version of the text series. Perpetua Light Titling is not well drawn either, but its proportions are better suited than those of any other series to large bookwork initials which at present are unfashionable.

Monotype Times

The Times was content for more than a hundred years with its Miller & Richard Modern Face, and with Monotype Modern Extended, the facsimile re-cutting which superseded it. This was not due to conservatism; *The Times* had more than once led the way towards new developments in printing technique, and had not failed to observe the introduction of new type-faces by the manufacturers of composition machinery. But in October 1932 the newspaper appeared in a completely new series and family of types throughout, more practical and more handsome than the old. The series

ory, as indeed had been done by Kit Dale,
pher Dale, who will appear as our squir
which he himself was then already burden

Figure 6-13c. Above, Monotype Times New Roman series 327 with long descenders, 10-point × 180%. Below, Lasercomp Times series 727 with lighter capitals drawn for the German market, 18-point.

pher Dale, who will appear as our squire
which he himself was then already burdened

was drawn by Victor Lardent of *The Times* and cut by the Monotype Corporation under the supervision of Stanley Morison, who was typographic adviser to both newspaper and corporation. A year later Times New Roman was offered to the printing trade in general, as Imprint had been twenty years before. In the next forty years, 'Times', as the series is generally known, became one of the most widely used of all type designs for bookwork in spite of its newspaper origins, and although it is useful rather than comely.

In structure the roman letters resemble those of Monotype Plantin; the first drawings by Lardent seem to have been based on a type-specimen sheet, perhaps from the Monotype Corporation. As in Plantin, the letters are unusually narrow and closely fitted, large on the body, and strongly drawn. The capitals are less than 150 per cent of the x-height, but still tend to mar the page by being 133 per cent of the lower-case in main-stroke thickness, and with lighter alternative capitals series 327 is known as series 727. The roman might be classified as an adaptation of an Old Face, but the italic is distinctly Modern in cut.

The larger founts demonstrate the design's lack of grace: the smaller, its clarity and economy. The range of sizes enlarges the usual boundaries, starting with $4\frac{1}{4}$-point, cut for small advertisements in *The Times*. The Times family of type faces contains more series than others. Times Condensed (series 724), Times New Roman Wide (427), and Times New Roman Book (627) are closely related to series 327. Times Bold (series 334) mates with series 327, Times Bold Condensed (725) with series 724, and Times New Roman Semi-bold (421) with nothing. Four of the titling series are evidently relatives, and a fifth (Hever Titling, series 355) appears to have been adopted rather than born into the family. There are Greek and Cyrillic series, the mathematical series 569 — a giant among the offspring — and constellations of special sorts for languages and sciences. This immense family has become a mainstay of routine typography, used by almost every kind of designer and printer for every imaginable kind of publication; and even those who favour variety in type selection tend to subside into Times.

Palatino (Stempel foundry)

Palatino was designed by Hermann Zapf for the Stempel foundry of Frankfurt in 1950, and takes its name from Giovanni Battista Palatino of Rossano in Calabria, an internationally famous calligrapher whose manual of handwriting was published in Rome in 1540. Adopted by Linotype & Machinery (at that time the name of the British Linotype company) first for metal and later for photo-composition, it has been gaining ground in British book production for some years. The design is almost immodestly idiosyncratic, but its success in a field strongly fenced in with traditions is deserved. The strongly-drawn letters are well suited to offset printing; their width makes them clear in small and medium sizes; and in large sizes they can be seen to be both unusual and graceful — although all sizes are reproduced from a single set of drawings.

This is the Stempel foundry's *Palatino*, set in 18-point with 2-point lead on a

Figure 6-13*d*. Mergenthaler Palatino, set by V-I-P photo-setter, 18-point.

Monophoto Univers

Adrian Frutiger designed the Univers family of type-faces in 1954 for Deberny & Peignot, the Parisian typefounders, and for the Lumitype photo-typesetter. Univers was the first design intended for both metal casting and photo-composition. It is still one of the most extensive families of related type designs; there are fourteen different forms of roman and italic in the Monophoto range, for instance. The Monotype Corporation introduced its metal version in the early 1960s, and the Monophoto versions were among the corporation's earliest photo-composition faces. The most familiar founts for book production are Univers Light, Medium, and Bold, Monotype series 685, 689, and 693.

Sans-serif types, of which the Univers family is a successful example, have been described as visually representative of the twentieth century, but their origins lie far back in the nineteenth. The first known type of this kind was shown in a type specimen of 1816 by William Caslon IV, who called it

People very often read books without the choice of type face or the many e
He won the first prize for very quick

Figure 6-13*e*. Monotype Univers Medium series 689, 14-point × 129%, above. Below, Lasercomp version of the same design and series in 18-point.

People very often read books without the choice of type face of the many e
He won the first prize for very quick and

'Egyptian'; the first references to the present title of this class of designs was by Blake & Stephenson in whose 1833 specimen it was entitled 'Sans Surryphs'. Thorowgood, who showed an example in 1832, called his 'Grotesque' and this term is still in use in Britain.

The sight of a page or more set in sans-serif types seems still to be unwelcome to readers in general, or at least is believed to be so by people who determine the selection of type-faces in books. The preferable legibility of types with serifs may be supported by theoretical factors, but tests have shown that a strongly-drawn sans-serif may be easier to read than a spindly type with serifs, if only because stout strokes are more easily observed than slender ones. Probably the soundest reason for not using sans-serif types for text in books is that people are not used to it, might not like it, and would not buy the books. If they would only get used to buying them, they would have little difficulty in reading them, if the right type-designs were presented in the right way. But until that happens, sans-serif types will usually be restricted to display and other abbreviated forms of wording.

Monophoto Apollo

In 1964 the Monotype Corporation commissioned the design of Apollo from Adrian Frutiger, who had designed Univers and adapted metal faces for photo-composition. This was one of the first types to be specifically designed for photo-typesetting, and part of the brief was that all sizes from 6 to 24-point were to be reproduced from a single set of type drawings. The series was designed to suit offset printing and does so. The lower-case letters are a little loosely fitted, and gain from the slight closing-up possible with some systems of photo-composition. In relation to the x-height the capitals are lower than those of any other series shown in this chapter, and Apollo is accordingly well suited to blank verse in which every line begins with a capital.

The first thing to remember about a film camera and that its function is to presen
The first thing to remember about a film

Figure 6-13*f*. Monophoto Apollo, 14-point × 129%.

Monophoto Photina

Introduced by the Monotype Corporation in 1972, Photina is an early example of a type-face originally designed for photo-composition. The designer, José Mendoza, was asked to bear in mind the requirements of metal composition and letterpress printing in case demand were to appear, but the series was particularly intended for photo-composition and offset printing. All too many types designed with such intentions combine too loosely into words, particularly with the close word-spacing possible in computer-aided photo-composition; Mendoza was invited to design a closely-fitted face, and he did so.

The Corporation's design brief was complex. The basic design was to be capable of adaptation in both weight and width into a family of related series. The first of the relatives was to be semi-bold rather than bold. The founts of medium weight were to lend themselves to composition with Univers Medium and to have the same set. And the minimal kerns seem to suggest that the design would be suitable for vertical scanning although such systems formed no part of Monotype plans.

The first thing to remember about a film camera and that its function is to present
The first thing to remember about a filmsetter

Figure 6-13*g*. 18-point Lasercomp Photina.

The resulting design has additional advantages. In proportion to the x-height, the *waist-line* or thickness of main-stroke is part-way between Plantin and Imprint, two early and sucessful essays in twentieth-century type design; most photo-composition faces are etiolated in comparison. The x-height of the 12-point fount is only fractionally smaller than those of Times and Plantin, largest on the body of all the types mentioned in this chapter; Mendoza has shortened the descenders of Photina with such skill that they are neither conspicuous nor ugly. The lower-case letters are not particularly narrow, but they are so closely fitted that the a–z length as a multiple of x-height is less even than that of Monotype Plantin. In waist-line the capitals and small capitals are similar to the lower-case, a characteristic unique to Photina among the type-faces described here, and the capitals are lower in proportion to the x-height even than those of Times. A page containing capitals, as almost any page is certain to do, and small capitals therefore presents an even and unspotted texture of strokes, but isolated in display the capitals and small capitals may look comparatively thin.

In small sizes the letters are more legible, and in large sizes they are more graceful, than those of most other photo-composition founts. No more valuable photo-composition series has yet appeared.

§ 6-14 Families, mates, and descendants

Typographic families consist of several series related to each other in design. Bertram Goodhue's Cheltenham, designed in 1896 for Mergenthaler Linotype, was an early example of such a family. Its offspring included Light, Medium, Bold, Condensed, Wide, Elongated, and Open versions of an unattractive design which became disproportionately popular as such inferior types do from time to time. The fourteen Monophoto forms of Univers already mentioned provide a better and more recent example.

The provision of special sorts to accompany the rest of a fount, even under a separate series number, is not usually understood to be the making of a family. Fournier, for example, includes series 185 as standard and series 285 with capitals reduced in height and weight, but a family is not usually made by two series alone; and even if it were, the latter series consists for the most part of sorts from the former. Again, the provision of a mated bold type to accompany a fount of medium weight does not usually warrant definition of the pair as a family. The term seems usually to be applied to groups of three or more series.

The addition of a titling series, where there is already a bold version, seems to justify the use of the term, but most families include more variations still. The variations themselves vary from family to family. Jan van Krimpen's Romulus, for example, designed in 1931 for Enschedé en Zonen of Haarlem, begot a semi-bold and a bold condensed, four weights of mated sans-serif, and a unique Cancellaresca Bastarda italic.

The use of a contrasting type, usually a bolder one such as the Fann Street Foundry's Clarendon of 1845, in the same line as the text face was an innovation of the nineteenth century, providing a compact and conspicuous form of sub-heading for works of information. The text face was normally unrelated to the bold, and quite often the two founts were out of vertical alignment with each other. By the end of the century such mates as Cheltenham and Cheltenham Bold could be set together in the same line, compatible in every way, and this may have been one reason for the mysterious popularity of the Cheltenham design. As machinery manufacturers took over the design and provision of types from foundries in the twentieth century, a mated bold became part of any generally useful fount of text size. Of the 26 designs examined in this chapter, only 6 lack a mated bold — Centaur, Poliphilus, Van Dijck, Janson, Fournier, and Scotch Roman — and all the rest are accompanied by a bold, semi-bold, heavy, or some other weighty companion of the kind. The earliest of such companions was Monotype Plantin Bold series 194, and the earliest pair to be cut together by the Corporation was Grotesque (sans-serif) series 215 and Grotesque Bold series 216 — these three all appeared in the later 1920s.

Bold italic founts are comparatively rare in metal type-setting, because of mechanical limitations on the number of alphabets deployed together in metal composition, but bold italic is becoming a commonplace in full-scale photo-composition founts. Bold founts, like sans-serif founts, usually consist of four alphabets — roman and italic upper and lower-case — and have no small capitals.

Punch-cutters, typefounders, and machinery manufacturers have been copying and adapting each others' designs since the European invention of typographic printing in the mid-fifteenth century. Machinery manufacturers, moving on from metal typesetting to photo-composition, copy and adapt their own designs for the new process, and as they develop new forms of photo-composition they copy and adapt their own copies. The Monotype Corporation, for example, having triumphed with Monotype Bembo, offered Monophoto Bembo in three forms — matrix-case A for 6 and 7-point, matrix-case B for 8 to 12-point, and matrix-case C for 14 to 24-point. Printers collectively preferred the economy of a single matrix-case for all sizes, and few invested in A or C. B became more commonly used, but among the discerning was considered to be an inadequate replacement of one of the best of all metal type-faces from any period. Accordingly the C type-drawings were adapted for digitization and Lasercomp output. But the title 'Lasercomp Bembo series 270' does not explain any of the differences between one Monotype version of Bembo and all the others, although the differences are quite conspicuous. In the era of metal typography, Monotype Bembo decently printed by one printer could be relied on to look almost exactly the same as Monotype Bembo decently printed on a similar paper by another printer, but there were differences in design between founts of large, medium, and small size. In photo-composition the founts of all sizes are exactly the same in design but the designer should no longer assume that the design is in all respects the one he used last time or saw used elsewhere, even when the same printer set a type of the same name on the same type-setter earlier in the same year. Type-face classification, identification, nomenclature, and series numbering deserve some study, but the best use of theoretical knowledge in practice is to inform the designer's constant and vigilant observation.

§ 6-15 Type design past and future

The selection of type designs for discussion in this chapter may appear to some extent irrelevant. Old Style series 2 and Modern Extended series 7, for instance, were beginning to be unfashionable when they were first cut in the 1900s; Caslon, Fournier, and Walbaum are ill suited to offset printing and rarely seen; Granjon and Janson have never been quite as popular among British designers as they deserved. Most of the designs were originally cut for metal casting by Monotype. The present and the future lie with photo-composition, many other types have been more recently designed, and Monotype and Linotype systems are no longer all but alone in book production.

But these types are not yet wholly out of date. In the British National Book League's annual exhibitions of book design and production in 1980 and 1981, all but four of the types were used in one or more books. The absentees were Bodoni, Janson, Modern, and Old Style. Only four type-faces omitted from this chapter made more than one appearance in those exhibitions — IBM Press Roman, Monotype Dante, Monophoto Ionic, and one or more inadequately identified designs by the prolific Fred Goudy. The dominance of Monotype systems is not now as complete as it was; of the 40 titles shown in the 1981 exhibition, 17 were set by Monophoto, 15 by Monotype, 2 by hand, 2 by Linotron, and 1 each by Linoterm, Linotype, Compugraphic, and VIP. Of the 20 typefaces used for these 40 titles, 16 are shown in this chapter; the other 4 were Gill Sans, Helvetica, Bulmer, and Melior. Typefaces first introduced in or before 1972 seem still to hold the middle ground in British book production. Newer designs have yet to prove equal popularity.

Every designer has to work out for himself the criteria by which he assesses any type-face, or the range of type-faces available with any type-setting machine, or the repertoire of any composing-room, but some criteria seem likely to be common ground. The general proportions of the characters are probably of greater importance than their details; one legibility study (Cheetham & Grimbly, 1964) indicates that Gill Sans may be easier to read than lighter type-faces with serifs. Some variety of proportions is useful; wide or narrow faces seem to suit wide or narrow measures respectively, and one of the wider faces, Baskerville, has been described as one of the most legible. A reasonably stout waist-line, combined with hair-lines distinctly thicker than a hair, has been welcome since 1913 when Plantin was introduced for twentieth-century use. Such designs, intended for the sharpest of reproduction on the glossiest of paper, are best for printing by offset on any paper. Spindly letters, which need to be thickened by a relief impression into soft paper, are less useful today. Very thin unbracketed serifs tend to disappear when photographically reduced to small sizes in photo-composition.

As long as all sizes of type are reproduced from a single set of type-drawings, designs intended for such a method, as Photina is, will be more successful than adaptations from the best of metal series. Unless such letters are closely fitted, they will tend to look scattered in large sizes, and even in text size may not combine well into words separated by narrow word-spacing. Unless the letters are strongly drawn with open counters, they will be difficult to read in small sizes. Close fit now tends to be preferred; display letters composed by *dry transfer* (§ 14-7) by designers rather than by compositors are often crowded more closely together than necessary.

Other characteristics are desirable in type-faces in general. They are not always found in revivals, because type-designs of previous centuries are not necessarily ideal in every way for today's books. The tall, bold capitals of Fournier series 185, for example, are too conspicuous on the page; the short, slender capitals of Photina are much to be preferred. The small capitals of Bell, thinner in stem than roman lower-case, are feeble in appearance; those of Janson may exceed the roman lower-case by rather too much but at least err in the right direction. The italics of Garamond and Caslon are more steeply and erratically inclined than usual, and in comparison with their romans are conspicuously thin-stemmed; it is a pity that Robert Granjon's more upright and more strongly drawn 'Scholasticalis' of 1565 has been neglected by revivalists. Ancient printing still offers starting-points for today's type-design, but the most enduring success has attended adaptations such as Imprint and Plantin. A pedigree is not always an assurance of satisfactory behaviour in all circumstances.

Details should never be allowed to intrude upon the reader's notice. A single unfamiliar or conspicuous character may be enough to cast doubt upon a fount; a clumsy asterisk or an eccentric set of arabic figures has before now marred an edition, and the Photina comma has distracted more than one author by its similarity in small size on a photocopy proof to a full point. Even departures from precedent, such as the use of separate characters instead of logotypes for some or all of the combinations fi ff ffi ffl fl *fi ff ffi ffl fl*, is questionable.

Type design and photo-composition are still developing, and may well continue to do so for years. The pace of investment and advance has become slower in a colder economic climate. First-class type design, like first-class typographic composition, cannot always be afforded. Until only the best is once again good enough for discerning publishers and printers, book designers will do well to form and preserve their own standards of typographic excellence, even in times when such standards cannot be applied to much of their work.

CHAPTER 7

Text pages

Success in book design depends mainly on success in the design of the text page. The most admirable solution of all other problems can do nothing whatever to redeem a failure here. Text design is the foundation of the typographer's craft, and can be the pinnacle of his achievement. Reputations may be made in more spectacular activities among illustrations and preliminary pages; it is here that the author and his reader can best be served.

One text page of any book is usually very much like another, and most books contain scores or hundreds of them. As though coaching a friend for a public speech, the typographer must make sure that the book, which is to address the reader at length, does so without either monotony or distracting gestures. Its manner must be persuasive enough to attract and to retain the attention of those who glance into the pages to see whether they wish to read or buy.

Text composition must not only *appear* to be legible, it must *be* legible. The casual glance must first be attracted to the printed words, and then must be invited to travel along the lines. Any oddity or ambiguity that may catch at the reader's eye or interrupt the rhythm of reading is better avoided.

A person equipped with determination and adequate eyesight can read almost anything without conscious difficulty; down to a certain level, bad printing is not incapable of being read. Printing which is bad below this level is too bad to read, and is therefore not worth carrying out. The limit is rarely if ever passed, but is all too often approached, in newspapers rather than in books, and in terms of frequency of mistakes in composition rather than of design. Reading without conscious difficulty is not the same as reading without strain. Type which is too small or too large, set in lines which are too close together or in words which are too far apart, blunts the pleasure of reading, and tends to discourage all but the enthusiast.

The unpopularity of even so small a change as the omission of serifs from a text type indicates that any printed detail may have its effect on the reader. The details of a page which is to be printed can be varied to a considerable degree: each type face has its peculiarities of serif, stress, weight, set, fit, and size on the body, and each fount can be set in a number of different measures and spaced differently both laterally and vertically. Ink of various degrees of black may be printed by different processes on papers of a variety of surfaces and shades, and there are other variables which are examined in other chapters. It is by a precise adjustment of all the variables of the text page to each other, under the guidance of a practised and sensitive eye, that legibility is achieved.

To devise a legible form of paragraph or page is not enough. Except at the very beginning or at the very end of an open book, two facing pages always present themselves, side by side. An *opening*, as these two pages are called, is the basis of every good design, which links and balances the two: they, and not a solitary page, are what the reader sees. If the pattern evolved for the opening is repeated throughout the text pages and indeed throughout the book, and if an appearance of intentional placing and alignment is imposed on every opening, the reader will develop, as he advances through the text, that confidence in the familiar and the expected which is the foundation of legibility.

A traditional part of this pattern is the even depth of full text pages throughout the book; also throughout the book, text area *backs* (is printed on the back of) text area, and as far as possible line backs line. The value of this tradition demonstrates itself when paper is inadequately opaque. Whether or not the position of the text area is or should be traditional in this way, if design is planning which governs the appearance of the edition, then part of design is the selection of a printer who will back the sheet in register with the front, and of a binder who will fold and cut squarely and accurately. If a plan is ruined by careless manufacture, the selection of suppliers may have been an inadequate part of the plan.

Even this is not enough. Clarity of presentation is essential. Any opening may contain diverse elements, which have to be combined with the main text and with each other in a coherent order and position. Extracts, tables with their headings and notes, footnotes, sub-headings, and illustrations need arrangement into a consistent succession of interruptions of the text, placed there in positions not only visually neat but textually apposite. The opening is a pattern to the eye, but to the mind it is a limited area of information. If a table is referred to at some point in the text, that table should if possible appear in the same opening as that point.

§ 7-1 Punctuation

Most punctuation is governed by convention, and is an editorial rather than a design matter. Author and editor may however agree with a typographer that some punctuation is inessential, and inessential punctuation is better omitted, in order to emphasize by contrast the presence of essential punctuation. While the clarity of the words is an editorial responsibility, the clarity of the typography is that of the designer, who is therefore entitled to propose that simplifying punctuation may contribute to legibility.

The simplification of displayed lines (chapter 9) is generally allowed to take precedence over the textual conventions of punctuation. The omission of full-points at the end of displayed lines is now a convention, and the typographer may reasonably claim some right to simplify punctuation more freely in display than in the text. Any contrary preference on the part of editor or author could without great expense be applied at proof stage, but simplification must not be allowed to obscure the meaning of displayed lines.

The full-point is an example of a mark which is apt to lose its force when used too often. Its chief purpose is to indicate the end of a sentence; when it appears within the sentence as well (as it may after an abbreviation), the impact of its presence at the end of the sentence is reduced. When there are many full-points within sentences, the text may appear to be so frequently interrupted as to be forbidding to the reader. The omission of full-points from *contractions* (abbreviations which include the last letter of the whole word), and from groups of initials such as USA, has now spread widely enough to be termed a convention. In assessing the need for certain uses of the full-point in a modern context, the book designer may bear in mind the probability that the author would have omitted them if he had realized that in his printed text they might convey a slightly old-fashioned appearance.

Before decisions about the style of punctuation become final, their effect on all instances in the text needs consideration. While they may for example suit abbreviations in the first few folios of copy, they might result in ambiguity if applied to abbreviations in later chapters. If they would, a specified exception or two may be seen as intentional rather than as clumsily inconsistent.

For most examples of quoted speech or writing, single quotation marks or *quotes* are now generally preferred. Double quotes are reserved for the minority of quotations which appear within a quotation. Many authors and typists use double quotes for the majority. This enhances the clarity of the typescript, and may do as much in an edition intended for young readers, but does not otherwise seem necessary in print, and can disfigure a page with a multiplicity of points. Apart from this, an author's punctuation is to be amended for typographic reasons only with discretion. Whether the publisher's contractual responsibility for the control of book production extends to punctuation does not seem to have been tested in law; under pressure, most publishers would probably concede an author's right to his own use of punctuation conventions. If an author were to reject punctuation in proof, the cost of rectification would be high.

That punctuation should as far as possible be so spaced as to maintain the text's appearance of close and even spacing is now generally agreed. The tendency of some nineteenth-century compositors was towards more than ample space — an em or even more between sentences, for instance, and proportionately wide spacing round some marks of punctuation. Today's practice tends to the other extreme, placing all punctuation hard up against the word it follows. This perhaps is rather too tight; in continuous setting, some punctuation marks may not be distinct enough from the rest of the text. Punctuation marks which rise from the main line as do : ; ? and ! may be clearer when separated from the preceding letter by a hair-space, and this will usually be possible (figure 7-1).

The forms in a simpler sentence are:—Exclamation: *What I have suffered!*; Question: *What have I not suffered?*; Exclamation with inversion: *What have I suffered!*; Confusion: *What have I not suffered!*

Figure 7-1. Above, punctuation close up against the words, in Lasercomp Plantin. Below, spaced punctuation, in Lasercomp Times series 727.

The forms in a simpler sentence are:—Exclamation: *What I have suffered!*; Question: *What have I not suffered?*; Exclamation with inversion: *What have I suffered!*; Confusion: *What have I not suffered!*

Conventions include the setting of an unspaced em rule—as here—'like parentheses or a pair of commas, before and after a parenthetical clause' (Oxford, 1981) among other uses. Without any space on each side, the rule appears to link rather than to separate words, like an elongated hyphen. A spaced en rule (–) or $\frac{3}{4}$-em rule may be preferable for the majority of purposes; one of these will be visible enough without being conspicuous. The em rule could then be used for the role assigned by Oxford to a two-em rule, 'to show that a sentence is left unfinished' (without space between rule and preceding word) and 'to denote the omission of a part or the whole of an undesirable word' (as in 'd— you, sir!', 'a b—y fool', and 'he called him a —', though such reticence is no longer in fashion).

To reduce the length of rules in this way, and to exchange double quotes for single, are among the few amendments considered permissible in extracts, which otherwise should be spelt and pointed in accordance with the original document.

§ 7-2 Text area

When the opening is designed as a whole, the dimensions of text area, margins, and leaf are selected to suit each other. If an earlier choice of leaf size is found to be incompatible with the intended text area and margins, the designer's plan will have to be changed at one of these points. The text area may be an early element in planning, particularly when it is to enclose such intractable items as tables, illustrations, or wide equations.

The selection of the measure is linked with that of the text fount; the width of one determines how many characters of the other there will be in an average line. Measures expressed in whole picas are convenient for metal composition, as the composing room is likely to possess spacing material cut or cast ready for such measures, but in Monotype keyboard and caster operation the measure is applied in units of set, and can be adjusted to the nearest unit. In the same way, photo-composition measures can be adjusted by fractional amounts, without reference to spacing material of fixed dimensions. In specification, measures are usually expressed in picas and fractions of a pica. The measure of this page is 26 picas; this contains on average 73.7 characters and spaces — more economical but less legible than fewer characters in a larger type or in a narrower measure.

The nature of the text may influence the choice of measure, as in verse. When the text is made up of many short paragraphs, as in an index or a dictionary, a narrower measure, whether in two columns or not, will reduce the frequency of white spaces within the text area.

The depth of the text area is governed by the number of lines of main text in a full page and by the body of the text fount including interlinear space. The clearest description of the text page's depth defines these elements and adds headline and folio, which increase the over-all depth; the depth of this page is 43 lines 10 on 12-point + headline + folio.

To use the same text area throughout has become a tradition of bookwork. The printer tends to adhere to this tradition in order to simplify imposition; the designer does so in order to provide the reader with an even pattern in opening after opening, and to emphasize the unity of the book as a whole. The text does not occupy the whole area on all pages, but on no page does it occupy more as a rule, even when this would enable the compositor to avoid a *widow* (a short line at the head of page or column) or to add a footnote linked to the last line of text. A similar tradition, not always honoured today, backs text area with text area; if the paper is inadequately opaque, part of the backing text area will otherwise show through the margins on the front of the page.

The compatibility of the main elements of the text opening can be verified by means of a diagram. This should indicate the cut edges of the opening, the outlines of text area and headlines as enclosed by the intended margins, perhaps the page number, and a line to represent the spine fold. This kind of verification should precede any final decision about dimensions or positions. If the diagram is started in pencil, lines which need adjustment can be rubbed out and re-drawn. Accurately drawn, perhaps on tracing paper which can be laid over a page of proof, such a diagram may prove a useful guide to the printer. The judgement of the designer's eye is paramount; the printed opening will be judged, and must be planned, by eye.

A diagram of this kind, representing the outlined elements of a page or an opening, photo-copied or printed in appropriate numbers, is known as a *grid*. Designers planning page make-up assemble proofs of text and illustration and paste them up into position on such grids to verify their plans and clarify their instructions.

Like those of the leaf itself, the proportions of the text area have their own aesthetic value. The appearance of the book and its legibility may be enhanced by a measure narrower than usual in proportion to the text depth.

§ 7-3 Founts and alphabets

Founts within a series may differ visibly from each other in appearance. In metal-cast series, two or three sizes of type may be cut from a single set of type drawings, but differences of detail and proportion between characters of small, medium, and large size may be quite easily seen. Monotype Caslon is an unusual example of a revival based on the founder's original variations between founts of different size, some of which are quite different in character from others. In photo-composition, when small founts are normally reproduced from the same matrices as large, the quality and general effect of one size may still differ from those of another. Careful choice of a series therefore entails examination and comparison of those of its founts which are to appear together on the page.

The set of the main text fount should be such that an acceptable number of characters will fit into each line. The a–z length of lower-case roman is not an accurate guide to the average width of characters in composition, as the proportion of wider characters is higher in the alphabet than in text; the lower-case roman a–w length is a more likely guide to the width of 26 composed characters. If the line is to contain fewer than 50 characters, close and even word-spacing, with comparatively few word-spaces to adjust, may induce an excess of word-division at line ends. More than 80 characters in the line may lead to discomfort in sustained reading (figure 7-3). There should be no such difficulty in measures containing the equivalent of two to three alphabets.

Small-type items such as extracts and footnotes may have too many characters in the line, by the standards applicable to the main text, but this should be tolerated when such items are rarely longer than a few lines. The type in which these words are set might have been too small for a work including long passages in smaller type still. Footnotes in a quarto or folio work may benefit from setting in double column, in order to reduce the number of characters per line.

There is, therefore, little excuse for thinking that conditions of labour today are very different from those that long preceded them; and it is important to realize that these conditions were all along factors, as they are now, in the problem of turning out good printing. Types and books reflect the state of the arts around them, because on one side typography is an art; but they are influenced by trade conditions, because it is also a trade. Not to face these two facts, or to neglect either one or the other, is merely to fool one's self!

DANIEL BERKELEY UPDIKE (1962)

Figure 7-3. About a hundred characters to the line is too many, particularly without interlinear space. This is 9-point Lasercomp Photina.

The old-fashioned habit of unnecessary capitalization persists, but the minimal use of capitals is just as characteristic of good modern composition as the avoidance of inessential punctuation. Convention requires a frequent use of capital letters which cannot be avoided, but the effect of these will be enhanced by using elsewhere, wherever possible, small capitals or lower-case. Even pronouns referring to the Deity are now set without special capitals. Sub-editorial care is necessary when an apparently intrusive capital indicates a specific meaning, as in 'this House', 'Foreign Secretary', and 'Primate'. Some oddities may result; '*Shorter Oxford English dictionary*' looks wrong, but the reduction of all but obligatory capitals in book titles and elsewhere still improves the general appearance and clarity of the page. Whole words or phrases set in capitals within the text, perhaps for emphasis, seem over-emphatic, and may look better in small capitals. The initials of honorifics may be well suited to small capitals rather than capitals, though attention is needed where a capital abbreviation includes a double initial as in Ph.D. Organizations are usually abbreviated in capitals. Acronyms are names made out of abbreviations, and these are sometimes set in capitals (NATO, UNESCO), sometimes in upper and lower-case (Fiat) and sometimes in lower-case only (radar, laser).

The combination of capitals and small capitals has respectable antecedents, contributes to clarity, and yet is somewhat out of favour, perhaps because even small capitals present a neat pattern. This may change, but the preference for aligning figures with capitals and hanging figures with small capitals is evidently well grounded, and likely to survive now that many founts include both kinds of figure.

Whether or not a chapter begins with an initial (§ 9-5), its first line commonly begins with a word or two in capitals or small capitals or both. When there is an initial ranging at the top with the first line, this may ensure that letters adjacent to the initial appear to align with it. This practice does however impart to a word or phrase more emphasis than the author may have intended it to bear, and the reader is likely to benefit from the use of upper and lower-case instead. To set the whole of the first line in any alphabet other than that of the text is doubtful practice; that line would then be disproportionately emphatic, and any word divided at its end would be completed in a different alphabet. If capitals or small capitals are to be used in the first line, they are best allocated to the first phrase, not merely to the first word which may be an insignificant one.

Small capitals, ranging as they do with non-extending lower-case letters and with hanging figures, fit well into text, and they too benefit from letter-spacing. When either capitals or small capitals are to appear frequently, the designer will do well to compare their appearance with that of the lower-case letters among which they will be set, to see how well they will combine. The best companions for lower-case will usually be those small capitals which come closest to it in x-height and weight of stroke.

Italic on the other hand is not so much a companion as an alternative. When italic is to be used for continuous reading, an italic of comparatively wide set, reduced slant, and minimal contrast of weight in comparison with roman will probably be easiest for the reader. Italic is more commonly used for exceptional phrases, and contrast of design will then provide difference and emphasis.

Underlining for emphasis is a convention of handwriting which has been excluded from typographic custom by the technical difficulty of setting underlines in metal. Even in photo-composition, underlines must either pass through the descending strokes or by avoiding them appear all but midway between adjacent lines of type. Where an underlined document is to be transliterated, the printer is entitled to italicize underlined words instead of setting underlines, but sub-editors will need to watch for textual references to underlines which should not appear in the edition.

The five alphabets of the standard bookwork fount supply enough variety for most forms of text composition. Bold and semi-bold are not well suited to occasional appearance in a text page designed for continuous reading; they appear to stand out of the page, distracting the eye away from the next words in lighter type. Bold and semi-bold may however be useful for sub-headings; for *lemmata* (head-words) in a dictionary or encyclopedia they are all but essential. If for some special purpose sentences or paragraphs are to be emphasized in this way, a semi-bold type is likely to mar the texture of the page less conspicuously than a bold of full weight.

§ 7-4 Lateral space

In metal composition, the justification of lines to fill an even measure is a mechanical necessity; the page must be rectangular to be locked up during imposition. The spacing out of words across the whole width of the line, on the other hand, is a convention of questionable value, essential to neither printer nor reader. The convention, however, is nearly always observed, even when wide and uneven word-spacing results, as though any other arrangement might disconcert the reader. Perhaps lines of even length assist the act of reading, and the to-and-fro of the eyes is comfortably rhythmic when the lines present an evenly bounded field for the attention. Certainly the familiarity of any typographic arrangement is likely to contribute to legibility, and the visibly rectangular page does distinguish prose text from verse and display lines. A full line also implies continuity; within the measure, a hiatus at a line end may appear to imply a pause, and perhaps a new beginning. This same convention, however, tends to discourage the use of narrower measures in bookwork and close and even word-spacing, practices which most typographers believe conducive to legibility.

Justification means spacing out the line to fill the measure, whether or not the line is filled with words; in metal, for instance, short lines at paragraph ends are justified. But there is no convenient, exact, and generally used term for composition in which the lines are all visually uneven in length, as in verse. This uneven style may in accordance with usage be called *unjustified*, as the term is unlikely to mislead, and as uneven lines are not in fact justified in photo-composition.

Major innovation in book typography is best introduced with subtlety and even stealth, inuring the reader gradually to new patterns of communication. Lines of visibly even length (apart from indented lines and short lines at paragraph ends) are still a prudent style for main text in the general run of books. Unjustified setting may be acceptable and indeed preferable when the measure narrows, perhaps to accommodate illustrations beside the text and within its area, and when the measure has to be narrow anyway, as on the jacket flaps (§ 17-1). Spencer (1969) finds unjustified setting no less legible than the more familiar squared-up page.

Whether the text is to be justified or not, word-spacing may be influenced and sometimes regulated by the designer. To ask a skilled and conscientious book-printer for close and even spacing, which is part of his ordinary stock-in-trade, may result in unusually tight setting; this will increase the cost of any insertions, which will cause the crowded lines to over-run into each other, and it may also lead to an excess of word division at the end of adjacent lines. If there is any doubt of a printer's spacing practice, preference for a stated maximum word-space may be included in the specification. In unjustified setting, word-spaces can be exactly even throughout, and the designer may well indicate what it is to be. In any kind of setting, he will need to do so in terms applicable to the typesetter which is to be used. If he intends to regulate the style of setting comprehensively, he must specify whether or not words are to be divided at line ends, and, if they are, how far short of the full measure the lines may be allowed to end. If words are not to be divided, the white space within the measure will depend on the width of the last word for which there was no room.

If word-spacing is too wide in several adjacent lines, rivers of white will meander down the page across the path of the reader's eye, breaking up the text into irregular islands of words. This is particularly likely when word-spacing is wider than the vertical space between short letters in adjacent lines. The space between sentences is now usually 'the space of the line' — the space between words in the same line; more is unnecessary, over-emphasizes the hiatus, and spoils the appearance of the setting.

The letter-spacing of displayed words in capitals or small capitals has become all but standard practice in British book design of the better kind. When such displayed words are so spaced, words in capitals or small capitals in the text will also need to be letter-spaced, if they are not to contrast disagreeably with the display lines. For economic reasons, letter-spacing in text should be mechanically even, not optically even, and in text sizes of type this should look neat enough. Usually medium letter-spacing is best for text; thin letter-spacing may be nearly imperceptible, and thick over-emphatic. In Monotype setting, for example, the unit-adding attachment can be adjusted to one, two, or three units for letter-spacing, and is nearly always set at two, or medium, equal to one-ninth of the set of the fount, which suits text composition well enough but might be a little too wide if the equivalent were used in display.

An *indent* is a specified space within the measure at either end of the line, and like other spacing within the line is best defined in terms of the set of the fount, usually an em or more than one. The space in a short line at the end of a paragraph is not specified and is not known as an indent.

The most common use of indents is to signal the start of a new paragraph; when the last line of the preceding paragraph fills the measure, there may be no other signal. The first line of a new paragraph is usually indented one em, perhaps the narrowest indent easily observed by the reader; in wider measures a wider indent may be clearer, and a two-em indent is sometimes preferred in measures over 26 picas. The first line of a chapter customarily begins without an indent, or *full left*, and this neat style also looks well under section headings. Lines set full left in this way maintain the rectangular pattern of the page, but pattern must not take precedence over clarity. When the start of a new paragraph is not indicated by an indent, some other signal will be needed, perhaps extra interlinear space between paragraphs, or a paragraph mark such as ¶.

Carefully regulated indents contribute to the clarity of the text and to its appearance of pattern and purpose. The length of any line or paragraph is determined by the text itself; all the designer can do, for items other than the main text, is to decide at what position on the page each is to begin, and how far full lines of such items are to travel along the main text measure. The first and last lines of each paragraph apart, indented items should as far as possible align with each other at left and right. One kind of indent other than paragraphs should be enough for any text. The simpler the pattern, the more clearly its purpose will be seen, and readers concerned with words rather than with their arrangement do not easily keep track of a logical structure indicated only by complex typography.

When indented passages are common throughout a book, indents at left and right may provide vertical axes on which other items may be aligned — headline, page number, section headings, and even chapter title. This may call for precise calculation and careful instructions; indents are usually described in ems of the indented fount, and when more than one fount is to be indented the same distance, each will be indented a different number of ems and units.

Indented passages may distort the pattern of successive pages. One example is an extract more than a page long, indented only at the left. When such a passage fills the page, a centred headline will appear to be off-centre, a verso page number at the outer margin will seem to be outside the measure, and the side margins will be out of proportion to each other.

Verse extracts, embedded in prose, would leave an ungainly space on the right if they started at the left of the measure. Without other instructions the printer is likely to centre each such extract visually. This may cause successive extracts on the same page to be indented a different distance, an untidy arrangement. The designer will do better to choose one indent for all verse extracts, so that at least they will align at the left with each other, if there is not too great a disparity of line length between extracts.

A hanging indent is one which reverses the normal pattern of the paragraph; the first line starts farther to the left than the rest, instead of farther to the right. This may be used to indicate a difference between the nature of the passage so treated and the rest of the text, or to emphasize paragraph numbers which stand clear of the rest of the paragraph, or to pick out for easy reference the first word. The last instance is useful in drama, when every speech begins with the name of its speaker. When such a paragraph beings with a number, the text after the number usually begins in alignment with subsequent lines.

§ 7-5 Interlinear space

Interlinear space is not a term in general use; in practice, people concerned with book production usually refer to *leading*, and their intention in doing so is clear enough. Leads are still in use, but in metal text composition most interlinear space is applied by casting on an increased body, and in any reference to photo-composition, leading is anomalous. Elsewhere references to *reverse leading* in photo-composition may continue, but here the unambiguous term is preferred.

Some interlinear space appears between all lines; extenders hold the lines some way apart, and the type itself provides a part of the space when the body is significantly deeper than the *gauge*, the vertical distance between the top of the highest letters and the bottom of the lowest. For the purposes of this book, interlinear space refers to space added to the body or nominal size of the fount.

The rigidity of metal composition imposes on designers limited scope for adjustment to interlinear space. If a page of forty lines of 11 on 12-point needs a point more interlinear space, either there must be three lines fewer, or type area, margins, and perhaps even format will have to be changed. If a resulting page of thirty-seven lines then looks too deeply spaced between the lines, half-point sizes would be a help but are not always available.

Finer adjustments are possible in the typography of a text page planned for some photo-composition systems; the designer can balance smaller increments not only in interlinear space but in body. Starting with forty lines of 11 on 12-point as before, and needing a little more interlinear space, he may find his printer could provide a specimen of $10\frac{1}{2}$ on 12-point. If that is still too close, the next step might be a specimen of $10\frac{1}{2}$ on $12\frac{1}{4}$-point, decreasing head and tail margins by 5 points each only.

In the days before type was leaded, text pages presented the appearance of a closely woven texture. Today's tendency is towards close word-spacing and some interlinear space, emphasizing the separate existence of each line while preserving the pattern of the page itself. A page which is either too open or too crowded may deter a prospective reader; familiarity is the surest inducement.

Once the reader has begun to read, the spaces between the lines have their own value. Legibility depends not only on being able to see the strokes which make up the letters, but on being able to see at a glance the positions of these strokes in relation to each other. If there is too much or too little space within or between the letters, the position and shapes of the strokes become confused, and the letters less easy to identify quickly and accurately. The same is true if the adjacent lines of letters, above and below that on which the eyes are focused, are too close; crowding into the area of attention, they distract the reader from the right object of his regard. If too much interlinear space is equally incompatible with legibility, it may be because the lines are too widely separated for rapid reading.

Concern for good book production today is not confined to publishers and printers. It extends to the general public, whose interest has been stimulated not only by the generally improved standard in the books they buy, but also by exhibitions and by an increasing literature on the subject. And a discriminating public is the best possible guarantee of the continuance of good printing. 			H. G. ALDIS

Figure 7-5. No type as large on the body as Monotype Photina looks easy to read when set without interlinear space.

The greater the number of characters in the line, the deeper the interlinear space should be. The reason appears to be that an extremely wide measure causes some difficulty in picking up the beginning of the next line at a glance, on returning from such a distance, and may lead to *doubling* or beginning the same line twice. Even if this is not a common experience, wide measures without proportionate interlinear space may well cause hesitation and a slowing down of reading.

While the amount of interlinear space is properly expressed as points or fractions of a point to be added to the type size, it is better assessed in the course of design as the vertical distance between short letters. One formula favours a distance roughly half as much again as the lower-case x-height; the distance in this book is fractionally greater than that. Whatever the formula, and whether a formula can be valid whatever the number of characters in the line, the distance between short letters does appear to the eye more significant than the distance between extenders. Founts with long extenders therefore need less interlinear space than those with short. The lower-case x-height of 11-point Monotype Times, for instance, is almost exactly the same as that of 14-point Centaur; when 14-point Centaur is set solid, the vertical distance between short letters is equivalent to that of 11-point Times on 14-point body or with 3-point leading.

Except perhaps when founts with the longest extenders are set in the narrowest possible measures, the reader of today is likely to prefer a point or more interlinear space in any text (figure 7-5). In some founts, extenders in adjacent lines may on occasion appear to touch each other if there is no such space, and this intrusion from the lines above and below may distract attention from the line being read. Page make-up in photo-composition is not yet mechanized in all systems, and when the printer is to make up paper or film images of the type by hand, cutting between the lines, two points of space between adjacent lines will help the compositor to avoid amputating the extremities of ascending and descending characters.

A loosely fitted fount seems to need extra space between the lines. Loose fitting is a fault in type, and fortunately not many designs suffer from it, but when it does appear the words should not be too closely spaced. The space between words must be visibly wider than that between letters; and if the page is to look like a series of lines rather than of irregular patches of letters, the space between lines must be visibly greater than the space between words.

Interlinear space in small-type items, such as extracts and footnotes, maintains an even texture in the page when it appears to be proportionate to that of the main text. To render it exactly proportionate is all but impossible in metal, as leads and moulds have normal size intervals of half a point at best; 9, 10, and 11-point type each with one point of interlinear space have interlinear spaces of about 11, 10, and 9 per cent of the body respectively. As the number of characters in the text measure increases when the type size is reduced, and as interlinear space should be proportionate to the number of characters in the line, this slight disproportion may represent an advantage to the reader.

§ 7-6 Extracts and tables

An extract in bookwork is usually a passage extracted from a source elsewhere. Since the words are not those of the author or editor of the edition, such passages are differentiated from the main text by typographic means. Extraction does not necessarily imply separation of part of a document from an unquoted remainder, as the whole of a document such as a letter may form an extract.

In tabular composition, each line is divided into two or more columns. Two-column setting is not tabular, since the measure contains two separate lines set side by side. Within the main text, tabular composition may be used without being designated as a table, but its style needs the designer's attention no less.

Extracts and tables, together with illustrations (chapter 14), provide the most frequent change in typographic style within the text. Footnotes (§ 7-7) by definition appear at the foot of the page and do not interrupt the text in the same way; the text should be coherent without them, but can hardly be so without the author's tables and extracts except when tables are gathered into an appendix.

Plans for extract setting should be based on the nature of all the extracts in the typescript, on their length, and on the frequency of their occurrence. The style of setting prose extracts may need adaptation to suit occasional verse extracts; letters, if they are to include addresses, dates, and other items may call for plans which will affect other kinds of extract. In long extracts legibility is particularly important; the reader may tolerate a few lines but not several pages of small type if it is really small. When several small-type extracts appear on many pages, the text may present an irritatingly fragmented look. The extracts may indeed be the most important part of the text, as in a collection of letters, and then linking passages by the editor might be better set in small type. In planning an edition which includes letters set in their entirety, the designer may do well to examine the value of following in typography the diagonal arrangement of address, date, and so on in the letters themselves. In other typographic contexts, separate items of this kind are usually rearranged more neatly, and not to do so when setting correspondence seems inconsistent.

Not all authors and typists observe typographic custom before preparing copy for extracts, and their arrangement in typescript is not always a reliable guide to their arrangement in type. If a designer follows the arrangement of a knowledgeably and consistently prepared typescript, he cannot go wrong. More often he will at least have to indicate the required style to the printer. The point at which the typescript is most likely to depart from printing custom, and influence the printer in the same direction unless the designer intervenes, is in breaking short extracts from the text, sometimes in a manner which suggests small type, when they would be better *run in with* (assimilated into) the text, differentiated only by their enclosure in quotes. When longer extracts are to be set in small type, indented, or otherwise differentiated from the rest of the text, this treatment is commonly reserved for extracts which will make some four or more lines of type.

For many editions there will be no need for any such differentiation. Quotes at the beginning and end of each extract, and at the beginning of new paragraphs within each, will be signal enough.

If the text type seems likely to suit the extracts but their difference from the main text appears to need emphasis, indents at left and perhaps also at right may be combined with extra spaces between extracts and adjacent text. This style brings with it the chance of apparent asymmetry in headline, folio, and margin, mentioned in § 7-4. The space between the text and any item set in a different way may have to be flexible; if the familiar half white line is called for and the extract begins but does not end on a page, the printer may have to set a whole white line in order to maintain an even depth of page. When text is leaded, extracts may alternatively be set in the text type without indent and with reduced interlinear space.

When extracts are conspicuously differentiated from the main text in this or any other way, they do not also need to be contained in quotes. Whether they should begin with the indent which signifies a new paragraph is doubtful; if they do, the indents suggest that each extract began with a new paragraph in the original document, which may not have been so. If the typescript seems to have followed the documents in indenting new paragraphs but otherwise starting level with subsequent lines, this may represent the intention of a meticulous author and had better be followed.

The widespread use of small type for extracts is questionable. Its popularity may be due more to designers than to authors and editors, as occasional passages in small type seem to offer some relief from the otherwise rigid pattern of the text pages, and this is a visual rather than a textual advantage. Extracts are normally an essential part of the text; to set them in small type seems to reduce their importance to that of an intermediate between the author's text and his footnotes. This may also cause a clumsy division between two parts of the extract, when it begins with a brief introductory sentence, followed by a sentence or more of the main text, all set in text type, before the extract continues in small type. The introductory sentence will then be in one size of type and the rest of the extract in another. The designer is likely to find that neither author nor editor has thought of setting all, from the beginning of the introductory sentence, in small type, enclosing in brackets the author's interpolated text. Small type extracts still have the disadvantage that shorter extracts, sometimes from the same source, are set in larger type.

For extract composition, type one point smaller than that of the main text will usually provide contrast enough. The distance between extract and text should be specified as minimum and maximum, as the space available will vary point by point according to the number of lines in each extract.

An author may choose to set out some verse extracts as prose, probably beginning each line of verse with a capital, and possibly indicating line ends with a *solidus* (oblique) or *modulus* (vertical) rule. This deprives the extract of the appearance of verse, and while the author may choose to do it the designer should not initiate it without the author's agreement.

More commonly verse is set in its own distinctive way, as separate lines, often grouped in stanzas. Where all or nearly all extracts are verse, and where there is no risk of the reader's supposing them to be the author's own words, this arrangement may be enough to distinguish verse extracts from text, without quotes or small type. Verse and prose extracts should, however, be treated in much the same way, as verse might otherwise appear not to be extracted at all; quotes round verse extracts tend to be slightly distracting, and small type may prove well suited to extracts which include verse and prose.

The possible advantage of a single indent for all verse extracts, to replace the random indents caused by centring on the longest line of each, has already been mentioned in § 7-4. In verse there are no paragraph indents, but line indents are sometimes used to emphasize a rhyme scheme not based on couplets. Whether to indent in this way, and whether to set a space between stanzas, must be for author and editor to decide; the designer may do well to make sure that indents do in fact follow and emphasize the rhyme scheme, and to specify the extent as well as the location of indents.

Verse divided into stanzas takes on the appearance of a single extract when the spaces above and below the extract are greater than the spaces between stanzas. Otherwise the first and last stanzas of the extract are closer to the adjacent text than to the intermediate stanzas.

Tables are usually but not always set in smaller type than the main text. Type size should if possible be such that all the tables in the book can be set within the designated area, but may be selected to suit the majority, smaller type being allocated to a minority of larger tables. This will involve calculation of the maximum number of characters in each column, and the space between columns (§ 18-1).

The best alignment for tables, as for any other item in the text, is upright, with the lines parallel with those of the text. When column headings are much wider than the rest of the column, they may be turned sideways; the tradition that all turned items should have their feet to the right ensures that different items in different parts of the book are turned in the same direction.

A table too wide to fit across the page may be turned in the same direction. This sideways position will look less peculiar if there are no upright items on the same page, not even headline or folio. If within such a table column headings are turned at right-angles to the lines, they should have their feet to the left, as they would otherwise be upside down on the page.

If a table consists of more columns than lines, it may be easier to arrange if the axis is reversed, changing lines into columns and vice versa. This should not even be proposed if it will reduce the clarity of the table or depart from the arrangement of neighbouring tables, and editorial agreement is obligatory.

Rules are useful for emphasizing the tabular nature of the setting and for separating the table from adjacent items. One simple arrangement for horizontal rules is to place a medium rule between table heading and table, another below the table, and a fine rule between column headings and columns. Such rules should match the width of the table, but not necessarily that of the text measure. Vertical or *down* rules are out of favour, partly because intersections with horizontal rules in metal hardly ever fit perfectly, and partly because they cross the path of the eye reading along the line.

The printer will often be unable to place a table next to its text reference, and the text may need to be sub-edited to guide the reader to a table elsewhere. When this is done, tables which share a page with text may be placed at the foot of the page, where they will not interrupt the text as much as they would if they divided it or appeared above its first line.

When a table is to spread across an opening, the reader will need some means of linking the verso part of each line with the second part on the recto, as the gap down the spine margin may disguise the alignment. The number or title of each line should then be repeated on the recto. A table too big even for an opening may have to be printed as a *folding plate* (§ 14-3).

A good composing room can translate almost any tabular copy into a reasonably clear and presentable example of tabular composition. The guidance of a skilled and conscientious designer will enable the same composing room to impart elegance even to these difficult items, and may then attract the reader's attention towards information he might otherwise disregard.

§ 7-7 Notes and references

References draw the reader's attention to another publication or document; notes expand upon the text from outside it. Notes and references may be treated alike; or references may appear within the text; or one treatment may be applied to notes and another to references. The designer's task is to preserve the text as far as he can from interruption by unsightly symbols, and from unseemly patterns of subsidiary composition.

When references are few, or in scientific works where they are often many, they may have been written into the main text. Scientific writers utilize an author-date system, which indicates a reference in the text by the author's surname and the year of publication; a list of references, usually after the main text, in alphabetical order of the same surnames, adds the author's initials and repeats the year of publication before completing the reference with the full title and the publication details. In texts concerned with the humanities, frequent references of this kind might be considered by the reader to be an unjustifiable distraction. The few references in this book are indicated by an author–date system.

Notes are always separate from the text. Most authors and readers appear to like footnotes, most printers and publishers prefer end-notes after the main text, at the end of chapter or book. Footnotes increase the cost and difficulty of make-up, reduce the text areas on facing pages to uneven depths, and tend to disarrange the orderly appearance of the page. The customary size for footnote type is about two points smaller than that of the main text. If superior figures are used for footnote indicators, as they usually are in works which make no use of them for mathematical purposes, the superiors in the footnotes themselves may be small and an ordinary arabic figure followed by a point will be more easily visible.

A printer experienced in academic work may be relied on to rearrange short footnotes (usually references or cross-references) into available spaces in short lines, and to avoid a tall narrow stack of such notes, but a neat and consistent treatment of this kind may depend on the guidance of an equally experienced typographer.

Footnote numbering is usually by page, but the custom of numbering from 1 to 9 and then starting again with 1, continuously throughout the chapter, dates back to the eighteenth century, and has the double advantage that the numbers do not have to be altered during page make-up, and that there will be no double superiors. The system works only when no pages include more than nine footnotes. To number footnotes serially by chapters without starting again after 9 also avoids alterations during make-up, but is likely to introduce double and even treble superiors, under which conspicuous spaces appear in the line.

The opening of a text in double column may look untidy if footnotes under each column reduce each or any column to a depth different from that of the three other columns in the opening. Some tidying up is possible by combining all the footnotes of each page into two columns of even depth, so that at least the text columns on each page are also even in depth. Under text set in a single wide measure, footnotes may gain in legibility if they are set in two columns.

Side-notes, set in the outer margin, are usually a guide to the different parts of the text, rather than an expansion or commentary upon it, and so may be described as a form of sub-heading. Notes written as footnotes could be set in this way, but only at the cost of a considerable area of marginal space left blank. The first line of each side-note is aligned beside the line of text to which the note refers, so the note needs no indicator in the text. Such notes are usually set unjustified in narrow measure.

End-notes are usually set about one point smaller than the main text, and this is in their favour as they are a little easier to read than the smaller type used in footnotes. As they are not for the most part on text pages, expensive rearrangement is usually considered unnecessary. The nature of the notes may indicate whether they should be at chapter ends or at the end of the whole text, but the possibility that most authors, reviewers, and readers prefer footnotes should influence book designers unaccustomed to reading annotated works.

§ 7-8 Sections

For the purpose of this chapter, a section is any paragraph or group of paragraphs differentiated from the adjacent text by typographic means.

The simplest form of section is most often seen in fiction. Within a chapter, the author wishes to indicate a new scene, as within an act of a play, in a manner more emphatic than that of the paragraph. In typescript, he is likely to leave a few white lines, perhaps with an asterisk or two in the resulting space. In the edition, the start of a new section may well be marked in much the same way, with a designated number of white lines. If asterisks or some other marks appear in the space between sections, that space and its purpose will be the more easily identified when it appears at the beginning or end of the page.

This and other kinds of section, with or without headings, may begin at the left, without a paragraph indent. This follows the custom of starting the chapter in the same way (§ 7-4), and emphasizes the difference between the first line of a new paragraph and the first line of a section. The absence of indent from the first line may however look clumsy when the second line begins a paragraph.

Writers seem to be more ready to introduce a new section than to show where it is to end. The reader will need some typographic indication of the end of the section, and editorial help may be needed in working out where it is to go. Often the end of one section is followed immediately by the heading which begins the next. Elsewhere, for example in a section of numbered paragraphs, there might be no sign to the reader that a series of special subjects comes to an end and that his reading now reverts to the main text. This is one advantage in setting hanging paragraphs when they are numbered.

Section headings can at times be difficult enough to deserve close attention and even experiment by means of specimen settings (§ 19-2). The style of each has to be planned to suit the longest and shortest headings in the series, and each has to be distinct in style from the headline (§ 7-9) and the chapter title. When there is to be more than one grade of section heading, each should not only be distinct from the others, but should appear to be either superior or subordinate to each.

Section headings are all subordinate to the chapter title, and should not approach too closely the emphasis of that superior heading. For roman or italic capitals, the text size will usually be big enough; italic upper and lower-case, and small capitals, may be a point or two larger, but a large size of small capitals may be all but indistinguishable from capitals in text size. If the typesetter cannot change type size during setting, section headings in anything other than the text fount may prove expensive. Roman upper and lower-case of the text series and body is rarely used, as it would contrast inadequately with the text itself. Bold type, whether of the text series or of another, tends to interrupt the text rather emphatically and to draw attention away from it, but may be useful when there are more grades of section heading than usual.

Section headings *broken off from* the text (set in a line separate from it) are usually either centred or placed at the left. If they are indented, variations in length may cause some of them to appear inaccurately centred. The centred position is reserved by custom for the superior grades of heading, but the status of each grade needs also to be signalled by the type in which it is set. *Run-in* headings, which precede text in the same line, should begin with a paragraph indent, as they may otherwise appear to continue a preceding paragraph when it ends with a full line.

The comparative status of different kinds of heading may also be indicated by the depth of extra space over and under each. This space must add up to a whole number of white lines, if the page is to be even in depth with pages which have no headings. If there is more space over the heading than under it, the displayed line will be closer to, and will therefore appear to refer to, the section under it rather than that above. Spaces commonly used for this purpose approximate to $\frac{3}{4}$ white line over and $\frac{1}{4}$ under (less might be imperceptible), $1\frac{1}{2}$ over and $\frac{1}{2}$ under, and 2 over and 1 under (more might be over-emphatic). If section headings of different grades are adjacent to each other at any point, the space between headings will also need allocation.

Some four grades of section heading are easily enough provided from the text fount. For example, grade A may be set in roman capitals, with 2 white lines over and 1 under: grade B in small capitals, with $1\frac{1}{2}$ white over and $\frac{1}{2}$ under: grade C broken off in italic upper and lower-case, with $\frac{3}{4}$ white over and $\frac{1}{4}$ under: and grade D in upper and lower-case italic indented one em, followed by a point, run-in with text, and with $\frac{1}{2}$ to 1 white line over the heading and at the end of the section.

Whether any text needs more than four grades of section is questionable. The relative status of each, and the purpose of setting it as a separate section, may be clearer to the writer's mind than to the reader's eye. Clarity of exposition and logical presentation do more to help the reader than do typographic divisions and sub-divisions. If no cross-reference is made to section numbers, they are superfluous; and points are always superfluous at the end of broken-off headings.

When section headings will make more than one line of type, the treatment of subsequent lines will need the designer's attention. Left to himself, the compositor may fill up the first line, turning over one word or part of a word; two lines would look better if more evenly divided. If some of the headings would otherwise fill the line, it is better to make two lines of them, in order to emphasize their presence by opening up space beside them.

§ 7-9 Headlines and footlines

The *headline* is a line at the head of the page, above the first line of text. It is not always occupied; when it is, the occupants are usually a repeated title of some kind, often the page number, and sometimes (as in this book) a section number, or (in historical works) the year in which the events below occurred. The repeated title is usually known as the headline, from the position it occupies, but has on occasion been placed in the *footline*, below the main text area.

In a book designed for sustained reading, headlines help the reader to find his place on taking up the book after an interruption. In a reference work they guide the reader to that part of the text he needs next. Headlines which do not fulfil either of these functions are unnecessary. Whether or not there are to be headlines seems quite often to be left to the designer.

Headlines do not introduce a new item of the text, and indeed most appear in mid-sentence, since few pages begin with a new sentence. Few authors write their own headlines; the copy is usually provided by publisher, designer, or printer. If headline copy will not fit into the available space, the designer may have to compile an abbreviated version. When headline copy differs from verso to recto, the verso headline is usually the superior item and the recto inferior, a reversal of the precedence usual at most other points in book production. In a symposium or anthology, the contributor's name may be set on the verso, the title of his contribution on the recto. A long title may be divided between verso and recto, preferably when the halves are drawn together by being ranged to the inner margin.

The content of a *running headline* runs unchanged throughout the book, and normally consists of the book's title. This does not assist the act of reading, may be of some help to printer and binder, and is often set only as a typographic relief from the otherwise rectangular pattern of the pages.

Section headlines refer to the titles of different parts of the book. The verso headline, for example, may repeat the part title, facing the chapter title on the recto. Such an arrangement will enable the printer to set the headlines at the same time as the text. *Page headlines* are those which refer to the contents of the page below them, and this may not be known until the pages are made up. In this book the verso headlines are chapter titles and the recto headlines are section titles with their numbers. A section headline in some books is likely to refer not to the section immediately adjacent to the headline but to a section title some way down the page, an old custom but not a particularly good one.

The type used for headlines should be distinct from that of the text, and from those of any section or other heading. If the type of headline and section heading is the same or similar, a difference of lateral position may prevent ambiguity. For example, headlines in italic may be indented inwards from the page number at the outer margin, to differentiate them from centred italic section headings, even when there is a difference of size.

Small capitals are popular for headlines because they are unobtrusive, and also because nearly all the letters are even in height and square along the top of the text area, parallel with the cut at the head. Capitals may be more suitable for quarto and two-column pages. Italic upper and lower-case will provide more room for long headlines, but should be avoided if there is much italic in the text. As a rule, the text size will suit the headlines; anything larger may be too obtrusive for this subordinate function, anything smaller may look inadequate. When the headlines on facing pages differ in content, they may also differ in alphabet; a contributor's name, for instance, may be set in small capitals, the title of his contribution in italic upper and lower-case. The lateral positions of headlines on facing pages usually reflect each other and so maintain the balance of the opening.

If the headline is too widely separated from the text, it may look isolated; if it is too close, it may appear to merge with the text. Six points of space will usually be enough, or nine points when interlinear space exceeds two points.

In the more fanciful kind of book, the typographer may be inclined to decorate the headline with a fleuron or rule or strip of border. Before he does so, he will do well to consider whether this ornament will retain its charm after scores or hundreds of repetitions.

A *folio* may be a format, a sheet of typescript, or a page number. The terminology of printing contains several ambiguities of this kind, which can be evaded by the use of unambiguous terms, as in this section.

Page numbers are usually set either in the headline or at the foot of the page. Probably the most convenient position is in the headline at the outer margin, where the reader can see them when rapidly turning over the pages. Whether at head or foot, page numbers are rarely farther from the fore-edge than the centre of the measure, as that might be inconvenient for the reader.

In some old books, page numbers in a line of their own were flanked by parentheses, brackets, or ornaments. These not only differentiated the number from other figures on the page but protected it against wear from rollers or inkballs whose whole weight would otherwise fall upon the isolated number. This custom survives to embellish a page which otherwise might appear somewhat stark.

Conventional methods of page numbering originate from book-production methods, rather than from the convenience of the reader. *Preliminary pages* (chapter 8), the number of which may have to be changed at the last moment before printing, are often numbered separately from the text in lower-case roman numerals, whether upright or inclined, starting with the first recto page on which anything is printed, or which precedes the first printed page if that page is a verso. Arabic numbering then begins on page 1, the first recto page of the main text. By custom recto pages are always odd numbers and verso pages even. The text series of page numbers usually excludes *plates* (§ 14-6), which have not passed through the press with the rest of the text.

Page numbers may be omitted from pages which do not need them, such as the title and other preliminary pages, full-page illustrations, and chapter openings, but the numbering continues as though the omitted number had been printed. The unnumbered pages are known as *blind folios*. The back pages of the book, such as appendixes, notes, and indexes, may like the preliminary pages be subject to last-minute alteration, but are numbered in the same series as the text, since to alter their page numbers would entail altering only those of the comparatively few following pages.

When page numbers are to be set in the headline, those pages such as chapter openings which have no wording in the headline may carry a page number in the footline. Alternatively, the page number is sometimes placed in the headline, between spaced brackets to emphasize its presence in an isolated position.

Page numbers are the feature that most printed pages have in common. The printer may have to rely on them in aligning page with page during imposition, when pages vary in depth, and in registering the print on the back of the sheet against that on the front; the binder may need to check his folding by their position. Page numbers should therefore be present on the great majority of pages and whether recto or verso should be an equal distance from the spine fold and the cut at the head. Any departure from such positions may deprive the designer of a reference point from which he can measure the vertical position of the rest of the page, and printer and binder of the means of making sure that all their margins are right.

The *footline* appears at the foot of the page, outside the text area, and contains the page number when that is at the foot. The first page of each section (§ 16-2) in which the book is sewn is signed with a letter or number in the footline to help the binder to identify it. *Signatures* in Britain are usually set in capitals or small capitals, in succession throughout the Latin alphabet which excludes J, V, and W, continuing if necessary to 2A or AA and so on. A itself is usually omitted, as it would disfigure the first of the preliminary pages, but signature A is understood to mean the preliminary pages or the first section which includes them, and the first signature to appear on a section is normally B. The American preference for numbered sections leads to a less conspicuous form of signature.

In order to help his own bindery to identify not only signatures but the edition to which they belong, a printer may also set in the footline the initials of the title or of the author or a works number, also on the first page of the section.

Collating marks are printed midway between the outermost pages of the section, so that after folding they are on the spine of the section and cannot be seen in the bound book. The mark on signature B is one step below that on signature A, and so on; when the sections are all gathered into the right order, the collating marks form a diagonal across the spine. To collate the book at this stage is to check the gathering of the sections by examining this diagonal line of marks; any interruption of the diagonal indicates an omission or a disorder.

The signature, with or without the initials or works number of the edition, may replace the collating mark on the spine, and will then also be invisible in the bound book. These items are all the preserve of printer and binder; only inaccurate folding, for example, is likely to bring the collating marks into view after binding. The designer on the other hand is entitled to regulate the setting of everything that is to appear on the text page, and may have to intervene to make sure that the footline is inconspicuous and appropriate to the style of the text page. This may need diligence, as signatures are not always placed on the page before imposition, which may follow the final stage of proof, so that signatures are not submitted for approval.

The ambiguities of printing terminology are in evidence here. A section may also be known as a signature or *sig*, so that the section has two names to begin with and *signature* has two meanings. The sections may also be known as the sheets or *quires*, the latter being a term once used to describe a number of sheets of unprinted paper.

§ 7-10 Chapters and parts

A writer who wishes to develop his theme by way of separate scenes or subjects usually divides his text into chapters. A reader who lacks the time or concentration to read the whole book in one session finds at the end of each chapter a pause in the narrative at which he can leave it for a while without losing his place. Publisher and printer, preparing a new edition, may be able to leave whole chapters all but undisturbed even when earlier chapters have been re-written to a different length, if each new chapter begins on a new page.

For economy, but without other advantage, a new chapter may begin on the same page as the closing lines of its predecessor. An awkward feature of this arrangement is that some chapters are likely to begin on new pages;

there will not always be room on the page before for the chapter heading and a minimal three or four lines of text. Usually the advantage lies with each chapter beginning on a new page.

When a short text has to be spread out as far as possible, the beginning of a new chapter may be dropped some way down the page. More than a few white lines may look intentionally wasteful, and if a new chapter faces a chapter's end of only a few lines there may be an inordinate area of space in the opening. Some space above the chapter heading may relieve the succession of full text pages, but in many books the heading of the new chapter starts at the level of first line of text on a full page.

The vertical level on the page of each line of the chapter heading — chapter number, chapter title, and first line of text, for example — usually repeats the pattern of other such headings in the book, and is repeated by the level of similar items in preliminary and *end pages* (chapter 8). When some chapter titles make one line only and others make more, the designer can choose whether to maintain approximately even spacing above and below the chapter title, or to start the text at the same level throughout, whatever the depth of the chapter title.

Even when the text needs to be extended rather than compressed, starting new chapters on recto pages does not tend to work well. Some chapters end recto and others verso, so the new chapter heading will sometimes face a blank verso and sometimes a full one, a variation which may puzzle the reader. Such an arrangement may, however, be worth while for an academic symposium, consisting of articles from various contributors who are to receive *offprints* of their contributions. Offprints consist of the pages of one such article only, preferably including no part of any other contribution, and are simplest to produce when each article begins and ends on leaves separate from adjacent articles.

The chapters of a consecutive text are worth numbering even when the text contains no cross-reference to the numbers; without numbers, they may appear to be separate items, as in a collection of short stories.

Roman numerals may prove a little awkward for chapters. Not all readers are familiar with their meaning, the larger numbers make an unhandy cross-reference, and they do not lend themselves to the tabular setting of a list of contents. Chapter numbers can be spelt out, but this again may make for a difficult list of contents, if numbers in the list are to be set in the same form as those in the text. Arabic figures are easiest of all. The word 'chapter' needs to precede the number only when the chapter number might otherwise be confused with a part or section number.

The chapters of a book may be grouped by the author into parts, to indicate a difference of treatment wider than the difference between the chapters within each part. Similarly, in a long and complex work, parts may be grouped into 'books'; the term 'volume' is better reserved for that part of the text which is gathered into one binding.

Book and part numbers may be in roman numerals or spelt out, to differentiate them from chapter numbers in arabic figures. This tends to cause no difficulty in the list of contents, as books and parts are comparatively few and are likely to be set differently from chapters in the list; nor do cross-references present much problem when the numbers are numerically small. If division of this kind is to mean anything to the reader, each book and part should have a title.

Book and part numbers and titles may be set over the heading of the first chapter in each. This, however, tends to obscure the similarity of chapter headings to each other, and when book, part, and chapter are all titled the number and variety of display lines may become too much for one page. Half-titles are usually preferable for book and part headings.

A *half-title* is usually a separate leaf, on the recto page of which a title is printed. Conventionally the text begins on the first recto after the half-title, so that the half-title's verso is blank. The rest of the half-title recto may however be useful for items which refer to the part as a whole, such as a quotation or a list of the part's contents, and the half-title verso is sometimes used for the same purpose or for a map.

Whether or not the text is divided into parts, it may start with a *repeat half-title*, the wording of which repeats that of the half-title or *bastard title* (§ 8-1) which precedes the preliminary pages. Half-titles are usually blind folios within the text numbering, so that when preliminary pages are numbered separately from the text a repeat half-title may occupy pages 1 and 2 and the text will then begin on 3.

When there are no parts, each chapter may begin with a half-title, but if there are more than a few of them the reader may begin to feel he is being offered more open space than he needs. More often, the half-title opens a new book or part. The setting of book and part numbers and titles, in position, alphabet, and size, should be related to chapter number and title in design, and are usually similar or more emphatic.

When there are part half-titles, the end pages are sometimes separated from the rest of the final part by a half-title of their own. When these include several items such as appendixes, list of references, notes, and index they will need a collective title for the half-title.

CHAPTER 8

Preliminary and end pages

T HE PRELIMINARY PAGES of a book, the *prelims*, are those which precede the main text. Those pages which are not part of the main text but follow it are known by various terms and here are called the *end pages*. Those intended for reference offer scope for a more inventive treatment than does the main text, and also for some latitude in the use of large and small type. Those intended for sustained reading are more like text pages, and may be classified as preliminary text.

Preliminary pages may be grouped and arranged in order in accordance with the function of each, with the conventions of book production, and with the preferences of author and editor. The first group introduces and explains (and may illustrate) the edition and the circumstances of its publication; it usually includes the half-title, any list of books by the same author, any frontispiece, the title-page, and the back or *verso* of the title-page, and there may also be lists of contents and of illustrations, acknowledgements, and a dedication. The second group, which may follow the first if that seems likely to be convenient for the reader, provides material for reference during reading, and may include glossary, list of abbreviations, bibliography, and errata — though this last is now rarely printed as part of the book. The third group calls for continuous reading rather for reference, and may include foreword, preface, and introduction, though an introduction may also form the first chapter of the text.

The second group defined above, apart from errata, may appear either in the prelims or in the end pages. The end pages may also include appendixes, end notes, index, and colophon.

The order of items in prelims and end pages is governed by convention with regard to half-title, title, title verso, and index, since the reader might otherwise not know where to find them, and since the half-title is the first printed page in the book; the publisher decides, or leaves editor or book designer to decide, about the rest, since the presentation of these pages may be a matter of book production which is usually assigned to the publisher by his contract with the author.

Prelims which are printed later than the text, for editorial reasons, are compiled as a multiple of four pages, in order to make a section for sewing (§ 16-4). The arrangements of end pages, and of prelims when they are printed with the text, may provide opportunity for alterations to the *extent*, the total number of pages in the book, in order to simplify the work of printer and binder. The extent may be increased by a blank leaf, preceding the half-title and excluded from the sequence of page numbers, and two blank leaves may follow the last printed page; preliminary items after the title verso may all begin on *recto* (right-hand) pages. If the extent is to be reduced, lists of contents and of illustrations and even acknowledgements may share pages, and any item may begin on an available verso page.

The convention by which the recto is preferred for the first page of most preliminary items still has some influence. In most books designed in a traditional style, two-page lists of contents and other preliminary items are commonly started on recto pages and completed overleaf, although the reader might find an opening more convenient for each. An appearance of reasonable economy, however, is generally preferred, and an excess of blank verso pages in prelims may appear lavish or inept. A blank recto may still be considered something of an oddity.

Every book is a problem in itself, and conventions are not laws. Imagination and common sense are more likely than habit to bring about a good result. No formula can be relied on unless the reasons behind it are known.

§ 8-1 The half-title

Perhaps because the terminology of book production seems fated to become more ambiguous year by year, the *bastard title* is now generally known as *the half-title*. This particular half-title differs from subsequent half-titles (§ 7-10) in appearing only on the first printed page, in consisting of the main title of the book rather than that of a part or other sub-division, and in being called *the* half-title rather than *a* half-title.

Whatever the original purpose of this first page of the book, its advantage now is that the endpaper is fixed to it and not to the title-page. More is said about endpapers in § 16-3; here it is enough to mention the usual method of *tipping on*, by which a narrow strip of the endpaper along the folded edge is pasted to the page which faces it. When the book is opened, the free leaf of the endpaper is turned to the left, and its tipped edge tends to draw up the leaf to which it is pasted and also to cover part of it; the quality and colour of the endpaper may also contrast with the text paper on the facing leaf. When there is no half-title, all this hardly interferes with the function of the title-page, but does detract from the appearance of the most important single page in the book.

The wording of the half-title is usually kept to a minimum; the title of book and of series if any are usually enough, and the author's name may be left out. The treatment should be inconspicuous, and related in style and position to the title-page from which the wording is derived. An elaborate, emphatic, or unusually placed setting could be mistaken for the title-page itself. The wording may include a series title and editor: the volume number and title, when the book is one of two or more volumes: or a synopsis of a novel, to assist library readers in their choice.

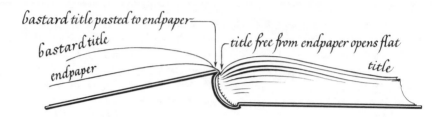

Figure 8-1. Part of the value of the bastard title is to enable the title-opening to open flat when the endpapers are tipped on to the outermost pages of the text.

Nothing is required by convention to appear on the half-title verso, and it may be left blank. It may on the other hand be used for a list of other works by the same author, whether or not they are relevant to the book in which they appear: for a series title, with the name of the editor and a list of other works in the series: for a dedication: or indeed for anything else the publisher may wish to appear there. Opposite, and in a sense balancing, the title-page, the position is conspicuous and emphatic, and should not be used for trivial items. The style of the setting should harmonize with that of the title-page opposite.

When a frontispiece is printed with the text, rather than as a *plate* (§ 14-6), it may well appear on this page. When there is to be nothing else on the half-title verso, the title-page setting may be designed as an opening, occupying both pages. In an illustrated book this may be taken a stage further: title and frontispiece can be combined as an opening, both extending together across both pages. When planning this or any other opening, the designer will need to remember that the spine of the opening will disappear into the *backing* (§ 16-6) and that after folding the facing pages may not align perfectly. If the frontispiece is a plate, the half-title verso will be concealed when the book is open at frontispiece and title, and an item of any importance may deserve a less obscure position.

§ 8-2 The title-page

The traditional purpose of the title-page is to announce the book's title, its author, its publisher, and the place and often the year of its publication.

Author and title usually appear together towards the head of the page, publisher and date towards the foot; in most books little else is added. The two groups of lines are usually separated by a substantial space, and the lines within each group are separated by spaces from each other, so that space tends to form a conspicuous feature of the page. The facing page, the half-title verso, is likely to be even more spacious; the wording which by custom appears on that page is not comparable in significance with the wording of the title-page, and accordingly should be distinctly subordinate in emphasis. In terms of pattern, these two pages may form the most important single opening in the book, but the traditional title-opening is dominated by unbalanced areas of space. The wording of the title-page gains emphasis from its isolation, but in breach of the doctrine that a book should be designed as a series of openings rather than of separate pages.

When display is centred throughout the book, a centred title-page provides an appropriate introduction to the arrangement of the pages which follow it. Some of the space in the centre of the title-page may be occupied by a heraldic or other device of author or publisher or series, by a small decorative illustration, or by a pattern of typographic ornaments. The large and magnificent devices or trade-marks to be seen in early printed books advertised the entrepreneur at the expense of the writer. Devices are now more modest in size, and should be equally modest in weight, constructed of lines no thicker than the main-stroke of the smaller type on an average title-page. Neither trade-mark nor decoration should compete in emphasis with the title itself. The balance of the opening can be adjusted by a suitable frontispiece (§ 14-6), particularly when illustration and caption are designed to face and balance the title-page.

The frontispiece may extend from its usual page to the title-page itself: a separate part of a drawn illustration, for example, may occupy the space in the centre of the title-page. When the display of the rest of the book is off-centre (§ 9-9), an off-centre title-page may be extended across the spine to the half-title verso. By such means the title can on occasion be designed as part of an opening, rather than as a separate page.

There is no evident utility in the invariable use of the recto for the title-page (or indeed for the first page of the text). If the force of bookwork tradition continues to wane, the publication details may as well be removed from the title-page verso to the half-title: the title may appear on the half-title verso, facing a short list of contents on the recto, and the text may begin on the verso of the contents list. So compact an arrangement might suit the economics of paperback production today, but might also puzzle the reader and cause him to wonder whether the printer has set his pages out of order. Similar innovations were introduced in the nineteenth century by William Morris at the Kelmscott Press, but may have to mature longer still before they are accepted as customary.

Meanwhile the typographer is challenged by the design problems and opportunities of the title-page. Since the opening is both significant and difficult to shape into a graceful pattern, his design will need imagination and detailed finish. He has comparatively few words to arrange, in rather more space than he needs, and contrast of emphasis between lines and groups of lines may require a variety of founts and alphabets. Propositions about the design of display, some of which apply to this task, are put forward in chapter 9; this section deals only with matters peculiar to the title and its facing page.

The conventional title-page, centred on a recto, may reasonably be seen as a single item of design, rather than as part of an opening, since that is how the reader is likely to see it. If in accordance with custom the setting is centred on the text measure, it may then appear to be slightly too far to the left. The remedy is to move the centre of the setting outwards to the right — a pica may be enough. When this is done, the setting on the half-title and its verso and on the title-page verso will similarly appear to be at rest if they are moved outwards in the same way. No part of any of these pages should however extend into the margins which contain the main text, in case the prelims appear to be part of another book. If on the other hand the setting of the title-page is substantially narrower than the text measure, its appearance on the page may benefit from deeper margins at head and tail, to balance with wider margins at each side. But there might then be a possibility that if the main title of the book is lower on the page than the chapter titles which follow it, it may appear to be less significant than they are.

In typographic style the title-page is usually related to display composition later in the book. The half-title introduces the title in the same way, and its setting may aptly be aligned at the top with that on the title-page. Any composition on the half-title verso should certainly align at the top with the title-page setting, to contribute at least an appearance of balance to the title opening. If the head and tail margins of the title-page are shared by its verso, the possibility of small type showing through from the back of the title to the front will be diminished.

Within the two main groups of lines on the title-page, convention permits some variety in the order of items. The title itself is often but not always first and top; the composition may instead start with the author. This follows the style of references, in which the author's name usually precedes the title of his work; it contributes to clarity when the name of translator or editor is also to appear. The lower group of lines usually includes the publisher's imprint, and the year and place of publication. Any name which identifies the publisher will do, and the briefer the better; the reader is not concerned at this point with the full legal title of the organization, and elaborate wording may distract attention from more important items on the same page. The order of publisher, place, and date is governed more by the publisher's preference than by printing custom.

The general treatment of the title-page depends on the typographer's opinion of its purpose. If it is intended to advertise or decorate, it may be elaborate in design, like the package of an expensive commodity. In this form it offers opportunities to creative imagination and technical skill, and may bear little resemblance to any other page in the book. If the title-page is intended to display a simple announcement with a discreet increase of emphasis, a plain treatment provides a subtle and limited opportunity. This latter attitude to the title-page is still quite widely held, but is not necessarily appropriate for all kinds of book.

§ 8-3 The title verso

A page is printed on one side of a leaf, which has a recto page on one side, the front, and a verso page on the other, the back. To be exact, a page has no back; but the back of a page is mentioned often enough in book production to have become a term in general use. For half-title and title, the term is apt enough, as for practical purposes both are leaves; where the printing of an item on the title-page is legally obligatory, that item is commonly printed on the back of the leaf, the title-page verso.

In accordance with Customs regulations in the United States, the country of origin should be stated on every title verso. The British Standards Institution (1971) has listed various other details which may also be printed on the same page. They may also be printed elsewhere, but they have their own importance, and when conventionally placed they are unlikely to be forgotten during production or difficult to find in use. Except on opaque paper, they tend to show through the title-page; but when they are modestly and neatly set, they will not be considered to mar it.

The full postal address of the publisher, and the International Standard Book Number, should be shown here, for the use of booksellers who wish to order a copy. For the protection of copyright, the copyright holder should be identified, with the year of publication applicable to the copyright notice. The publishing history of the book is intended to explain to the reader what version of the text is to follow, and what editions if any have been published before it. The British Library or Library of Congress Cataloguing in Publication Data are required to be set out in a manner incompatible with almost any seemly typographic arrangement of the rest of the page; when set in this way, the Data rarely appear to be a natural part of the book. Some publishers add a paragraph limiting uses of the book which might be defined as re-publication after sale. In Great Britain the printer's imprint must appear in the book, and is commonly combined with the statement of the country of origin on the title verso, but may also be printed as a *colophon* (§ 8-11).

The title verso may therefore contain more different items than any other page in the book — different in both purpose and appearance. To impose typographic unity on such a mixture is far from easy, but the page deserves effort; it is likely to be consulted before purchase, and should offer the discriminating buyer clear statements of fact without inelegance.

The type should as a rule be smaller than that of the text, as the information it conveys is not part of the author's message. Although the different parts of the wording are disparate, the appearance of the page may benefit from the use of one fount only. The wording is often an uncomfortable mixture of single lines and continuous sentences, and is best treated as display and arranged in a manner conformable with display composition elsewhere in the book. A white line is enough space between items; more may cause them to straggle down the page. There is no need to place the last item at the foot of the page, unless this seems the best place for it.

When the page is used for publication data of these various kinds, other kinds of item are better placed elsewhere, as they might otherwise be overlooked by a reader not concerned with such matters as the Standard Book Number. Certainly this gathering of administrative trivia is not the milieu for a dedication.

§ 8-4 Lists of contents and illustrations

Book titles today are often brief and sometimes cryptic, and they do not all define the nature of the book very clearly. In general, the title-page at least refers to the book's theme; if the chapter titles have been well drafted, the list of contents may provide a useful enlargement on the scope of the text. The list of contents may also be consulted by the reader more often than other preliminary and end pages, and so may best appear in a position where it will be easy to find. For these reasons, it may well be placed on the first recto after the title-page. When chapters have numbers but no titles, there is no apparent need to list them anywhere.

The headings of preliminary items should generally be linked in style with the headings of chapters, and may well be set in much the same way and at exactly the same height on the page as chapter titles. Headings do not need to include 'list of', when the way in which a list is set is evidence of its nature. If the drop to the first text line of preliminary items matches the drop to the first text line of the first chapter, and if the text type is used throughout the preliminaries, they will present a tidy introduction to the typography of the main text. A more modest style of heading and a smaller text type, on the other hand, would emphasize their subordination to the main text, and may be preferable when prelims extend to many pages.

A list of contents which includes chapter numbers is usually set as a three-column table, with chapter numbers on the left, chapter titles in the centre, and page numbers on the right. This convention is so well established that there is no need to identify the outer columns with the words 'chapter' and 'page'. If the list is to include part numbers or section titles or both with page numbers, the two main columns of numbers will look neatest when they contain one size of figure only, as the different set of different sizes of figure would give the column an uneven outline; part numbers may be centred, there may be no need to show page numbers for parts, and the page numbers of sections may be run on among the titles (as in the list of contents in this book) or indented into a separate column. While the text measure is often best for preliminary items in general, a wide space between chapter title and page number may render them difficult for the reader to link, and the whole list may benefit from being indented at left and right to close up this space. The use of *leaders* (…) to draw the eye from title to number complicates the page and is out of fashion. When the wording will make more than a line, the designer may do well to specify the point at which each line is to be broken, how far if at all the turnover is to be indented, and whether or not chapter titles may extend into the column of page numbers. Lacking such guidance, the printer might resort to an arrangement neither incorrect nor obscure but lacking in grace.

The customary tabular arrangement of this kind is clear and familiar but not obligatory. A short and simple list of contents in particular may lend itself to some other style; each chapter title may for example be centred in a continuous line starting with chapter number and ending with page number. When the chapter titles represent a sequence of ideas, they are perhaps better not mingled with numbers in this way, and the style of the list of contents may need to be adaptable to the list of illustrations.

Chapter titles in the contents list are often set in the same way as the chapter headings though in smaller size, so that the pattern of the text is introduced in the prelims. The titles of items in preliminary and end pages may, however, be set in a contrasting style to indicate their subordination to the chapter titles. The wording should be the same in list and heading, and so should any combination of capitals with smaller letters, and any punctuation.

The list of illustrations is in effect a supplementary list of contents, and is perhaps best placed immediately after the list of contents and set in much the same way. The list of illustrations is often much shorter than the list of contents, and may reasonably be run on after it, or printed on its back. Illustrations printed with the text are usually listed separately from plates.

This is partly because when illustrations are printed with the text their page numbers in the list refer to the pages on which they appear; the numbers of the plates usually form a different sequence, divorced from text page numbers, and their positions are likely to be indicated by reference to the number of the text page facing each, or to the numbers of the two text pages between which more than two plates appear.

Long and detailed captions may be needed by the illustrations themselves, but should be abbreviated before being set in the list, which provides a guide to their position rather than explanation or description. The list of illustrations is an apt position for any acknowledgements assigned to illustrations, as an acknowledgement set under the illustration itself may tend to distract attention from picture and caption.

Other preliminary lists may include a glossary and a list of abbreviations. The preliminary pages are likely to be the best place for these, as the reader can run his eye down them to identify unfamiliar terms before he starts reading the text. When preliminary items are to begin on the next available page, whether verso or recto, there is something to be said for beginning such lists on a verso; from farther on in the book, the left-hand page is easier to consult than that on the right, and, if the list or lists make more than one page, this may reduce the number of openings to which the reader will refer. The position in the prelims of the lists depends on their nature; a long list, likely to call for repeated reference, should be easy to find, and might therefore precede the preliminary text; a short list, which the reader may be able to memorize, might face the first page of text, following the preliminary text.

The lists may alternatively appear in the end pages. Some authors choose to combine index, glossary, and list of abbreviations, to save space, and to reduce the number of different parts of the book to which the reader is invited to refer. Lists are usually set either as tables of two columns or as *hanging* paragraphs (§ 7-4). Tabular setting suits short *lemmata* (the *lemma* is the term glossed or indexed, the abbreviation extended, or the subject-title of a note) and short explanations; the columns are compact, and the page better balanced than if all the type is crowded to the left. Longer lemmata and explanations are better as hanging paragraphs, which reduce the number of turned-over lines.

The use of roman or italic, and of upper or lower-case, should follow the textual presentation of the lemma, for clarity; lemma and definition may be separated in the list by punctuation, to avoid introducing a change of alphabet. Unless economy requires smaller type, the text fount is likely to be suitable.

§ 8-5 Preliminary text

Items of preliminary text are constructed of continuous sentences, not of lists or tables or display. Since they are continuous, they may be set in the same style as the main text, unless they are long and their difference from and subordination to the main text calls for typographic expression. The various items may well be grouped; to switch from text to list and back again might seem restless. Foreword, acknowledgements, preface, and introduction, in that order, are the most commonly used terms, but some books also have an author's preface, prefatory note, preface to the second edition, and so on.

The order of the items is influenced by their different functions, which have been clearly defined by Burbidge (1969):

... the preface is the author's account of the scope and purpose of his book, how it came to be written, and to whom he is obliged for assistance. A foreword, which may be written by the author or by someone else, is a send-off, something which touches on the subject-matter in a very general way, and perhaps relates it to other works in the same field, thereby giving a sort of fillip or stimulus to the more serious matter to come. An introduction deals wholly with the subject-matter of the book. If it serves merely as a general preamble to the text then it should, of course, form part of the prelims; but if it is vital to the comprehension of the text (being thus an integral part of the book), it should be moved out of the prelims, imposed as the beginning of the text, and perhaps even numbered as chapter 1.

In accordance with this definition, acknowledgements may form part of the preface; if the style of the preface is unsuitable for this, or if the acknowledgements are very long, they may be set as a separate item, perhaps before or after the preface. Since this book was first published in 1956, its acknowledgements have appeared among the end pages, for the reader to see after studying the work and visualizing what help the author may have needed.

§ 8-6 Dedication and corrections

Dedications are less fashionable than they used to be, but authors are entitled to make them. However brief, a dedication deserves a page to itself; it may represent a memorial, or speak for strong feeling on the part of the book's creator. The recto facing the title-page verso is unsuitable, as the dedication should not have to share an opening with such routine items as the Standard Book Number. Elsewhere within the prelims, the dedication might appear to interrupt the sequence of information. There is much to be said for the half-title verso, or for the last page of the prelims, if either page is needed for nothing else.

A very short dedication may look insignificant if set in anything smaller than capitals of the text fount, but it should not appear to compete in emphasis with chapter titles. If it is aligned with the topmost display line on the facing page, or with the first line of text when it faces the start of the first chapter, it will contribute to the balance of the opening.

Unwelcome afterthoughts used to appear all too often in prelims, under such titles as addenda, errata, or corrigenda — or, if followed by one amendment only, addendum, erratum, or corrigendum. There is no need for Latin, and 'addition' or 'correction' will do. In the past, the main text of any book might be printed some time before the prelims and end pages, and the intervening period (and particularly the compilation of the index) was all too likely to reveal omissions or other mistakes. Today most books are printed from beginning to end within a much shorter period; one day of two shifts, for example, may be enough for 5,000 copies of a book of 256 pages. A reputable printer moreover is more likely to print a *cancel* (a replacement leaf, leaves, section, or even sheet) than to release to the binder and the public mistakes which might appear to be his fault or that of the author. If mistakes have to be published, and if they are such that the reader might be misled, he must certainly be warned against them; mistakes unlikely to mislead are better left buried in the text, to be disinterred only by the most attentive.

§ 8-7 The text

The first page of the main text may be considered the most important single page after the title-page; if the title-page has convinced the reader that the book is worth his while, it is on the first text page that he begins his work in earnest. The convention that the text begins on a recto page is so familiar to readers that it should be broken only with obvious intent, and preferably in the course of a generally unconventional treatment. When the text does begin on a verso, the page numbering may need care if arabic numbering begins with the text, as the first page of text should then not be page 1. When conventions are cast aside, those which guide printer and binder are better retained, and page numbering should always begin on a recto page.

A *repeat half-title*, usually repeating the words and perhaps also the design of the half-title which begins the book, may precede the first page of text, whether or not there are other half-titles further on. This half-title is often numbered in with the text, which then begins on page 3, but may also form part of the prelims. Whichever numbering applies, recto and verso of any half-title leaf are normally *blind folios*, on which no number is printed.

§ 8-8 Appendixes and notes

The end pages are a part of the book similar in importance to the prelims, and may therefore be separated from the text by beginning a new leaf, on the recto page. When there are half-titles within the text, the first item in the end pages may be preceded by a half-title.

Oxford (1981) and Hart (1978) now agree that the plural *appendixes* is dropping out of use; earlier editions of both authorities preferred it for bookwork, restricting *appendices* to scientific use. But Oxford and Hart still prefer *indexes* to *indices* as parts of a book, and since the end pages may on occasion include two or more appendixes and two or more indexes, the plural *appendixes* appears the more compatible. Perhaps it will not drop out of use among publishers who seek to apply sub-editorial polish to their editions.

Appended to the main text, appendixes usually appear immediately after it. The difference in importance between appendix and chapter may be indicated by a change of numbering method (chapter 19, for example, being followed by appendix A), by a style of appendix heading more modest than that of the chapter (perhaps similar to headings in prelims), and by the use of a smaller text type. Emerging from the main text, the reader may not otherwise observe at first that he has entered an appended text. Particularly when appendixes are short, they may run on after each other, even if the preceding chapters begin on new pages.

The reason for setting appendixes in smaller type than the main text applies also to the notes, which may follow the appendixes but in their absence follow the main text. When notes are collected at the end of the book in this way, rather than placed at the foot of the text page, a fount one point smaller than that of the text is small enough as this is easier for continuous reading than the smaller type usual in footnotes. To begin each footnote with a reference to a text page number is convenient for the reader but expensive for the publisher. If to avoid this expense the notes are numbered continuously throughout each chapter, the text is likely to be interrupted by double and even treble superiors. When many of the notes make the equivalent of an average paragraph or more, economy will be served by starting each with a paragraph indent; when most are shorter, clarity will be gained from setting each with hanging indents.

§ 8-9 List of other works

Two titles are common, *references* and *bibliography*. The purpose of a reference is specific; in his text, the author refers to another work, and in the list he describes it in such a way that his reader can trace it if he wishes. The purpose of a bibliography is inadequately defined by its title, which may appear at the head of a list of references, of works consulted by the author, of books commended for further reading, or of works written by the subject of a biography. If such lists can be separated from each other in accordance with the function of each, and given a more explanatory title, their purposes will be clearer to the reader.

The type size appropriate for appendixes and notes is likely to suit such lists. Most book titles occupy one or two lines only, and may therefore be set as hanging paragraphs. Precedence is usually in alphabetical order of authors' surnames, with which each entry begins. Small capitals, with or without initial capitals, are often used for authors' names, but this can look clumsy; a combination of capitals and small capitals is out of fashion, and if small capitals are letterspaced as they should be, the names appear to be over-emphasized at the expense of the titles. Roman upper and lower-case will prove clear enough for names; and if each name or group of names is followed by the year of publication, as it often is in references within the text, that will separate and distinguish name from title. References are sure to contain a high proportion of initial capitals for proper names; if capitals are also used indiscriminately within book titles, the page is likely to look indigestible. For scientific works in particular, convention prescribes bold figures for the volume number in a reference to a periodical; specialist readers prefer such conventions to be respected, but there may be lists in which the page can be healed of a rash of bold figures by use of the term *volume* or its abbreviation *vol.*

§ 8-10 The index

The index is usually the last part of the book to be compiled. Appendixes, notes, and so on may appear in proof at the first stage, but until the pages have been made up and a page proof submitted, an index based on page numbers cannot be assembled. For this reason the index is almost invariably the last major item of text in the book. In that position it is easy enough to find during reading, and may be taken to catalogue all references which have preceded it. When appendixes and notes are set in a fount a point or so smaller than that of the main text, and the index follows in a smaller fount still, in order to fit into a two-column measure, the cadence of type-size extends to the end of the book.

In a book of verse, an index of first lines may be best set as a table, like a list of contents. Most indexes consist of a single word or phrase followed by one or more page numbers and perhaps by some analysis of references. When a proportion of entries makes less than a line of the text measure, as in most indexes, two or more columns save space and balance the page. To help the reader in identifying entries, the index is usually set as hanging paragraphs, with additional indents to indicate subordinate entries. A white line appears between the letters of the alphabet in all but the shortest index; in long indexes, the new initial may be set as display, centred in a deeper space.

§ 8-11 Colophon and imprint

The colophon was the first means by which the printer-publisher proclaimed his part in the technical process, secured his moral and legal rights — such as his rivals were prepared to acknowledge — in the commercial exploitation of the finished article, and vouchsafed any further information that might stimulate its sales. As the name implies, the colophon is the end-part or tail-piece of the book. It should therefore be applied only to statements which appear after the text, and its loose use for the publisher's name or device on the title-page is to be deprecated (Steinberg, 1974).

In a trade edition, the colophon of today is likely to consist only of the printer's imprint. British law requires this to appear in the book, American law and custom regrettably permit it to be left out. The end of the book may prove an awkward position for the imprint; there may be no page to spare at the end, and the book may end with a section or more of full-page plates. The imprint is therefore usually printed on the title-page verso, and combined with the country of origin which has to be printed there. The printer is entitled to some say in the choice of position, and has a duty to provide the correct wording of his own imprint; his legal liability might be affected by incorrect wording.

Some publishers like to name the text type-face in the imprint. There is no harm in this practice, and little good either; it may stimulate in authors opinions about book design which are influential rather than valuable, and is unlikely to provide the general reader with information he can use. If the typeface is to be named, it should be described precisely (§ 6-7).

To add more data to imprint or colophon in a trade edition might now be considered pretentious, and in practice might prove inaccurate. A last-minute change of production plans could vitiate any description of paper or bookbinder. Author, publisher, and printer have taken a major share in the production of the edition, and have different reasons for printing their names in it; the designer's part is minor in comparison, and he should seek to add his own name only if the publisher concedes that his work is a major feature of the edition. A freelance designer or a design organization may reasonably claim that no equally effective form of advertisement is available. In a show-piece of book design and production, or in a limited edition, an elaborate colophon may be acceptable; otherwise, the reader may well be left to observe for himself that a designer among others has invested care and skill in the edition.

CHAPTER 9

Display and ornament

Dısplay ın books is traditionally understood to mean composition in type sizes larger than those in which text is usually set. In Monotype work, such type is normally set by hand; in any work, selection of a display series different from the text series may still necessitate separate composition. In this sense, display is traditionally a composition process separate from that of the text, carried out with display founts differing in constitution from text founts.

However, photo-composition can now provide large versions of the entire text fount, together with the opportunity of setting them by means of the text composition process. In this chapter, display is separated from text composition in terms of design, and is treated as composition requiring design decisions other than those which regulate the continuous paragraphs or equivalent material of the author's text.

Text composition is designed for sustained reading, display composition for reading at a glance. The selection and arrangement of type for display are governed less strictly by convention than is text design because of this difference of function. The very term *display* suggests an element of show, of decoration, and even of advertisement; and the best display work is that which fulfils the primary object of all typography, communication, and in addition is touched (but not too heavily) with creative originality. In display, ornament, which is little used on the text pages — where it might interfere with their function, however slightly — also has its place.

§ 9-1 Composition in metal

Monotype composition casters normally produce composed type in sizes up to 14-point. With the large-type composition attachment, this facility extends up to 24-point, but the number of large-type lines in any book is likely to be so small a proportion of the whole setting that its use for this purpose is rare. Founts of large type are more likely to be produced on the *supercaster* (which casts type, leads, and ornaments, but not in composed order), distributed into case, and set by hand.

Line-cast composition may be set by hand or machine; founder's type (§ 4-1) is set by hand. Monotype, line-cast, and founder's type can all be combined for display with any form of text composition, but the printer's task will be eased if incompatible metals are not too extensively mixed. Characters and ornaments not available as typographic sorts can be printed from line-blocks (§ 11-3) locked up with the type.

Most metal-cast founts above 14-point have a different constitution from that of text founts, lacking small capitals, inclined and other alternative arabic figures, special sorts other than the most familiar of accents, and sometimes even italic characters. The Monotype range of such display founts within text series is extensive, and is supplemented by separate display series, not designed for continuous reading, nor available in composition matrices, nor for the most part cut in text sizes. There are also *titling* founts, within or separate from text series and display series (figure 9-1*a*). Each of these usually consists of one alphabet of capitals, with figures, ligatures, ampersand, sterling and dollar signs, and punctuation. The capitals occupy almost the whole body, leaving only a vestigial beard, so that they can be set as *initials* (§ 9-5). In size, titling letters are often usefully intermediate in capital x-height between the text founts, and much of their value is provided by variation in weight from that of the standard capitals.

WEISS INITIALS I

WEISS INITIALS II

BEMBO TITLING

PERPETUA LIGHT

Figure 9-1*a*. Four titling series. E. R. Weiss's *Weiss Initials, Series* I *and* II (Bauer); *Bembo Titling* (Monotype); and Eric Gill's *Perpetua Light Titling* (Monotype).

Open display letters, within the outline of which white space appears, and particularly light titling or other display founts, are relieved of that boldness which prevents large sizes of solid letters from blending in weight with smaller founts (figure 9-1b). Decorative founts are those in which decoration has been added to the letters, or in which the letters are distorted into decorative patterns. Such letters are designed for use singly as ornamental units, rather than in combination to produce words, though the plainer specimens of the style may provide readable composition in small doses.

CASTELLAR

LUTETIA OPEN

OPEN CAPITALS

Chisel

Figure 9-1b. Four open types. John Peters's *Castellar* (Monotype); Jan van Krimpen's *Lutetia Open* and *Open Capitals* (both Enschedé); and *Chisel*, adapted from a late 19th-century Latin Bold Condensed (Stephenson, Blake). The first three types have capitals only.

The metal shanks of large characters were trimmed at the beard to fit them into spaces which would otherwise have been vertically inadequate, or mortised on the right to bring smaller type into the area of the shank and close to the initial itself. These cutting operations concealed the difficulties of fitting large characters neatly among small ones. In photo-composition there would be similar difficulties if such decorative initials were in fashion, and no doubt equal skill and other methods will be found in time to fit photo-composed initials and text together.

Founder's type was supplied by typefounders to printers in the form of sorts, for composition by hand. During the twentieth century, composing and casting machines provided printers with their own foundries, and the typefounding industry began to close down. Few of the historic typefoundries now survive, but some of them are rich in the possession of ancient punches and matrices and in the work of type-designers of our own time. Not all the old material is accessible, and some of it is interesting rather than admirable, but there are still matrices of previous centuries from which display types could be cast to enliven books of today (figure 9-1c). Even the best of this material has yet to appear in photo-composition form, and since it is absent from current typography there is no demand for it, but these inventive letter-forms have been recorded and deserve to be remembered again one day.

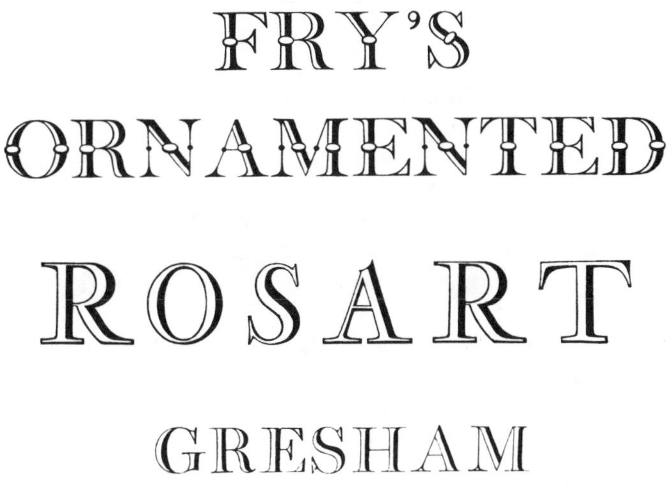

FRY'S

ORNAMENTED

ROSART

GRESHAM

Union Pearl

Figure 9-1c. Four old display types. *Fry's Ornamented*, a design of the late 18th century (Stephenson, Blake); *Rosart*, a design of the first half of the 18th century by J. F. Rosart (Enschedé); *Gresham*, late 18th-century (Stevens Shanks); and *Union Pearl*, a late 17th-century design which is the oldest English decorated type (Stephenson, Blake). The first three were originally designed in capitals only; *Gresham* has a more recently designed lower-case.

From time to time in photo-composition as in metal type the intervals of size between display founts pose a design problem when one fount is too small and the next too large for a specific purpose or space. The standard progression of metal founts is 14, 18, 24, 30, 36, 48, 60, and 72-point; occasional additions include 16, 20, 22, and 42-point; other sizes are rare. Some photo-composition systems can adjust the size of the typographic image to the space available, others are limited to metal gradations and may not be able to encompass all of them. Since most books contain few lines of this kind, the usual remedy is to set the largest available size and adjust its scale by camera.

Didot bodies in type are those of continental Europe, and are slightly larger point for point than Anglo-American bodies. A Didot fount cast on Anglo-American point body is likely to require extra depth, so that 14-point Didot type (14D) is cast on 16-point Anglo-American, 16D on 18-point, 20D on 22-point, 24D on 30-point, and so on. The same series may for instance include a 14-point (Anglo-American) and a 14-point (Didot) fount which differ visibly in gauge; using Didot sizes, designers should specify both the Didot size and the Anglo-American body, to avoid ambiguity.

§ 9-2 Photo-composition

Most photo-typesetters can in a single operation set most of the type sizes required by most books, following the size gradations of metal founts. The Monophoto 400/8, for example, provides sixteen sizes from 5 to 24-point; only an extensive metal series would include more.

For sizes over 24-point, such a machine can be supplemented by a display photo-typesetter. This kind of unit is designed for comparatively slow composition with large type, sometimes in a single continuous line which has to be cut and pasted up into position to form a succession of display lines. Most of these units can set type up to 72-point; some, perhaps designed for newspaper use, can set it much larger. Many enable the operator to adjust the fit of the letters while setting, an advantage when letter images, drastically enlarged from a smaller matrix, look too loosely fitted. Some provide opportunities for playing tricks with letters, changing their shapes, applying a *mechanical tint* (§ 9-11), and even setting them in a circle instead of a straight line. Typographic ornaments cast in metal for hand composition can be copied and set by such machines as these, and several of them enable typographers to reproduce characters or ornaments of their own design.

Some photo-typesetters offer much more variety of size than anybody need expect of metal, and may, for example, be capable of setting any size from 4 to 72-point in $\frac{1}{2}$-point steps. This enables the designer to approach the layout of display from a different direction. An increment of $\frac{1}{2}$-point in body on a 24-point fount, for example, will increase the width of a 20-pica line by only 5 points. Lines can therefore be set to fit a designated width almost exactly; in the standard progression of fount sizes, the difference in width between the lower-case roman alphabets of the 18 and 24-point founts of a typical series is more than 5 picas. In apparent size, the gradations of photo-composition display founts can be infinitely subtle; the difference in apparent size between 18 and 24-point metal letters is coarse in comparison.

Because their characteristics are plain to view, big letters need close examination before use. A good text fount, enlarged in the course of photo-composition to double, treble, or quadruple its text size, will not necessarily appear to be a good display fount. Any deficiency in sharpness of outline, in length of extenders, in detailed finish, and in fit may be so accentuated by size that a reader may feel that something is amiss, some quality he cannot define is absent. Adjustments to fit, which are possible on some photo-typesetters, may reduce the impact of such faults.

Display founts in photo-composition, when reproduced from the same matrices as text founts, offer the same array of characters, including large sizes of small capitals, italic figures, accents, special sorts, and bold type which in metal are often omitted from display founts. In those systems which rely on a separate fount of matrices for each size of type, such constituents of text founts which do not usually appear in display may be provided in text sizes only.

§ 9-3 Type selection

The choice of a text fount is usually limited by the equipment of the chosen printer, who cannot usually afford to buy a complete set of matrices for use in one or two books only. Metal display founts, on the other hand, are more flexible. The Monotype Corporation provides many of its display matrices on hire so that printers can cast and lay cases of one or more founts at short notice without high cost. Typefounders may be willing to supply small quantities of cast type for a few display lines. Display is likely to form a small proportion of the composition cost of any edition, and special arrangements may be worth while when they are not inordinately expensive.

Dry-transfer lettering copied from type-faces can be composed by the designer, particularly when the book is to be printed by offset, and fixed in place by the printer. The letters are supplied on the pressure-adhesive side of a sheet of transparent film, with register marks to show alignment and fit. The film is laid in position on white card, and the character is transferred to the card by rubbing across it through the film. Adjustments of fit are laborious and far from easy to apply evenly but are still possible; this facility is all too often exploited by cramming letters too closely together.

Whatever the source of display type, the existence of each required alphabet needs to be verified before its use is specified. Even the vast repertory of Monotype display founts does not extend up to 72-point in all series. Even when a 72-point upper and lower-case is to be had, this and some smaller founts may lack italic and are sure to lack bold. By no means all bold founts include bold italic. In photo-composition the sizes of fount available are determined by the system, not by the matrices, and a designer who wishes for sizes intermediate between 18 and 24-point for example may have to set a few display lines in 24-point and reduce them by camera.

In the absence of instructions, a printer will usually set large display lines from founts of the series used for the text, if they are in stock. Close relations do not, however, always look alike, and in some metal series — Caslon and Walbaum are examples — there may be a distinct difference of appearance between small sizes and large. Even when text and display are reproduced from the same photo-matrices, enlargement may cause some incongruity.

The designer may prefer display founts of a series other than that used for the text. Established customs of bookwork have been questioned during this century; some survive, others may yet revive, and the question itself has brought into being some comparatively new answers. One typographic theory exalted sans-serif type as the proper medium for a contemporary message — a type implying modernity, its outlines stripped of old-fashioned ornament, its appearance matching in starkness the architecture of the day. This theory has not been widely accepted as applicable to text composition, but that does not disprove it. The use of sans-serif type in text is certainly a custom of the twentieth century, and while such type may or may not be too unfamiliar in mass to be truly legible for sustained reading, it is clearly legible enough for display. The letter-forms may remind the reader of newspapers and magazines, posters and other advertisement, rather than of books, but there is not always evident harm in this. The tradition that mixed series should be related to each other in appearance and in period of origin is no longer very influential and is often breached, for example, by the use of sans-serif in display. When mixing series, however, some typographers still prefer to use for display a type which in origin or design or both is compatible with that used for the text.

The meaning of *display* for the purposes of this chapter embraces smaller type as well as larger. To use the text fount, or another fount of the same series, for such apparatus as headlines, page numbers, and captions is a custom which has its convenience for the printer but is not always economically essential. Subtle differences between two founts of similar size may be discernible by the reader, but only far enough for him to feel the wrong type seems to have put in an appearance. If the difference between two small founts is marked and purposeful, the mixture of series is likely to be effective.

Decorative letters are less widely used in bookwork than they were at one time, partly because few successful founts of this kind include lower-case, and partly because such letters are often seen at their best when they are almost too large and too bold for the book page. The charm of fanciful letter-design begins to pall when it is exposed on too many pages. Such founts sometimes appear only on the title-page, but a decorative fount which includes lower-case, if not too exotic in design, is sometimes used for chapter titles as well. The only kind of decorative letter which deserves no place in books is one which includes characters difficult to recognize at a glance.

The fount used for display should not be so big that it appears to be uncomfortably large when the book is held at a distance suitable for reading the text. Letters which are too big may be just as unpleasant to look at as those which are too small, and may seem even less attractive. The emphasis of particularly big display may also overwhelm any text which appears in the same opening, and distract the reader from the smaller type. Within the text area, even the capitals of the text fount may look rather too big when displayed, particularly if they are tall and bold in proportion to the short letters, and those of a smaller fount may be preferred. In metal composition, any change of fount within the text area may prove costly if it is frequent. If the series is one in which the capitals are no heavier than their lower-case, the stem of the capitals of the smaller fount may be thinner than that of the text lower-case, which, combined with letter-spacing, will give the heading an appearance of contrasting lightness. The heading may then seem to be subsidiary to the text, an effect which may be useful.

The proportion in size between display, text, and page dimensions is a matter of taste. As a very rough guide, twice the size of the text fount will usually be quite big enough for the largest display on the same page as part of the text. On the other hand, when small type is set in double column on a big page, the size of the display should be related to that of the page rather than to that of the text fount.

Relative emphasis can be adjusted by the use of different founts, different series, and different alphabets. Bold letters sometimes look smaller than their equivalent in medium weight, because the counters of bold letters tend to be smaller. Emphasis depends partly on the waist-line of the letters, partly on the alphabet used, and partly on apparent size. In large sizes, capitals which are equal in x-height to the lower-case of a still larger fount seem more or less equal to it in apparent size and sometimes appear to be more emphatic. Letter-spacing capitals and small capitals seems to enhance their emphasis, particularly within the text. Probably most readers consider capitals to be more emphatic than small capitals, small capitals than italic upper and lower-case, and italic upper and lower-case than roman.

In large sizes, when one line is set in the capitals of a display fount, and the next in upper and lower-case of the same fount, the lower-case is curiously apt to look slightly too big, particularly if the capitals are not large in proportion; the lower-case of a fount somewhat smaller is likely to clarify the intended difference in emphasis. Where emphasis is to be graded, the difference in size between displayed lines or groups of lines should be obvious; if, for example, the main title-line is in upper and lower-case, and the sub-title is in capitals of a smaller fount, the x-height of the capitals in the sub-title should not be greater than that of the lower-case in the title. The diminishing thickness of waist-line in smaller founts must be remembered; if the capitals of two different founts are set in one line, the disparity of waist-line is likely to be uncomfortably obvious, which is one reason for the unpopularity of the practice. Another reason is traditional; in metal the alignment of two founts on the main line requires time-consuming handwork. Above all, mixing founts in a line does not make for neat display.

Massed capitals or small capitals are less easy to read than lower-case, and therefore tend to be reserved for picking out and emphasizing an occasional phrase, or a group of short phrases; long titles and sequences will be much more legible in upper and lower-case. Capitals or small capitals are sometimes used rather ineptly for a list of contents in which the chapter titles are composed of several words; the appearance of a list set in this way may be quite discouraging to the reader. Lower-case also has the advantage of its narrower set; it fits more words into one line before running into a second. Laterally, lower-case italic is the most economical alphabet of all.

§ 9-4 The display line

In text composition, the setting of each line starts at the left of the measure (apart from such indented instances as the first line of a paragraph) and continues until there is not enough room left for another word or part of a word. At this point the words are usually spread out to justify the line, which may end with part of a word and the hyphen which connects it with its remainder in the next line. Given a measure and fount, the printer is entitled to treat display in the same manner. This may do well enough for display items in text type or smaller; for display in larger founts, it is all too likely to look clumsy. The designer's specification should regulate the division of all divided lines in display, whether by means of a general instruction or by marking the copy in each instance, if he is to make sure of a result he will be pleased to see in proof. To have the type set first and then rearrange it in proof, thereby increasing the bill for alterations to type, was never a good practice, and in photo-composition is likely to turn out prohibitively expensive.

The point at which every large-type line is to be divided should be marked on copy by the designer, as the rhythm, the emphasis, and even the meaning of the heading may be affected by line division. Large type is better left unjustified; the reduced number of words in the line tends to result in excessively uneven word-spacing. In the absence of justification, the division of words between lines can be avoided.

In the copy for any set of chapter titles, captions, or other forms of display, variations of length are to be expected and must be allowed for. When all the chapter titles but one consist of a word or two only, and that one exception consists of a dozen words or more, the design of the chapter titles must suit the long as well as the short. The apparent width of every line should be planned. The designer may for example decide that no caption is to be apparently wider than the illustration above it. In practice, captions are likely to be set in a single measure; the visual width of each is adjusted to that of its illustration by indents at left or right or both, or by line-division marked on the copy to produce a similar effect, to avoid changes of measure during composition.

Large-type display lines tend to look better when there is some extra space beside them, as well as above and below; even when such lines are to be centred, they may benefit from indents at left and right, to prevent a long line from filling the measure. Similarly, space beside a sub-heading helps to distinguish it from the text.

When a display item of any kind will make more than one line, its meaning will be easy enough to assimilate if it is broken into two or more lines roughly equal in apparent width. Without instructions to this effect, the printer is likely to fill up the first line and *turn over* (insert in the next line) a single word only, an arrangement which not only lacks grace but presents phrase or sentence in a way nobody would choose to speak it. If a headline is allowed to fill the measure, the reader may at first glance confuse it with the text, and even a narrower setting may come too close to a page number in the same line. If typographic economy cannot prevent this, or if a headline cannot be fitted into a single line, it will have to be re-written. When display lines are ranged to the left, rather than centred, the printer should be instructed to *leave* such lines *short* (of the end of the measure), as he may otherwise suppose that any which comes close to filling the measure should be widely word-spaced in justification.

In text composition, word-spacing and letter-spacing are usually mechanical, and in a well-designed book set by a capable printer this is generally acceptable for the type-sizes usual in text. Display such as headlines, normally in text or similar size, is spaced in the same way, and indeed any other method is likely to be uneconomic. When large-type display is set by hand in metal, it should be word-spaced and letter-spaced with apparent evenness, which differs from mechanical evenness. In lower-case, for example, a word-space between two short letters should be slightly narrower than one between two ascending strokes such as those of d and b. In the same way, the letter-spacing between V and A should be less than that between M and N. Single-type composition in metal has accustomed designers to this refinement, which is part of the high polish a perceptive eye expects of any first-class printing.

Optically even word-spacing and letter-spacing is never going to be easy for keyboard operators, unless facsimiles of the typographic image and its spacing can be shown in enlarged size on the screen of a visual display unit. David Kindersley and Dr Neil Wiseman (1976) have devised a software solution for computer-centred systems, but it would be distinctly expensive to a printer and machinery manufacturers appear to lack interest in a facility for which they observe no demand.

In text composition the designer can, if he feels it is necessary, designate the inner and outer limits of word-spacing, probably in terms of units, and he can specify thin, medium, or thick letter-spacing. In unjustified display, he will do well to regulate all such spacing, as printers differ in their practices. All spacing should suit the fit and other characteristics of the types to be set, but as a general guide a middle or mid space, 4 to the em, will usually be close enough for lower-case word-spacing; in some closely-fitted display founts, thin spaces, 5 to the em, may not be too narrow.

The letter-spacing of display sizes tends to be proportionately narrower than that in text (§ 5-7); in previous centuries, such spacing was sometimes so wide that words became almost incomprehensible, but today's style favours compact setting. When the unit system allows for fine adjustment at this point, spaces between 12 and 16 to the em are likely to suit display sizes of capitals and small capitals without looking too narrow for use in text.

§ 9-5 Initials

The beginning of the first paragraph of a chapter is sometimes marked by a large initial letter. This is almost entirely a matter of custom and decoration; the initial does not assist the act of reading or reference, except for instance in prayer-books, where initials distinguish new items — prayers, responses, and so on — from their neighbours.

An appearance of purpose is essential to the successful placing of initials; this begins with the alignment of the initial with the neighbouring text. The main line of the initial letter should continue the main line of the type in one of the adjacent lines of type, and the next text line should extend immediately under the initial letter. In metal, a capital letter used as an initial has to have its beard trimmed to make this possible, as there would otherwise be an unsightly white space below the initial; a titling letter usually fits without trimming.

A drop initial aligns at the head with the head of the capitals or small capitals of the first line of text, and drops to align at the foot of the letter with a subsequent line; a five-line drop initial, for instance, extends from the first to the fifth line. Exact alignment cannot be left to chance or the printer; initial, leading, and capital or small capital x-height of the text fount must be matched in depth to make a fit. If the head of a drop initial is to appear to range with the head of the first line of text, the adjacent text will need an evenness of height which lower-case letters collectively do not possess. This evenness is usually achieved by the use of either capitals or small capitals for the rest of the word which contains the initial, for the rest of the phrase, or even for the rest of the line. This tends unduly to emphasize a random part of the text, and is one of the arguments against the use of drop initials. Another argument, against the use of initials in general, is that in a book designed in the centred style they provide emphasis at an off-centre point.

The main line of a raised initial continues the main line of the first text line, and rises beyond that line; and an initial may also both rise and drop.

The lateral placing of the initial needs as much care as its vertical alignment. In an off-centre design, a raised initial can be indented some way from the left margin, perhaps aligning with a similarly indented chapter title. Any initial which drops is best set at the left margin, as one or more text lines at the chapter opening would otherwise be short. The bigger the initial, the more important its appearance of alignment at the left with the left margin or whatever indent it is ranged on. The vertical main-stroke of a big T, for example, should align at the left with the left margin, leaving the left half of the cross-bar to project into the margin, and Hart (1978) shows other examples. This may entail extra time and cost during make-up, but the initial will look all the better for it.

The text to right of the initial needs equal care. If a large capital L is used as a drop initial, and if the text which follows it is set against the right-hand edge of the letter's shank, the word of which L is a part will be interrupted by a wide gap. The rest of the word must be set close up against the vertical main-stroke, even if a metal initial has to be mortised to receive it. As for the succeeding lines, they may reasonably be set against the shank; any other arrangement is likely to demand too much time of printer and designer.

An initial of four or more lines tends to be so bold that it seems to stare at the reader out of the page. An open or an extra light fount is particularly apt for this purpose, as the letters will be more compatible in weight with the text in which the initial appears.

§ 9-6 Elements other than letters

The terms used in this section to describe various kinds of display material are not generally accepted as definitive. Here they are assigned to the meanings they are intended to bear throughout this book, in which other book-production terms are treated in the same way.

A *rule*, in the sense of printing material, is a straight line of limited or of variable length and of unvarying thickness. In metal, rules are type-high strips of brass or type-metal; in photo-composition, a rule is the line such material would print. Two punctuation rules, the en and the em rule, are now more widely used in text than others, including the $\frac{3}{4}$-em and the 2-em rule. The oblique and vertical rules are also sometimes borrowed from mathematical notation, for display rather than for text, usually with some space on each side. Display rules in metal may be cut to a required length, and are described in terms not of length but of design. A fine rule is as thin as can be effectively printed from a relief surface. A medium rule is $\frac{1}{2}$-point across the printing surface, and thicker plain rules may be described in terms of the same measurement. Multiple rules print more than one line in parallel; among the more popular of these are a double fine rule and a thick and thin rule. There are various forms of dotted and broken rules, consisting respectively of dots and dashes of various sizes at various intervals. An ornamental rule contains within its outline a repeated white pattern, or has a patterned outline, or both.

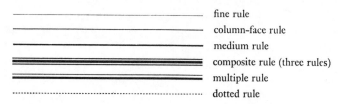

fine rule
column-face rule
medium rule
composite rule (three rules)
multiple rule
dotted rule

Figure 9-6. Some rules.

A short rule of some six picas is sometimes centred to indicate a division between two passages of text or display, or placed at the left between text and the balance of a footnote continued from the previous page and therefore not identified by a footnote number.

A *dash* varies in thickness along its length, and each is designed for printing in one width only. A *Bodoni dash*, so called because of its frequent use by that printer, is swelled and tapered. A *cut-off dash* has a diamond-shaped centre tapering into a thin plain rule on each side. Ornamental dashes of many kinds are available in metal, some of them unobtrusive and graceful; intricate Victorian designs are still to be seen, and modern artists have drawn or engraved dashes to the order of individual presses or publishers.

A *fleuron*, or printer's flower, is an ornamental unit designed for use alone but sometimes capable of assembly with other units of the same or differing design to form an area of decoration. A *border unit* is an ornament designed for combination with other units of similar or different design. Striking results can spring from combinations of fleurons, border units, and rules; the planning of such combinations benefits from the designer's use of sorts, or proofs of them, which can be arranged and rearranged in search of successful patterns.

Rules, dashes, fleurons, and border units, alone or in combination, can be used in display to indicate division and to provide emphasis, shape, and decoration. For these purposes they can be arranged and sometimes combined as spots, patches, lines, or frames. The possibilities of experiment are alluring, but reticence is to be advised; decoration by means of such standard printing material now tends to be seen as hackneyed, obtrusive, and incompatible with most kinds of text. Such ornaments are still found suitable for endpapers and binding cases; the higher flights of decoration in books are reserved for drawing or engraving commissioned for a specific edition.

§ 9-7 Pattern

Pattern in display may be defined as the visual effect of printed elements purposefully arranged. The most effective kind of pattern lends itself to the act of reading; the various elements are not scattered apparently at random, but are presented in due order to the eye. In this sense, an arrangement of display may be ineffectual if its pattern is too subtle to be observed by the kind of reader for whom the book is intended; the arrangement may be purposeful, but its visual effect is not that of a pattern.

To design a typographic book is to design a succession of openings. Most readers start at the beginning of this succession and work their way through to the end; some return from the text to scan or refer to the prelims or other pages; none can see all the openings at once. Since the reader's view of any opening is separated in time from his view of any other, the openings should be unified in design. The arrangement of display in each opening can and should remind the reader of other openings in the book, and of the identity of the edition before him.

Pattern is designed for the eye, and should be planned in visual rather than in mechanical terms. The left alignment of the drop initial, already mentioned, is an example; the initial should appear to be aligned, even when it has to be moved out of typographic alignment to do so. In the same way, successive lines of large type aligned at the left with each other, particularly when each begins with a capital, may need differing indents if they are to look evenly indented. If a chapter title in large type is preceded in the same line by a number and an em space, the title may look too far to the right, particularly if the rest of it is centred in the next line.

The edition need not be unique in the arrangement of display. A design which suits one text may do very well for another. For an edition to be internally consistent is enough. If the general style of its display is similar to that of other titles in the same series, that may serve to remind the reader of related titles. In this area, however, the typographer is likely to find scope for innovation and for deployment of the creative imagination.

§ 9-8 Centred display

The centred style of display is older than books as we know them, and can be sen in lapidary inscriptions of imperial Rome. By the time the first title-pages were composed in the early sixteenth century, printers had adopted this simple formula for the lateral placing of display. The tradition lived on, it has survived in a period of radical change in book-production techniques, and it remains valid. Centred display is easy enough in application to spare the typographer's time, and all too many books have to be designed quickly if they are to be designed at all. It is a vernacular, familiar enough in use to be common ground between publisher, designer, and printer; all understand its conventions and find it easy to apply. The tradition is ancient but not stale; within it there is room for excellence, when every detail is meticulously planned, and the style of display rises beyond routine tidiness to clarity and grace. Convention is the basis of communication, and the centred lines reflect the rigid formality of the text pages.

When major display is centred on the text measure, items of minor display set in the text fount and in smaller founts are quite often placed off-centre. These may include headlines and sub-headings, perhaps at the left margin, and page numbers at the outer margin. This too is an established convention, but the scrupulous typographer may prefer a more compatible arrangement.

The centre of the text measure is normally the point on which display is centred. A succession of display lines can be centred on some other point, in a departure from tradition. Two or more lines, centred on each other, indicate the position of an alternative centre where one line might not. This other centre should be well clear of the conventional centre, if there is not to be an insidious effect of slipshod work rather than of unconventional design. Such a departure is far from easy when the measure is narrow or when the wording does not lend itself to it.

The centred style seems to be more subject than other styles to inept line division. Such particles as 'the' and 'of' are spoken with less emphasis than nouns and verbs and other parts of speech. If such a particle is broken away from the rest of a passage and set alone in a line, it seems at the same time to be over-emphasized by its isolation and to be removed from the natural path of the reader's eye.

§ 9-9 Off-centre display

When a major item of display, such as a chapter title, is not to be centred, the typographer determines the position of the beginning of the line. The point at which the line ends is determined by its length, and in a design sense is a random position. In this, display reflects the placing of text lines; all begin either at the left margin or at a planned distance from it.

In the simplest off-centre arrangement, major display begins at the left margin. When chapter titles are short, this may leave an uncomfortably wide space to the right of the display. If the title-page is treated in the same way, it may appear to be receding from the centre into the spine; better perhaps to centre the main title, align all other lines on the page with it at the left, and range half-title left and title-page verso right to the same distance from the inner margin.

Display can be extended into the margin at left or right. This increases the cost of make-up, at least in metal, and is not very convenient for all photo-composition systems, but it can be done. If such an unusual position is to be taken up, it should be emphatically occupied, perhaps by part of a single line in large type or by a large figure; anything set in a more modest fount may look as though it has strayed out of its field. Too deep an incision into the spine margin may cause part of the display to disappear into the *backing* (§ 16-6).

When display is indented at left or right, an indent too narrow to be conspicuous may look merely untidy. Longer lines indented in this way sometimes appear to be inaccurately centred; the designer should always have some idea, however approximate, of the point at which every display line will end, and so how it is likely to look on the page. There may be more than one width of indent, for different elements of display, but this introduces the danger that display will seem to be scattered rather than assembled.

When chapter titles are indented at the left, the first text line of the chapter, perhaps picked out with a rising initial, may be indented to align with them, instead of starting at the left margin.

When display is ranged to the right, at the right-hand margin or at a distance from it, the position on the page at which each line begins will be random, being determined by the length of the line. In a single line of large type this is likely to be acceptable; a group of lines arranged in this way, each starting at a differing distance from the left, might irritate the reader, particularly if a very short line draws his eye right across the page, away from the left-hand edge where he is accustomed to begin.

Display ranged to the outer margin — to the left on the verso, to the right on the recto page — lies near the fore-edge, accessible to a reader leafing through the book in search of a specific passage. Page numbers are commonly treated in this way, for the same reason. Ranging to the outer margin can, however, increase the cost of composition if many lines other than headlines are involved, since the display cannot as a rule be correctly placed until make-up assigns each page to its verso or recto position; some of any display set with the text will have to be re-aligned. Off-centre display has become familiar, but the opening still tends to be symmetrical; display ranged outwards is usually either at or equidistant from each outer margin.

§ 9-10 Vertical position

A typical text page, together with the various items on the page of a chapter opening, establishes different vertical levels with which other items on other pages may well be aligned. The headline provides a level normally used for little else. The first line of text on a full page lies on a second level. The chapter number makes a third, when it occupies a line of its own lower than the first line of text on a full page. The first line of a chapter title is then likely to make a fourth level; the first text line of the chapter makes a fifth, the last text line a sixth, and the footline, when it contains a page number, a seventh.

Throughout the book, each example of a specific typographic item of display should as far as possible appear at the same level on the page as other examples of that item. If a particularly deep table rises into the headline, the headline is better omitted than raised. The first and last text lines normally contain the text area, though extra depth (always equal in both pages of the opening) is sometimes employed to ease a problem of make-up. The chapter title and the first text line under it provide levels for the headings and first text line of preliminary and end-page lists and text. The title on the title-page is rarely placed higher than the first text line on a full page or lower than the chapter titles at chapter openings, to which it might otherwise appear subordinate. The last line of the title-page may rest on the position of the last line of the text area; if the first line of the title-page is dropped a certain distance from the head of the text area, the last line of the title-page may appropriately be raised an equal distance from the foot of the text area.

Alignments of this kind may be extended to the placing of illustration. The head of the picture may be raised to the level of the headline or to the head of the text area; the last line of the caption may be dropped to the foot of the text area or to the footline, leaving varying spaces between pictures and captions.

The purpose of this kind of alignment and spatial relationship is to present an even pattern from opening to opening, and to assert the connection between different pages and items of the same text. A kind of symmetry is maintained by the use of a limited number of levels for all display and for the first line of each kind of list or text, and by the introduction of no item at any other level, particularly on adjacent pages.

The vertical position of any space or item on the page lower than the first line of text on a full page can be specified in terms of the text line or lines that would have occupied the position described. At the opening of a chapter, for example, the text may begin on the eighth line, and would then be *dropped* to that line. The *drop* to a text line must be defined in whole lines, unless any fraction is complementary to another fraction elsewhere on the same page; the foot of the page would otherwise be out of alignment with that of a full text page. In the same example, the chapter number may be centred on the first line, the chapter title on the third, and the first line of text under the title on the fifth. The drop to display lines may be measured and specified otherwise, but reference to lines is convenient for the compositor, and facilitates the checking of alignments in imposed proof.

The vertical intervals between display lines benefit from the same kind of attention as interlinear space in text composition. Adjacent display lines should not look more crowded than those of the text, except to produce a special effect; usually a proportionate addition of interlinear space will be enough for a single item set as two or more lines of upper and lower-case. The vertical distance between two lines of capitals should not usually be less than the x-height of the capitals. Associated items should be closer to each other than dissociated items, and display lines in a centred layout are usually closer to each other than to any other element on the page such as the text, except for sub-headings which in this sense are more a part of the text than, for instance, the chapter title. If one item is to look vertically centred between two other items, slightly more apparent space will be needed under it than over.

Apparent space is likely to include the distance between the top of the short letters of lower-case and the top of the capitals; this has to be judged by eye. Certainly if two lines of different size are to seem to be printed at the same level on facing pages, they must be aligned with each other at the mean line.

Space confers emphasis, and vertical space in particular. A chapter title may even look effective in the text fount if it is generously separated by space from everything else. Too much space, for example between headline and text, may result in too much emphasis.

§ 9-11 Mechanical tints

By shading, cross-hatching, or stippling with a pen and black ink, an artist can simulate a shade or tone of black, a form of grey. Artist or printer can instead apply, in a single operation, a similar form of shading to an area, by using a mechanical tint. This term does not indicate any real variety of colour, unless the tint is printed separately from the rest of the page; a tint of this kind appears to the eye as an area of reduced intensity, usually even in pattern. A simple form of tint can be reproduced by printer or engraver by varied applications of a half-tone screen (§ 10-2); a 5 per cent tint, for example, consists of a plain light area of half-tone in which the size of each dot is 5 per cent of the distance between dots from centre to centre.

An artist may apply a mechanical tint to his drawing by dry transfer, and the variety of transfers available includes dotted, ruled, grained, and decorative patterns. A common use for a tint of this kind is to differentiate sea from land in a map to be printed in one colour. If the tint is not cut away around and behind small type or lettering, the words may be indistinguishable from the background. When the size of the drawing is to be reduced before printing, a fine pattern, the elements of which are very small or very close together or both, may not reproduce well.

In display, tint panels may be used to emphasize headings, as a reproduction of type (preferably large and reasonably bold, if it is to be clear) can be combined with the tint to print black on a grey background; or type can be *reversed* (§ 10-2) into the tint, to print white on a grey background. Tints may also be printed in a second colour as a background for black type or illustration, not only in display but for example in tables or chess diagrams.

CHAPTER 10

Camera and film

Photography is the formation of a visible and stable image on a chemically sensitized surface, by radiations of light or other forms of energy. Different kinds of photography are embodied in various preparatory processes in printing, and consist of different stages and different exploitations of the photographic principle. Photo-composition, examined in chapter 5, is one example; electronic scanning, not dealt with in this book for reasons explained in the preface, is another. The purpose of this chapter is to introduce some photographic methods of text and illustration reproduction, and to equip the designer for further study of processes which, in more senses than one, continue to develop.

The impact of photography on printing, and indeed on publishing, has been revolutionary. Any text which has ever been printed can be reprinted in almost exactly the same form, without type-setting, though the type itself may have been melted down centuries ago. New composition can be electronically set and rearranged, and can pass from composing room to press without the weight, rectangularity, and fragility of composed metal. Illustration, much of which in the nineteenth century depended on wood-engraver as well as artist, proliferates in the twentieth century from photographs and photographic processes of reproduction. The possibilities of publication in book form have been immensely widened; the illustrated book of today does not merely look different from those of a century ago — it may be a radically different kind of book, subordinating text to illustration and combining the two in a manner made possible only by photography.

In the first part of the twentieth century, the attention of publishers and printers was attracted by letter-forms, while the emergent type-setting machinery came to terms with the traditions of printing and with the possibilities of innovation. That was where attention was needed then, and if tomorrow type design is neglected it will deteriorate. But the typographer's scope will be gravely limited unless he now takes an equal interest in the photographic aspects of printing. Unless high standards of photographic reproduction are constantly demanded and paid for, in spite of the cost of first-class work, they will give way under economic pressure, and the second-rate will swamp them.

Opportunities for good printing, offered by developments in type design, illustration, presswork, and paper will be lost if photographic advances do not continue, in pursuit of excellence; there are all too many opportunities for bad printing.

Photographic techniques are exacting. When the designer holds in his hand a copy of the image he intends to reproduce in a book, he may form in his mind's eye the kind of reproduction of that image he believes possible and appropriate. Between copy and reproduction lie mazes of physics and chemistry, ranks of elaborate and delicate machinery, varieties of sensitive and unstable substances, and vagaries of temperature and humidity and light. The printer makes his way through all this as best he can, not always as effectively as he would wish, but sometimes with triumphant success. The more closely his problems and methods are studied by the designer, the more likely are illustrations in particular to start off in the right direction on their journey towards the printed image.

§ 10-1 Camera and copy

A camera constructed for reproduction processes in printing is known as a *process camera*. Special purposes apart, the smallest of these has to be big enough to reproduce the largest of book pages. Since cameras designed for text production deal with editions consisting of many pages, most of these are capable of reproducing at each exposure a number of pages or a number of illustrations. For this reason, and because the whole apparatus — including copy board, lights, and movable lens — is mounted on a single bed, a process camera is usually a large one. Because of its size, it is static, and copy comes to the camera; unlike a portable camera but like other industrial machinery, the camera is not transported to its subject. If the subject itself cannot be brought to the camera, it will have to be represented by a photograph as *camera copy*.

In preparation for process photography, camera copy is placed on the *copy board* of the camera. In most process cameras, focus is shallow; depth has been relinquished in favour of sharpness, and copy has therefore to present a plane surface to the lens. The pages of a bound book, for example, are likely to prove difficult as copy; in some books every opening can be made flat, however laboriously, but in others the curve of the leaf into the spine distorts the printed area, and such books may have to be dismembered to make camera copy. The focus may be so shallow that correction lines pasted on top of repro proof or printed copy are fractionally out of focus. The best camera copy is flexible, thin, and flat; the best copy board is a vacuum frame in which such copy is flattened under a sheet of plate glass.

Copy on a transparent enough paper may benefit from reproduction by transmitted light, which shines through the copy into the camera. Any correction in white paint prevents this, as the paint is sure to be opaque by transmitted light and to be seen by the camera as black. The opacity of any image also becomes more important than its apparent blackness by reflected light.

The photographic image in the camera is generated by the radiation of light on to a chemically sensitized emulsion which has solidified as the coating of a film base. The reaction of this emulsion varies according to the intensity of the incident light, and in negative film the image which is to be printed represents reduced intensity or an absence of incident light where the copy is dark or black. The image, in short, is defined by its background. If a set of copy, such as the pages of one or more printed books, presents its images on one or more different shades of paper, different exposures may be necessary, and some unevenness of image definition is likely to result. If any one page is presented on two shades of paper, some irregularity is all but certain. The best paper background for any text or illustration copy is a brilliant white, of a colour even throughout the set. The reflective intensity of photographs as copy is increased by a glossy surface.

A photographic record of colours and intensities tends to differ from that of the human eye. When a photograph as copy has been touched up with apparently white paint of the wrong kind, the paint tends to be reproduced as grey. Coloured photographs or transparencies, to be reproduced in a single colour, generate an imbalance of monochrome tones, with blue too pale and red too dark, unless filters are used. Such originals are better translated into monochrome prints and approved by author or editor before reproduction begins.

In the same way, any original too big for the copy board, or requiring drastic reduction or enlargement, should be presented to the camera in reduced or enlarged form. Reproduction scale from a single exposure is limited; typical limits are reduction to one-fifth of, or enlargements to five times, the original size. If an original is to be reduced or enlarged beyond such limits, the operator may have to make a positive opaque photographic print or *bromide print*, with maximum reduction or enlargement, and apply a further reduction or enlargement to that instead of to the original. The quality of detail and of tonal balance may suffer from this indirect treatment. The distances between copy, lens, and emulsion are adjustable, and regulate the scale of reproduction, so that it can be larger or smaller than, or the same size as, the original copy. On many cameras, scale is adjustable to the nearest 1 per cent, and originals marked for a percentage reduction are more likely to be exact in reproduction than those marked with the dimensions of reproduced size.

§ 10-2 Line and half-tone

Line subjects are those which can be effectively reproduced without adaptation. Reproduction pulls from metal type, paper output from a photo-typesetter, and diagrams or pictures drawn in black on white paper without intermediate shades are typical subjects for line treatment. Old steel engravings are sometimes too faded, or the fine lines are too thin or too close to each other or both, particularly when they are to be reduced in size.

Half-tone subjects are those in which the reproduction is to include continuous tone, shading for instance from black through greys of varying intensity to white. Each impression of a printing press normally conveys to the paper one kind of ink, of one colour and one degree of intensity. Since the number of impressions economically available for any page is severely limited, and indeed for most work a single impression only is possible, continuous tone subjects, such as photographs of people and scenery, simulate the gradation of tonal intensity. This can be done by analysing the tones of the original into a fine pattern of dots, which vary in size according to the intensity of the tone. The distance between the centres of all dots is constant; the darker parts of the subject are represented by larger dots, and the lighter parts by smaller dots. In the darkest areas, the dots meet and join until the printing surface is either continuously solid or interrupted only by white dots. In the lightest areas or *highlights*, the dots may become almost imperceptible; some may not print at all, so that there is no continuing pattern; and in some areas there may be no dots. The brightness of such highlights depends on the shade of paper on which they are printed.

At one time the most widely used method of image analysis was to pass the image through a transparent screen ruled with a rectangular grid of opaque lines; such a screen, half transparent and half opaque, provides the name for the half-tone process. The opaque lines are equal in thickness to the spaces between them. The screen is rigid, and is used at a distance in front of the emulsion. For some reason unknown to most printers, light passing through each rectangular orifice generates a circular dot in the emulsion. The ruled screen may be expected to produce a particularly delicate gradation of tones; other kinds of screen tend to produce different kinds of result. A *contact* (half-tone) *screen* is marked with a rectangular pattern of dots which vary in opacity from centre to circumference, and is used in close contact with the emulsion. This too analyses the image into circular dots, and provides an effective medium for emphatic contrast between tones. Whatever the type of screen, the lines of dots are normally diagonal in reproduction, to render the pattern inconspicuous.

The *gauge* of the screen is expressed as the number of lines to the linear inch or centimetre. Choice of screen gauge is limited by paper and printing process, and affects the appearance of the printed image. A screen of 100 lines to the inch is coarse enough to print on all but the roughest of paper, but the dot pattern will be all too visible. If the paper surface is smooth and hard enough, the pattern of a 120-line screen will be less obvious but still apparent. On a *coated paper* (§ 15-4) a 133-line screen pattern will be visible rather than obvious; a 150-line screen, easier to print by *offset* than by *letterpress* (§§ 11-5 and 11-1), can hardly be seen. Screens of 200 or 300 lines and even finer, capable of offset reproduction by a printer experienced in such work, effectively simulate continuous tone but diminish tonal contrast. Screens up to 150 lines tend to break up and blur fine detail in illustrations, and any screen too fine for the printing circumstances results in inferior production with spotty highlights and congested shadows.

Photographically, half-tone reproduction is an imperfect medium. Carefully utilized, it is capable of conveying most information accurately enough. But any reader is entitled to relief from the two-dimensional nature of pictures, in a variety of tonal emphasis and contrast which may provide an illusion of depth in a third dimension, and which helps his imagination to suspend disbelief. Author, publisher, designer, and indeed all who handle the original illustrations would like any printing process to do justice to the general appearance of the pictures, which may range in tone from almost pure white to the most intense black, all brightened by the glossy and even glazed surface on which they have been developed. Since photographic processes are known to be capable of tonal adjustment, to seek for more than justice might not seem unreasonable, in search of a printed image which improves on the original. Such effects may be achieved by *make-ready* (§ 12-2) in letterpress printing or by photographic retouching in offset when such work can be afforded. The general run of half-tone reproduction, on the other hand, represents a compromise between economy and quality. The designer's part is to understand the process well enough to identify and obtain the best results within economic reach.

A camera designed primarily for text reproduction in line is likely to have capacity for several pages at each exposure. Line illustrations for such a camera can be effectively grouped for a single exposure as long as they are similar in background shade and have scale in common. The adjustment of exposure for half-tones depends on the nature of the image and the tonal balance required of the reproduction of each. There may be scope for grouped exposure, when the same scale applies to similar subjects in a set, but the best reproduction, in offset half-tone at least, depends on individual exposure.

The relationship between the tones of the original and the dot pattern of the printed version can be regulated by exposure, among other means. If everything in every original is to be reproduced, including the palest shades in the highlights and detail all but concealed in the shadows, there will have to be a dot pattern over the whole printed area, and the dots must not be allowed to expand far enough to present a solid printing surface. Such an effect may be described as a pattern of 5 to 95 per cent, for instance; in the highlights, the diameter of the dots will be 5 per cent of the distance between their centres, and in the shadows 95 per cent. If, however, more contrast is needed, the highlight dot can be reduced to nil; while a few dots may appear in such areas, they are minute, and there is no continuous pattern. The darkest areas of the original will then tend to close up to 100 per cent, or solid black, while intermediate tones towards either extreme become paler or darker. This effect seems to be more popular than a faithful rendering of all shades in a reproduction of less dramatic contrasts. By the same means, the background of a crayon or wash illustration can be *dropped out* of reproduction although the picture itself is reproduced by half-tone.

Half-tone screens can be used not only for analysing the tones of photographs but for panels and similar areas which are to appear grey instead of black, sometimes to form a frame and a background for line reproductions. The gauge of the screen has to suit printing conditions, but the designer can indicate variety of tone by variations in dot pattern between about 10 and 90 per cent, equivalent to very light grey and very dark. If anything is to be printed over the panel, a 30 per cent pattern will probably prove dark enough; if an image is to be *reversed* (§ 10-3) into it, the image is likely to be clear enough on a screen of 40 per cent or more.

An illustration which has already been analysed into a half-tone pattern is not a good subject for half-tone reproduction. The screen patterns combine to produce an extraneous third pattern known as *moiré*, even when they differ in angle and gauge. A half-tone print can, however, be treated as a line subject, to be photographed without a screen, in the hope that the dot pattern can be directly reproduced. This may work well enough when one of the coarser screens was used for the original print, as in a newspaper, but such *dot-for-dot* reproduction of finer screen needs experiment. The legibility of type is diminished by screening except in the larger display sizes, as the outline loses definition with all but the finest screens.

§ 10-3 Film

Before exposure, film coated with light-sensitive emulsion is fixed into the back of the camera facing the lens, and through the lens the copy board. During exposure, light reflected from (or transmitted through) the copy passes through the lens on to the emulsion, and locally hardens it and renders it insoluble; where the black parts of the copy reflect no light, the emulsion remains soluble. Between the soluble and insoluble conditions there is no gradation. The exposed negative is removed from the camera;

the soluble parts of the emulsion, corresponding to the black parts of the copy, are washed away, leaving the transparent base clear and bare, while the insoluble parts of the emulsion are blackened, rendered opaque, and chemically fixed.

Photographic exposure usually generates in light-sensitive emulsion a negative image, reversed from black in the copy to transparent in the negative and from white to opaque. If the original image is transparent or negative, as in most photo-matrices, the image formed in the emulsion will be opaque, and is described as positive although it is in fact a negative image of a negative image. The image presented to a camera for monochrome reproduction is normally positive, the emulsion image negative. If positive film is needed, neither lens nor screen need be used; the developed negative is placed in close contact with a second sensitized film, and exposure to light through the negative produces a *contact* positive. Alternatively, a positive-reversal emulsion, insoluble until exposed to light, is capable of developing a positive film image from positive copy.

Line and half-tone can be combined in a single emulsion by the interposition of a half-tone screen between two exposures of one emulsion with one original. In the lighter areas of a half-tone picture, lettering and indicating lines may be added in positive line; in the darker areas, such items may be reversed in negative line. In negative film, the half-tone pattern can be eliminated locally with opaque paint, to increase contrast or to extract an illustration in silhouette from its background.

In the absence of specific instructions, the printer is likely to square up most half-tones into rectangles, leaving the dot pattern to show over the whole area. This is the simplest and least expensive form of half-tone reproduction, and a designer whose methods and requirements are un-familiar to a printer may do well to assume extra work will be done only if it has been requested. The rectangular background of a *squared-up* half-tone is interrupted, sometimes in an unsightly way, when dropped-out highlights are crossed by one or more of the edges. When text is to be reproduced from film, half-tone illustrations are separately exposed and assembled with the text film. In negative assembly, the exact dimensions and positions of half-tone illustrations are represented on the camera copy by black rectangles of paper, which generate transparent rectangles in the negative. The half-tones are developed on film much thinner than that used for the text, so that they can be sprayed with transparent adhesive and mounted in position on the text film. Half-tone film mounted separately in this way has to be thin, or its edges may distort the text film under pressure applied to them both. In positive assembly no black rectangles of paper are necessary, and this technique lends itself to *cut-out half-tones*, in which the edges of the illustrations correspond with the edges of the image on the camera copy, or of that part of it which is to be reproduced. When a page of positive film consists not of a single piece of film but of two or more pieces joined together, special precautions have to be taken to prevent the join-lines from being reproduced, however faintly, on the printing plate; once they have appeared there, they become difficult to eradicate.

The purpose of negative or positive film in printing is to transfer an image to a printing surface, or plate, an operation described in chapter 11. The plate has first to be *bichromated* or coated with a light-sensitive substance. The film is laid on it with the film emulsion next to the bichromate; contact has to be close over the whole surface, to ensure accurate reproduction at all points, and this is usually effected in a vacuum frame with plate glass over the whole exposed area. The bichromated plate surface is exposed to brilliant light through the film, for a precisely calculated period of exposure, and the printing image is developed on the plate in much the same way as any other photographic image. Printing plates are not easy to correct before printing, and after printing most of them are too expensive to store unused; techniques for generating printing images on plates without the interposition of film are beginning to appear, but for correction and storage in book production the advantages of film seem likely to be preferred for some time to come.

CHAPTER 11

Printing processes

WHEN as a publisher's assistant I started work on book production in 1947, typographic composition was invariably metal — photo-composition was in the development stage. Nearly all newly published books were printed from metal type; big editions were likely to merit metal stereoplates, sometimes with electrotyped half-tones let into them, and usually flat for flat-bed printing; and while rotary photogravure was competitive in terms of cost and quality for jackets and coloured illustrations, most colour-printing was done by four-colour half-tone on flat-bed letterpress machines. Offset photolithography was competitive for music scores and similar specialities, and for reprints for which neither type nor stereos could be provided, but was too uneven in colour and too pale in half-tone reproduction to be much used for new books. Photographs and paintings were sometimes reproduced by flat-bed bitumen-grain photo-gravure; and for subjects of this kind, and for short-run illustrations in which extreme accuracy and delicacy were required, there was also collotype in monochrome or colours.

Each of these printing processes could be assigned to one of three classes. Letterpress printing is a *relief* process, in which the *printing surface*, that part of the whole surface presented to the paper which conveys ink, is raised above the non-printing part of the surface. Offset and collotype are *plane* processes, sometimes classified as *planographic*; printing and non-printing surfaces lie in the same plane, but only part of the surface conveys ink, or some parts convey more ink than others. Gravure is a *recess* or *intaglio* process, in which the ink is contained in recesses cut into and below the non-printing surface. These classifications help to remind us how the processes work; but the proper study of the book designer is the processes themselves, not the classifications among which they may be divided.

Each of the processes was itself a kind of classification of different printing techniques. The recess principle apart, for example, flat-bed gravure and rotary gravure shared differences rather than similarities. In terms of the kind of paper required, the economics of production, the pace of output, and the general appearance of the printed image, these two forms of gravure might reasonably be considered to be different processes and even members of different classes of process. Unless each was studied as separate from and independent of the other, use of either might lead to disappointment. In more widely used processes, inadequate discrimination between different aspects of one process could lead to loss of goodwill, time, and money. Unsatisfactory editions are the evidence, but to identify examples might be damaging to reputations. Designers who know how to prevent any possibility of this kind of disaster are well placed to make the best use of the most appropriate process.

Printing and other processes have changed since 1947, and the balance between them is now wholly different; almost everything is or could be printed by offset, for instance. But methods of process assessment and principles of process selection seem to have changed very little. More than one kind of press, operating in accordance with whatever process, may still offer more than one kind of cost, of production timing, and of visible nature in the printed image. The probable effect of any method of printing, as of other book-production procedures, on whatever is to be reproduced and printed has to be foreseen while plans are being laid. Paper specification has to be influenced by printing as by binding processes.

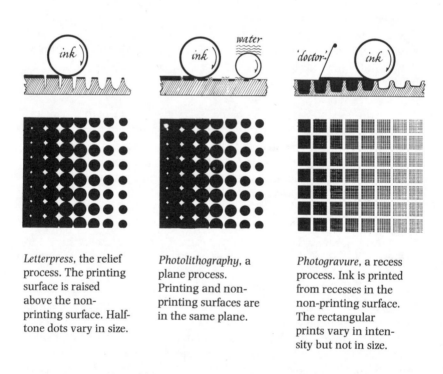

Letterpress, the relief process. The printing surface is raised above the non-printing surface. Half-tone dots vary in size.

Photolithography, a plane process. Printing and non-printing surfaces are in the same plane.

Photogravure, a recess process. Ink is printed from recesses in the non-printing surface. The rectangular prints vary in intensity but not in size.

Figure 11. Working principles of the three main classifications of printing process, with a comparison of the kinds of image by which they simulate continuous tone.

Industrial machinery is most economically used when it is filled to capacity. The cost of printing the biggest possible sheet on a specific press is much the same as the cost of printing a smaller sheet on the same press, whatever marginal variation in price the printer may allow. Industrial design comprises design for industrial production as well as for public use, and the economics of a choice of sheet size and hence of page size will benefit from the designer's knowledge of the various dimensions of a press's maximum capacity. A reduction of a fraction of an inch or a few millimetres in page size may all but halve the printer's price per page for presswork.

In terms of machine time and paper wastage, preparatory operations are likely to cost several times as much as running a thousand sheets through the press. The cost of such operations is therefore more important than the cost of running the press, until the number to print rises to a point at which the latter cost equals or exceeds the former. And on this preparatory work or *make-ready* much of the quality of printing depends.

In terms of design and of quality, presswork differs from composition. The submission of specimen pages and proofs enables the composing-room to rectify mistakes during the preparatory stages, and to prove the composition correct before it is printed. Presswork quality cannot be proved in advance. The designer's choice of paper, process, and printing surface will be influential but may not be decisive. Presswork methods and quality standards may vary not only between printers but between presses under the same roof. Any printer has to do his best with the task assigned to him, often on paper he did not choose, occasionally on paper of a kind he has not printed before, and sometimes in conditions which are adverse in some other way. Presswork and paper are the economic base of publication, since without both there can be no edition, but neither is mechanically infallible, and there are times when they are incompatible. Most presswork represents some compromise over quality; the collective opinion of book designers (though not necessarily that of page designers) has and should always have its influence on presswork standards in book production.

Success in composition, or in illustration reproduction, or in both combined, and success in presswork are interdependent. The pressman can do nothing to remedy repeated and conspicuous faults of type-casting, for example, and the quality of his presswork from so undeserving a surface cannot be satisfactory.

§ 11-1 Letterpress

Text, as distinct from illustration, is sometimes referred to as the 'letterpress' of a book. Here, and indeed generally in printing, *letterpress* means the relief process of printing.

The purpose of any printing process is to lay ink on certain parts of the printed surface and not on other parts. The ink comes from the printing surface; those parts of the paper on which no ink is laid correspond in position to the non-printing surface. In letterpress printing, the printing surface is raised in relief above the non-printing surface; the former is coated with ink from inking rollers and is impressed upon the paper, the latter is reserved by its lower plane from contact with either paper or rollers.

Letterpress is the simplest, the most ancient, and until recently the most widespread of printing processes. Letterpress is giving way to *offset*, but is not obsolete yet; short runs of fine editions, long runs of paperbacks, and enormous numbers of national newspapers are still printed from the relief surface. The composition of metal type itself provides a printing surface; photo-composition provides paper or film from which a printing surface can be made. Most editions published between the European invention of moveable type until about 1970 were printed from type on flat-bed presses (§ 12-4), and some of the once familiar and long appreciated characteristics of this kind of printing will be difficult to replace. Letterpress ink is undiluted, and for comparatively slow flat-bed presswork it is stiff in consistency; the thick film of ink laid on the printed surface can impart to the image a density of tone difficult to match by other means. The pressure of the relief surface against any paper forces ink outwards to the edge of the image and quite often beyond. The edge is consequently reproduced with emphasis, and the image as a whole may appear to have been strengthened, by an effect known as *squash* or *ink-squash*. On a soft-surfaced paper, the same pressure may indent the image as a whole into the paper, generating what has aptly been called the third dimension of letterpress printing. This third dimension, an often minute appearance of depth as well as height and width, has an appeal which is difficult to analyse; it reminds the observer, perhaps, of the permanency of letters engraved in stone, or of printing as it was in the days of the hand-press. To bibliophiles this evidence of impression is one of the elements of *fine* printing as distinct from good printing.

A relief printing surface is an imperfect medium for the exact transmission of an image. First to the inking rollers and then to the paper, the surface presents an irregular mesh of areas and edges and dots. The larger areas would benefit from a heavier impression than the smaller, but make-ready which provides some differences of this kind cannot provide all. Any part of the printing surface which has no other printing element close by on either side — a page number centred at the foot of the page is an example — supports in isolation the weight of the inking rollers, tends to collect and convey too much ink, and is likely to suffer disproportionate wear. Each element of the printing surface is separated from its neighbours by non-printing depressions, varying in depth according to their width; the narrower and shallower of these are all too likely to receive surplus ink, pressed outwards over the edge of the printing surface, and to transmit this ink to the surface of any paper flexible and uneven enough to touch it.

§ 11-2 Printing from type

Once composed and corrected, metal type becomes the least expensive of printing surfaces. Processes other than letterpress derive their printing surfaces from plate-making and perhaps from photography, the cost of which is added to that of composition. *Make-ready* (§ 12-2), during which the press produces no saleable sheets, takes longer in letterpress than in other processes, wastes more paper, and inflicts more wear on the type than printing an edition of moderate size except on a hard abrasive paper. Type locked up in chase for printing is heavy and rectangular; the forward impression movement of bed and chase can be reversed only by massive reciprocating mechanisms, and the speeds normal in rotary printing from curved surfaces are unattainable by the flat-bed press.

The principal constituent of type alloy is lead, not the hardest of metals. Type is vulnerable to mis-handling before printing, to wear in the course of make-ready, to repeated pressure and abrasion during presswork, and to scrubbing after use. Pulling proofs from type *off its feet* (vertically askew), or with too heavy an impression, or from galleys with floors of uneven thickness may cause visible wear. An incautious movement with a metal tool may scratch several adjacent lines of type; to rectify such damage is the printer's duty, but watchful designers still see every page at every stage of proof to make sure that such accidents have been marked. Line-casting alloy is less hard than single-type alloy, and should be subjected to a minimum of proof-pulling and a maximum of examination in proof.

Composed metal type can provide camera copy for reproduction and for printing by processes other than letterpress. *Reproduction* or *repro proofs* are intended not for reading but for the camera, and are pulled after completion of all the stages of proof-reading and revision which usually precede printing. In sharpness of profile, in intensity of blackness, and in evenness of colour and impression from page to page, repro proofs should match or exceed the best of continuous printing on smooth paper, but for economic reasons should be pulled without prolonged make-ready. When the first edition is to be printed from type, *repro sheets* may be pulled unbacked from the production press after make-ready, and kept as camera copy for future reprints.

Once type has begun to wear, however slightly, reprints with amend-ments, and new editions reproduced mainly from the original type, may become difficult to print well. Words or passages newly set after the previous printing may stand out in sharp contrast to the part-worn types around them, distracting the reader's attention and bearing witness to shoddy work. The printing quality of type cannot be maintained indefinitely throughout a series of reprints and revised editions without some form of reserve, which will provide a replacement surface when type begins to deteriorate; stereo moulds (§ 11-4) may be enough.

The appearance of the image in letterpress printing is visibly influenced by the nature of the paper surface to a greater extent than in other processes. Metal type-faces were designed to subject themselves to this influence. On a very hard smooth paper, for example, sharp serifs, fine hair-lines, and slender stems combine into an unwelcome effect of dazzle, yet on a softer and more elastic surface the same type may even look mellow. On the latter kind of paper, stoutly built characters may appear gross, though they would be well suited to the former. The designs of metal types deserve to be carefully and skilfully matched to the most appropriate papers. On such paper, letterpress printing from type is more capable than other processes of a black and glossy impression, when these qualities are required, and of exact representation of the characters as they were intended by their designer to appear.

Evenness of the image's colour and profile throughout a book is one of the principal qualities of good printing. Unevenness of text printing may be conspicuous enough to distract the reader; even a layman is likely to have formed in his mind a clear picture of the characters he is reading, and to notice a change in their appearance which may seem to imply a change of emphasis. Letterpress is sensitive to variations in paper thickness and in paper surface. But when the paper itself is even in every way, and when ink, skill, time, and a good press contribute to the best work, an impression which continues evenly from page to page from the beginning of the book to its end affirms the unity of the book and the relationship between page and page.

§ 11-3 Blocks

A relief printing surface from which an illustration is to be printed is known as a *block*. The term is derived from the wood blocks in which illustrations have been cut or engraved for reproduction since the early days of printing. Then whole books including the text were sometimes printed from blocks. Today a block is understood to provide illustration; when a process by which such blocks can be made is applied to the preparation of text as well, the printing surface is known as a *plate*. The making of blocks is known as *process engraving*, although a process block is a comparatively thin plate and is not engraved.

A plate of suitable metal is coated with a light-sensitive emulsion, and exposed to light through a negative of the printing image. The non-printing surface of the metal, unprotected by the developed image, is etched and drilled away to well below type-height. The metal block is mounted on a base of wood, composition, or metal for printing, and being flat and rigid cannot be used with a *rotary press* (§ 12-5) which requires a curved printing surface. If rotary printing is intended, a *duplicate plate* (§ 11-4) will be necessary. Etching takes time, and the expensive materials are irrecoverable; prices for process engraving rise steeply with any increase in area.

Though engraved in zinc, copper, or magnesium — metals harder than the alloy in which type is cast — line-blocks are equally vulnerable to damage and wear. They do not provide the sharpest reproduction of delicate detail, or the most solid body of colour when there are uninterrupted areas of ink to be printed, particularly when the paper surface is fibrous rather than mineral. A fibrous paper may be too uneven for complete contact with all parts of a solid printing surface; it tends to receive too little ink and too little pressure, and to present a starved and mottled appearance. The printed surface has to be flattened; if the paper is rough, the printing surface will have to flatten it. Forcing its way through the superficial fibres, the image transmits a thickened reproduction with an irregular outline.

Half-tone blocks for bookwork are usually etched in copper, which stands up well against wear, but the slightest scratch or other damage is emphasized by the screen pattern. Replacing a damaged half-tone block without serious loss of quality may require either the negative or the original. Half-tone originals should always be kept in hand, at least until the first printing has been approved. When repeated reprints are expected, some form of reserve material may be the only guarantee of quality maintenance in successive editions; the assortment of original illustrations from which the blocks were made is all too likely to have been dispersed after first publication, and process engravers do not store negatives for long.

Half-tone blocks and paper have to be matched to each other with particular care. The closer the dots to each other, the shallower the non-printing depressions between them. The extra ink and pressure applied to printing the dots effectively on a fibrous surface tend to fill the depressions and to force paper against the surplus ink there, causing dark and middle tones to deteriorate into solids, and marring lighter tones with spots and local filling in of the same kind. A *machine-finished* paper (§ 15-3) may be smooth enough for a 100-line screen; a *supercalendered* paper (§ 15-4) will do for screens up to 120; and finer screens are best reproduced on a *coated* paper (§ 15-4), preferably with a glossy surface. Most printers find 133-screen half-tones fine enough to reproduce detail on smooth coated paper — finer screens still are possible but may be difficult to print well by letterpress.

In the printing of text from type, evenness of impression is one of the first requirements. In the printing of half-tone blocks, a selective variety of pressure is applied. When the pressure on the darker tones is progressively increased, and that on the lighter tones proportionately abated, the reproduction gains a contrast not necessarily present in the original. The hard edge of the half-tone dot on glossy paper, the dense film of ink applied in flat-bed letterpress printing, and the intensified difference between highlight and dark tone impart to this kind of reproduction a sharpness and contrast difficult to match by other means. The consequent appearance of depth, perspective, and almost of a third dimension — as though the reader were looking at reality rather than at a reproduction — is very welcome to authors and readers; I have even heard complaints of its absence from electron micrographs, scientific photographs so enormously enlarged as to look even less terrestrial than a lunar landscape. The economics of book production have made monochrome illustration familiar, but the human eye still prefers the contrast and illusion of depth conveyed by colours.

Letterpress half-tones are best printed separately from all text other than captions. The glossy white paper which provides the most favourable colour and surface for these illustrations does justice to a limited number of type-faces only, and returns a dazzling reflection from bright light into the reader's eyes; text is better printed on toned paper with a matt surface. For these reasons, text and half-tone illustrations, which can be printed together by any process, have long been printed separately from each other in letterpress work of good quality. This has given rise to a convention that half-tone illustration need not appear at a relevant point in the text, but may be placed in a textually random position which suits the economics of binding but not the convenience of the reader.

§ 11-4 Letterpress plates

A letterpress plate is a relief printing surface, normally made as a single piece, from which text or illustration or both can be reproduced. In newspaper production an entire forme may be duplicated as a single curved plate for rotary presses; in bookwork, economy in material and weight usually requires single-page plates.

In flat-bed printing, plates provide an alternative printing surface to replace or supplement type and blocks which might otherwise wear out in extensive and repeated use. For rotary work, in which the printing surface is curved round a cylinder, plates are essential, as type and blocks are rectangular. Photo-composition can be printed by letterpress from a relief plate.

The *stereo-mould*, from which the printing surface of the *stereo-plate* receives its shape, is capable of becoming soft enough to accept under pressure an indented impression from type and blocks, and of becoming hard enough to withstand the heat and pressure of casting. The stereo-plate is made by casting. The material of the plate is capable of becoming liquid for casting, and of subsequently becoming solid and hard enough for printing. The oldest form of stereo, cast in an alloy similar to that of type-metal, originated in the eighteenth century, and today rubber and various kinds of plastic are also in use.

The moulding process is not precise enough to duplicate the half-tone pattern of fine-screen blocks, which consists of minute dots and of narrow, shallow depressions between them. For this purpose *electro-plates* or *electros*, a more costly and exact kind of duplicate plate, are to be preferred. A plastic *electro-mould* is rendered electrically conductive by being sprayed with silver compound, and is placed in an electrolytic bath. On the recessed image of the printing surface, a film of copper is grown by electro-deposition. The copper film, retaining its shape, is stripped from the mould, laid face down, and filled or backed with molten stereo metal, to give it body and strength.

Photo-composition can be converted into a relief surface for letterpress printing by means of *photo-polymer plates*, which in large areas are less expensive than metal blocks, and which can be curved for rotary printing.

At their best, stereos and other forms of duplicate plate provide for monochrome printing an impression all but indistinguishable from that of the original printing surface. As in all printing, however, an image which is not printed directly, as from type, but is converted into another printing surface, tends to lose some of its accuracy in the process.

§ 11-5 Offset

Offset, litho, and *photolitho* are abbreviated terms which today usually indicate *offset photo-lithography*.

Lithography takes its name from the fine-grained limestone which in the eighteenth century provided the first lithographic printing surface. This surface is flat; printing and non-printing areas share it, in contrast with letterpress printing areas which stand up in relief above the non-printing areas. In lithography, the difference between printing and non-printing areas is that the former accept grease which repels water, and the latter accept water which repels grease; the process is based on the mutual antipathy of grease and water.

If a picture is drawn in greasy ink on smooth limestone, a sheet of paper pressed against the stone will naturally receive a print of the drawing. Much of the ink will have been transferred from stone to paper, and a second sheet would receive a much fainter print as a result. But a second print and many others, each as good as the first, can be pulled if the ink is replenished between impressions, and this can be effected without redrawing. The whole surface of the stone is damped; water slides off the greasy picture, but elsewhere dwells on the clean porous surface. Greasy ink, applied also to the whole surface of the stone, is repelled by the damp areas, and replenishes the picture which is greasy but free of water. The printing image is now fully inked again, and the non-printing areas are clean though damp. A second pull is likely to receive as much ink as the first.

Limestone is still in use, but not in industrial book production; it presents a good surface for the lithographic artist, but it is expensive, heavy, fragile, limited in size and endurance, and laborious to prepare. The lithographic surface of today is provided by a thin metal plate. The metal is not porous, but when suitably roughened or *grained* its surface can retain an even film of damp until it evaporates. The plate is coated with a light-sensitive emulsion, and exposed through negative or positive film. The coating over the printing part of the surface is hardened and rendered insoluble by light, that over the non-printing part remains soluble. The unexposed emulsion is washed away, the hardened image reinforced; the clean metal is now ready to receive water and reject ink, the image to receive ink and reject water. Since the image is photographically generated, the process is termed *photo-lithography*, and since it consists only of a thin film of grease and other ingredients it is too delicate to endure heavy pressure and the abrasion of a fibrous or mineral paper surface. The image is therefore printed from the plate on to a cylinder covered with a smooth rubber *blanket*, which receives it without more than the lightest pressure. The blanket in turn, flexibly sympathetic to any unevenness in the paper surface, transfers the image to paper. This purposefully indirect form of printing is known as *offset*, and is

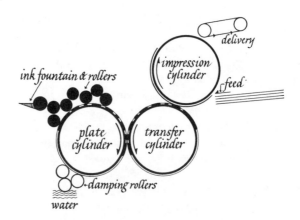

FIGURE 11-5. The principle of an offset lithography press.

not to be confused with unintentional indirect printing, *set-off*, of which an example is damp ink transferring itself from one sheet to another within a stack of sheets or within a bound book. The offset principle is mainly exploited in photolithography, but can also be used with a relief surface, and even in photo-lithography the use of water may yet be rendered unnecessary by the development of ink-rejecting plate surfaces.

The early development of offset photo-lithography in book production was applied to reprints of editions for which type was no longer available. Even when the first printing of a book is from type and no illustration is involved, a first reprint by offset is now economically possible, and subsequent reprints based on existing negatives economically favourable. During the last hundred years, photographs have largely superseded other forms of illustration, and have brought into being new kinds of publication in which illustration may occupy more space than text. Offset has proved its affinity with photographic illustration, and for a majority of illustrated books, in their first printing as well as in reprints, is the preferred process. The accelerating pace of development in photo-composition, which can be reproduced by other processes, has favoured offset, and supports its growing domination in book production as in other fields of printing and publication.

An image printed by offset is not indented into the paper but is laid on its surface. There is no ink-squash; any local thickening is likely to be due not to pressure or inking but to imperfect exposure of negative or plate. On a fibrous paper surface, the outline of the offset image tends to be more faithful to that of the original than a letterpress image on the same paper, and even colour illustration can be printed by offset on bookbinder's cloth. On the other hand, the film of damp protecting the clean areas of the plate against ink is partly transferred to the paper, which at once begins to stretch, particularly across its machine direction or *grain* (§ 15-3); this may give rise to presswork difficulties if the grain of the paper lies round the printing cylinder instead of across it, and the grain which suits the press may prevent the book from opening flat (§ 16-2).

Ink and water are applied successively to the plate before each impression, and although they tend to repel each other some mingling is to be expected, so that offset ink is partly emulsified. The balance of ink and water tends to stray, and the plane printing areas have a limited capacity for ink, so that the film of ink tends to be thinner than in flat-bed letterpress. In letterpress printing the ink film is split between printing surface and printed surface at the moment of impression; in offset it is split twice — first between plate and blanket, then between blanket and paper. The offset image accordingly is all too often less dense and less even from sheet to sheet than printer and designer would like. Improvements in paper, ink, and presswork technique are eradicating much of this weakness from the process. On fibrous paper surfaces in particular, offset deals better than letterpress with large areas of flat colour and with half-tones in which accuracy has more value than contrast. Half-tone screens used in offset are usually finer than those used in letterpress; 133 will do for any reasonable surface, 150 is common, and when the subject and the paper deserve it and the printing will allow it, 175, 200, or even finer screens may be used. Coated papers help offset to produce its best results, particularly from half-tones, but the finish of the paper can be more matt than the letterpress process would require for work of equal quality.

§ 11-6 Offset film and plates

Film is easy to store. The material cannot be re-used, and its weight and bulk are negligible; type kept in store represents part of the printer's working capital, reserved for future use in shelves which occupy valuable space. Some kinds of film deteriorate in store, and to rely on the indefinite availability of film may prove unwise. Replacement of damaged or deteriorated film may be possible by photographing a printed copy of the book, but some loss of sharpness is to be expected, and half-tones will have to be reproduced from continuous-tone originals if acceptable quality is to be maintained. Pages of film can be stored in imposed formation, which relieves reprints of the cost of re-imposition.

Film can be transferred from one printer to another, and indeed from one country to another. Duplicate film can be made by contact (§ 10-3) for another printer's use. The transfer of film from one printer to another needs careful handling; contact film tends to lack some of the quality of the film exposed to the original, the printer who made the film may have based his price for it on the expectation of reprints, and the printer who receives it may have difficulty in extracting from it the best possible result. Film exposure and make-up are far from standard in procedure throughout any printing industry. Even in less subtle areas of technique, one printer's material is likely to provide another printer's problem.

Film can be altered by stripping correction film into the existing negative of a page, or by amending the original page itself and photographing it again. Here as elsewhere in printing, however, a succession of printing images prepared together at one stage of production should always be even throughout the book; but images prepared at different stages, and assembled together within a page or as a succession of pages, should not be expected always to appear perfectly even. The result of amendment may be reasonably inconspicuous if a whole page, rather than part of it, can be reproduced at one time, and if the facing page is treated in the same way; contrast between two openings is less obvious than contrast between facing pages.

The image is transferred (*printed-down*) to the offset plate through negative or positive film. Negative print-down is easier than positive, since any imperfection in the opaque or non-printing areas of the film can be deleted with opaque paint. If the transparent or non-printing areas of a positive are imperfect, for example where one piece of film is joined to another, an unwelcome mark may appear on the plate, and may be far from easy to delete. Most photo-composition on film is positive, and as techniques continue to develop, the single-piece positive — a whole corrected and made-up page, including display and illustration on one piece of film — prevents any trouble with join-lines.

Printing-down through positive film reverses part of the procedure described in § 11-5 when a positive-reversal emulsion is not used. Light passing through the positive partly hardens the coating over the non-printing areas, and leaves soluble the coating over the printing image, from which it is easily washed away. The whole plate is covered with water-resistant greasy ink. When the plate is soaked for a while, the light-hardened coating softens and dissolves under the ink, so that both depart to leave the non-printing areas bare and clean. Over the printing areas, the ink adheres directly to the plate, and dwells there to form the base of the printing image. If the exposed metal in the printing areas is etched before the first application of greasy ink and while still unprotected, the image is fractionally deepened, and this increases the thickness of the ink printed by offset on to the paper. Plates made in this way are said to be *deep-etched*, though the etch is in fact shallow.

If the emulsion side of the film is not firmly in contact with the plate across the whole area of the image, or if there is over-exposure in the course of printing-down, there may be a slight spread of light outwards from the transparent areas of the film. A negative image may be slightly thickened as a result; in positive film, it is the non-printing areas which tend to thicken, at the expense of the printing image, which is accordingly sharpened and even diminished. Any distortion of the image is a fault, and identifiable distortion should be rare; but positive film still tends to produce a sharper image than negative film.

Some correction to plates is possible, but a new plate is usually preferable to the possibility of trouble developing in the correction area during the run. Plates can be stored, but for most work new plates are made for each printing; this affords an opportunity to rectify any faults before reprinting. The image on an ordinary plate may begin to deteriorate after some 25,000 impressions, but a deep-etched image should endure for 50,000 or more.

Precision of image profile, density, and endurance are all available when two or three metals are used for plate-making. One method of making bi-metallic plates is to build up a resinous reproduction of the printing image by photographic means on a copper base. A chromium coating is then formed by electro-deposition on the base and round the resinous image. The resin is washed out and the plate is ready for printing. The hardness of the chromium and the fineness of its grain make possible the use of extremely fine screens, while the depth of the image contains a particularly rich film of ink. A plate of this kind is capable of an emphasis, contrast, and detail not easily achieved by other means. On the white background of a good quality paper, the clear, intensely black letters appear to stand up out of the page. Logically this should lend itself to legibility better than the old-fashioned style of printing in which each letter is seen at the bottom of a dent in the paper. Such plate-making however is extremely costly and is economically possible only when a very long run is planned.

§ 11-7 Offset text and illustration

The photo-composition image is usually though not invariably positive, whether on transparent film or on opaque paper from which the camera translates it into negative film. The best quality is to be expected of photo-composition film positive, as it is the most direct process; the typesetter image itself, rather than a reproduction of it, is in contact with the plate during print-down. But cameras, film, and camera technique now enable competent printers to print down a true and consistent typographic image from camera copy.

Offset tends to do justice to very fine line illustrations, with lines tapering to nothing, and close cross-hatching in shadowed areas. In letterpress, the waning thickness of the line often breaks into nothing towards the end; at the edge, where the raised surface takes the weight of the inking rollers without adjacent support, the line may be thickened by squash, which is also likely to fill minute interstices between lines with ink, except on a very glossy paper. An old steel-engraving, for example, might require a half-tone screen for letterpress printing, but might be a line subject for offset. In large areas of black or of some other colour, the offset ink film is less dense but also more even than that of letterpress, an advantage generally exploited in the printing of coloured panels on jackets and covers.

The screen pattern in an offset half-tone can be induced by exposure variation to dwindle into smaller and smaller dots until there is no pattern left but only a few random dots or none at all. The offset process has no difficulty in reproducing this effect, which hardly ever works properly in letterpress, and drawings which include shading in chalk or wash can be faithfully printed in this way. Highlights in photographs, similarly treated, add brightness and contrast to offset half-tones which may otherwise seem a little flat and pale by comparison with letterpress half-tone. In the darkest shadows, where dots expand and join up into solids interrupted only by pinpoints of white, the pinpoints, which indicate different intensities of shadow, are more likely to remain open in offset printing than in letterpress which tends to fill them with surplus ink.

The sharpest and most emphatic offset half-tones are likely to be printed on a glossy and indeed shiny paper of a brilliant white. Such a surface, as a border round the illustration, as a background showing through the highlights, and as a base for a glossy and therefore intensified film of ink, goes some way towards compensating for offset's lack of contrast in half-tones. This kind of paper, however, provides the reader with an unsympathetic background for the text. A mineral paper surface absorbs less ink than a fibrous surface, and is usually to be preferred for half-tone work of good quality. A matt paper with a slightly creamy tint offers less vivid illustration but a more mellow background for sustained reading. For these reasons, the convention of printing half-tones separately as plates is not yet obsolete even in offset work.

The quality of letterpress half-tone at its best can be matched and even exceeded by a form of *duotone* offset printing in monochrome illustration. Duotone means more than one kind of printing, but this technique consists of printing two different images, both black, on the same illustration. The second black image, printed over a complete picture, reinforces the darkest areas only, perhaps with a glossier ink and with a different screen angle, and adds nothing to the medium and lightest tones.

§ 11-8 Collotype, gravure, and screen

Earlier sections of this chapter deal with design factors in those printing processes most widely used in industrial book production. Other processes, developed and deployed in other forms of printing, also contribute to books, though the contribution may be no more than a jacket or some pages of illustration. Three of these processes, available to publishers but unlikely to be found under the roof of a bookwork printer, are outlined in this section.

Collotype is one of the most subtle and delicate of all printing processes, used mainly for short runs of continuous-tone illustration. A base-plate of glass or metal is coated with a light-sensitive emulsion of gelatine, and the coating is dried in a dark oven. The image is printed down on to this coating through a continuous-tone negative. The plate is soaked in a solution of glycerine and water, and this permeates each part of the coating differentially, according to the degree to which each part has been hardened by exposure to light. When greasy ink is rolled over the surface, the drier areas accept more ink and the damper areas less, in analogy with the principle of offset. The gelatine printing surface has a limited endurance; in the finest work, a few hundred impressions may be enough to cause deterioration, and greater resistance to wear can usually be achieved only at the expense of quality.

The gelatine tends to transmit to the printed image a fine, irregular, and all but invisible reticulation; apart from this, collotype is the only process which prints in continuous tone. Even in the darkest areas, the film of ink is thin and matt. The process is capable of the finest detail and of the most gradual transition from one tone to another, but not of emphatic contrast in illustration or of adequate emphasis in the reproduction of type. Coated paper provides no advantage.

Letterpress printing and printing from a relief surface are synonymous. Offset and collotype are both *plane* or *planographic processes*, in which the printing and the non-printing areas lie in the same plane, on the same surface. *Rotary photogravure*, generally known as *gravure*, is a *recess* or *intaglio process*; the ink is retained in recesses engraved into the non-printing surface. Unlike collotype, gravure is a major industrial process, used for the high-speed printing of magazines, often in colour, and usually with illustrations. In book production it is now largely superseded by offset; colour printing by gravure can be brilliant and intense, but the reproduction of type lacks sharpness, and the cost of plate-making is usually too high for any but a large edition. Photogravure itself superseded techniques of hand engraving when photography began to supplement drawn illustration; early photogravure was printed from a flat plate, but means of printing down on to a curved plate were found towards the end of the nineteenth century, and rotary presses are now standard.

The printing image consists of a pattern of minute square cells, etched into a cylindrical copper surface, with a frequency of 150 to 400 cells to the inch. The coarser screens provide contrast in illustration; the finer screens sharpen the typographic outline to some extent. More than one gauge of

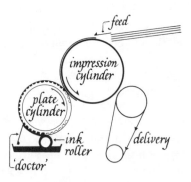

FIGURE 11-8. The principle of a sheet-fed rotary gravure press.

screen can be used on the same cylinder. Where a light tone is to be printed, the cells are shallow and the film of ink thin and transparent: darker tones are printed from deeper cells, which convey a thicker and more opaque film of ink to the paper: where nothing is to be printed, there are no cells in the surface against which the paper is pressed (figure 11).

In gravure printing the forme is a single plate, shaped as the curved surface of part of a cylinder, and once it has been made it will have to be replaced if alterations or corrections are necessary.

In printing, the cylinder revolves in contact with ink rollers which convey to it a thin spirit ink. As the inked surface revolves, the non-printing areas and the top of the cell walls are scraped clean by a sharp edge known as a *doctor* (figure 11-8). When the paper is brought into contact with the printing area, there is next to no ink on the surface which touches the paper, but the ink is drawn out of the cells by the suction of paper pressure and removal. Particularly in fast printing, some ink may overlap the cell walls, and all the ink may be flung towards one side or one corner of each cell, so that the printed image presents a mottled rather than a cellular pattern.

The best gravure printing has the merit of reproducing photographs with emphatic contrast between dark tones and light, and also with faithful reproduction of detail. Glossy paper is unnecessary, but quality benefits from a matt coated paper surface. Even the finest screens roughen the edges of printed letters with their pattern, though not to an extent unacceptable in short captions and illustration numbers. The film of ink in the dark tones is dense but dull in reflection.

Screen printing is a stencil process, in book production restricted almost entirely to covers and jackets. Ink or paint is squeezed through a stencil supported on a screen, which used to be silk but is now usually of wire; the process is still sometimes known as silk screen. The stencil can be prepared photographically, and is capable of reproducing the larger sizes of text type without undue loss of legibility, and a reasonably coarse half-tone screen without undue debasement of its pattern. The value of the process lies in the thickness, and, where required, the opacity of the ink or paint; unlike other processes, this one can easily print white on a black background. An effective jacket can be produced by printing light-coloured lettering on a darker-coloured paper of individual texture.

CHAPTER 12

Imposition and presswork

MPOSITION is the arrangement of pages in a rectangular formation for printing. The pages consist of whatever is to appear on the appropriate side of each leaf, whether anything is to be printed there or not. In letterpress imposition, for instance, a blank page is imposed not as a space in the *forme* (§ 4-9) but as a rectangular element lower than type-height, against which adjacent pages of type or blocks can be braced for locking up. The position of each page in the formation is determined by the manner in which the sheet is to be folded; this in turn is determined by the capabilities of the designated folding machine, by the number of pages assigned to each *section* (§ 16-2), and by other factors. One of the standard formation or imposition schemes diagramatically explained in *Book impositions* (1962), or occasionally a non-standard scheme, has to be agreed between printer and binder for every set of sheets and for any *oddments* (§ 12-1) which are to be folded. The correct application of an appropriate scheme ensures the right order of the pages in the folded section. Imposition may have to be repeated whenever the book is reprinted, and in this sense it is an essential preliminary to presswork rather than a culmination of composition.

Presswork is the work of the printing press, as distinct from combinations of presswork with composition and other preparatory tasks of the printer which are collectively known as *printing*. The term *machining* is also used in place of *presswork*, but tends in other contexts to mean machine-work as distinct from hand-work.

§ 12-1 Imposition

The number of pages in the sections is one of the factors which contribute to the structure and appearance of any book. Apart from this, the imposition of many editions hardly concerns the designer as long as it is done well. When a book contains much the same sort of text and illustration, presented in much the same way, as any number of other books being produced at the time, printer and binder may usually be left to agree about the imposition scheme. But the book designer is entitled to say what sort of scheme will best suit his plans, in terms of the number of pages to the section and in other such terms.

Industrial design is a matter of planning not only a product but the application of industrial processes to its manufacture. Industrial book design involves planning a page not only as an item to be composed but as one of a number of similar areas to be printed on one side or the other of a sheet, and as a position in the sequence of pages of the section into which the sheet is to be folded. To do this, the book designer may have to calculate, after make-up into page and before imposition, which pages are likely to appear at which positions on the sheet and within the section, and whether those positions will suit his arrangement of the text and illustrations.

Planning sheets and sections is a matter of more than imposition. The relationship between sheet and leaf size, between the dimensions of the page and those of the maximum printed area of the press, and between sheet size and the press's maximum sheet capacity, have been considered in §§ 3-1 and 3-3. Illustrations which extend beyond the area of page or leaf, or across the facing pages of an opening, and those which are printed as plates, separately from the rest of the book, are dealt with in §§ 14-3, 14-5, 14-6, and 16-3, and § 16-2 is concerned with folding in general.

The assembly of a sewn book is based on a unit of two joined leaves, each of two pages, with a stitch through the central fold at the spine. Sheets printed on both sides, and sections, therefore make four pages or a multiple of four. Most sheets are folded into halves at each fold at the centre of alternate edges of the cut or folded sheets, or in a different sequence which has the same effect. A *broadside* (§ 3-1), printed on both sides for folding in the widely used way, accommodates 16 octavo pages, a double broadside 32 pages, a quad sheet 64, and a double-quad sheet 128. Sheets of the same sizes would have room for half these numbers of quarto pages.

The outermost pages of any section are printed from the *outer forme*, which also prints alternate openings throughout the section. In a section of 16 pages, for example, pages 1 and 16 (the outer pages), 4 and 5, 8 and 9 (the *centre spread*, § 14-3), and 12 and 13 are printed from the outer forme; pages 2 and 3, 6 and 7, 10 and 11, and 14 and 15 are printed from the *inner forme*. When for example colour is to be economically used throughout a book but not on all pages, coloured illustrations may well be restricted to the inner or the outer forme, and hence to alternate openings. The centre opening or spread of the section is the only opening in which the facing pages are imposed side-by-side with the spine between, so that the horizontal alignment of any illustration or table which occupies a whole opening cannot be impaired by an inaccuracy in folding if it is printed across the spread.

The superficial area of the sheet is not always halved at each fold. If the first two folds are so placed as to reduce the area to two-thirds and one-third respectively, subsequent halving folds will produce *duodecimo* (12mo) or *sexto* (6to) pages (§ 3-1), which are smaller in area than octavo and quarto pages when all are based on a broadside of the same dimensions. A broadside, printed on both sides for folding in this comparatively unusual way, accommodates 24 12mo pages, a double broadside 48, a quad sheet 96, and a double-quad sheet 192. Sheets of the same sizes would have room for half these numbers of 6to pages. If the first two off-centre folds turn the outer edges inwards on the same side of the sheet, there will be some openings in the section in which the verso page was printed on one side of the sheet and the recto on the other.

In 8vo or 12mo folding the maximum number of pages to the section is likely to be 32 or 48, but there may be 64 or 96 pages on the quad sheet, which may therefore contain several sections. Within each sheet the sections can be arranged at will, subject to appropriate relative positions for sections which are to be *inset* together (§ 16-2).

The standard sheet of an edition is the size of sheet on which most of the pages are printed. The standard imposition for this sheet is usually *work-and-back*, or *sheet-work*; each sheet is *worked* (printed on one side) from one forme or offset plate, and *backed* (printed on the second side) from a second forme or plate. On a *single-cylinder* press (§ 12-4), a *work-and-turn* or *half-sheet* imposition may be used to print a standard sheet from a single forme. The sheet is worked from this single forme, turned over, and backed from the same forme, which has been so imposed that the product is two half-sheets with an identical printed image. The number printed from the single forme is half the number printed from work-and-back formes.

A *printing oddment* is a sheet printed from a forme — or, on a perfector (§ 12-4), from a pair of formes — containing a half or a smaller fraction of the number of different pages in the standard forme or formes. The sheet itself may be of standard size, but the forme may include pages duplicated, perhaps more than once, to provide two or more identical sections from one sheet; or a perfector may be run on a single cylinder only, with a work-and-turn imposition. One way or another, the presswork cost per page in an oddment is higher than in a standard sheet. A *binding oddment* is any section which has to be folded into fewer pages than the standard section. A book which includes printing and binding oddments is likely to cost more to produce, not only per page but in total, than a somewhat longer book made up of *even workings* for both printer and binder.

If economy were all, books would always be designed, and adjusted in preparation, to make these even workings. But the ways in which texts are assembled by author and editor, the nature of the text itself, and the reader's requirements of the printed page, often combine to make this difficult or impossible. Publishers and their designers have not generally assumed responsibility for *casting-off* (§ 18-1), and in the absence of a laborious and exact survey of the copy, design plans result in even workings only by chance. After the text is in type, alterations to the preliminary pages, additional illustrations, new appendixes, an index of unexpected length, or any one of various unforeseeable chances, may alter the extent of the book. Diligence and determination on the part of publisher, designer, printer's estimator, and sometimes author and illustrator, can, however, contribute to the possibility of an exact fit. The problem, in the initial stages of design, is to decide whether it is possible to aim at fitting, and, if it is, whether it is worth while.

Certain authors habitually rewrite extensively when the book is in proof. Technical and scholarly works, and highly topical publications, are likely to require additions and deletions during all stages until the formes actually go into the press-room. If a series style has to be adhered to, the designer has little opportunity for manœuvring the extent one way or the other. For books of this kind, even workings are usually not worth attempting.

When the book is a long one, the extra cost of oddments may add so small a percentage to the cost of printing and binding as to be negligible. If the run is short, the extra cost may not mount up to a considerable sum. When large numbers of a book of short or medium length are to be printed, an attempt at even workings is worth while.

No standard procedure determines the stage of composition at which metal type is imposed. If pages are imposed as soon as they are made up, before they are sent out in proof, any subsequent amendment which affects page numbering or necessitates the transfer of type from page to page is likely to be expensive. This rarely happens with film, which is inconvenient to show in imposed proof. For academic works, for others in which major amendment may have to be admitted, and for those in which an even working is essential, the most economical procedure is to number preliminary pages separately from the text, and to impose after author and editor have returned the final proof. This has the additional advantage that the preliminary pages can be imposed either with the text or an an oddment, perhaps with the index, in the event of late additions. Some publications are shown to the book trade in the form of *book proofs* (§ 19-3), for which a simplified and temporary imposition of metal type and lock-up in the bed of a proofing press may be enough. The printing imposition itself, together with its margins and with alterations and corrections marked on the author's final proofs in page, should always be checked by the printer's reader in imposed proof or *press revise*, whatever the printing process.

When pages are made up in film or paper instead of metal, the intended position of the image on the page may not be evident during imposition. In metal, when a half-title setting, for example, is to appear under three white lines, the compositor makes up metal spacing material for those three lines over the setting. In the course of imposition, the metal head of the page can then be aligned with the head of other pages. Full text pages present no problem, as the first and last lines of text define the corners of the text area, and the page number serves as a point of reference. When the first and last lines of the page are not occupied by text and there is no page number, the *stone-hand* or imposition compositor may have no sure guide to position, and his work will be the more reliable if the position of all such pages is marked on the final proof, or indicated by corner-marks defining the upper and outer corners of the otherwise empty headline or first line. In the same way, the intended margins of illustrations printed as plates should be made clear to the binder by means of an imposed proof or a marked sheet.

Sensitized film for pages to be printed by offset is an expensive material, and is often used only for the printing area. Negative film may be mounted on a card template, pierced by rectangles to accommodate the pages; positive film may be sprayed with a transparent adhesive and mounted on a base of insensitive transparent film. Film is more likely than metal to be stored in imposed form; *chases*, with their *furniture* and *quoins* (§ 4-9), are costly equipment designed for printing rather than for storage, but light-weight duplicate plates for letterpress printing may be held imposed in much the same way as film. Re-imposition for reprints is an economic factor which should have no effect on quality.

§ 12-2 Make-ready

Make-ready is that stage of preparation for printing which involves the press. The purpose of make-ready is adjustment — of register, of impression, and of ink supply.

Register is the relationship of position between the images produced in the course of two or more printings on one sheet, whether the images are on the same side of the sheet or on opposite sides. The printings may be simultaneous, as on an offset perfector: or successive, as in a single pass through a four-colour press: or separate, as when a sheet is worked in a first pass through the press and backed in a second. In offset, the alignment of pages is checked before the plate is printed-down; once on the plate, the position of each page is fixed. In letterpress, the pages are separate from each other, and their alignment can still be corrected during make-ready. When a sheet is perfected with simultaneous impressions, register between its two sides is invariable, and if pages and plates have been precisely adjusted in relation to each other and to the sheet, register will also be correct. In successive impressions, some variation is possible, but if all alignments have been corrected during or before make-ready, errors of register will be rare. Separate impressions increase the likelihood of register variation; moisture absorption and mechanical stretch affect the dimensions of the sheet between impressions. In colour half-tone, register needs to be accurate to a twentieth or smaller part of the percentage to which a sheet of paper can expand in damp conditions. For most book printing which does not include colour, division of the paper into sheets at the paper mill is rapid and inexpensive rather than exact; unevenness in squareness or in size is likely to result in register variation, and also in variation of distance between the printed image and the edges of the sheet. This latter form of inaccuracy does not necessarily throw out the register between forme and forme, but is all too likely to result in erratic folding. In order to minimize the chances of any such effect, and particularly for colour printing, paper can be guillotine-trimmed before printing, but this is usually considered to be uneconomic for a single colour. Even in a single colour, however, register from front to back of the sheet deserves the printer's efforts, particularly if the paper is somewhat transparent. If the lines of type on each side of the leaf are inaccurately registered, the type printed on one side of the leaf will discolour spaces between and beside the lines on the other.

Register between separate impressions depends on various factors including the size of the sheet. No paper is entirely stable in size, and the larger the sheet the larger its expansion and contraction in varying conditions of humidity and under the pull of the cylinder. A quad sheet is likely to prove too big for accurate colour register in separate impressions, unless part only of the sheet is to be printed in colour and the colour pages can be grouped close to lay-edge and gripper-edge. The *lay-edge* of the sheet is that edge which buts against the side-lay of the press, and so establishes the relative position of sheet and forme.

Dimensional instability of a different kind may become troublesome if close-register work is attempted with stereos, because of uneven shrinkage in moulding and casting which can render register not difficult but impossible.

The appearance of any image printed by letterpress is governed by the nature of the paper, the condition of the printing surface, the quality and supply of ink, and the weight of the impression. *Impression*, in this context, is pressure between paper and printing surface, and also the depth in the paper surface to which this pressure is allowed to force the relief printing surface. Heavier impression suits paper with a rough surface, worn type, and thin inking; lighter impression is more apt for richer inking, a fresh printing surface, and smooth paper. Some type series still in use are based on letters designed for a comparatively heavy impression; their thin main-strokes and very thin hair-lines are most clearly seen when widened by being forced deeply into the paper. Such presswork is associated with the typography of the past and of the private press, with the types of long ago, with paper of a rough individual surface, and so with much that is great in printing. This association is the main reason for the popularity, now in decline, of heavy impression, 'the third dimension of letterpress printing'. Impression of this kind, however, is an anachronism; it tends to distort the reproduction of the printing surface, to accelerate its deterioration by wear, and to cause an unsightly corrugation of the printed area unless the paper is thick. If a book of many pages is printed heavily on thin paper the impression, thickening the text area but not the margins, may cause difficulty in the bindery. The depth of the dent in which each letter appears may be held by some to enhance the decorative value of the page, but can hardly be supposed to improve its legibility, and may even distract attention from the letter-forms themselves. The reasonably smooth, hard, and even machine-made papers now generally used in book production, and the precision-built presses of today, combine to ensure a solid print without obvious indention or distortion, and the majority of text types now in use needs no more.

§ 12-3 Ink and inking

The selection and preparation of ink are almost entirely a matter for the printer. But the printed image is made of ink, and a designer who seeks quality in presswork cannot afford to shed all responsibility for ink and for the way in which it is applied to paper. Not only ink and paper, but the press, the printing process, and in letterpress the depth of impression have all to be adjusted to each other. The designer can bring his influence to bear on this adjustment without usurping the printer's right to run his presses by methods and with materials he has found effective. The scope of such influence, however, is limited by the circumstances of presswork. Inferior paper, for example, may demand the printer's utmost skill, but does not lend itself to a superior result; and intense colour of image cannot be expected of high-speed printing which may be economically necessary.

Intense colour is not necessarily a virtue. Colour printing is discussed in chapter 13; this section is concerned with black. Not much is now heard of the axiom, once influential in book production, that emphatic contrast between type and paper may be spectacular but tends to dazzle the reader. The proliferation of half-tones printed by offset with text is only one of the factors which has promoted whiteness in book papers. Intensely black type on these white surfaces, less mellow than the creamy esparto papers popular in the first half of the century, may attract the reader's attention but may also distract him in the course of sustained reading. Small type, particularly in the more delicately drawn designs, may benefit from contrast of this kind; larger and bolder founts seem to stare back at the eye, instead of awaiting examination.

Offset half-tone, on the other hand, needs all the ink the printing surface can carry. If it is too heavily loaded, the darker tones will begin to close up into solids, but up to this point there can hardly be too much colour. With a comparatively thin film of ink at best, the process has to compete in depth of contrast with letterpress half-tone and with the deep tones of the photograph from which the image was taken. Any monochrome half-tone is a shadowy representation of three-dimensional reality, to which colours contribute a simulation of depth. Until a full range of colour becomes economically possible for all photographs used as illustration, designers and printers are bound to maintain the development of offset half-tone towards clarity, detail, and an appearance of depth derived from tone contrast. The dropped-out highlight, as white as the paper surface showing through it, provides the whitest possible tone; the contrasting blackness of ink is certain to be increased by continuing development in ink-making and presswork. Meanwhile, if demand for greater intensity causes any printing surface to be overloaded with ink, the printed image will be thickened, and dot patterns will begin to close up.

The intensity of any colour, including black, is increased by a glossy surface which reflects white light. This effect can be observed by comparing two copies of the same book-jacket, one before and one after *laminating* (§ 17-4). A similar effect results from mixing varnish into ink, but varnish tends to sink into absorbent paper rather than remaining on top of the printed image. Since a glossy image therefore tends to appear only on a glossy paper, the reflective value of the printed page is even over the whole of its area. A glossy image is unlikely to prove easier to read than a matt one, but may well be more popular.

Ink varies in transparency according to the nature of the mixture and the thickness of the ink film. Too thin a film of black on white paper looks grey because the paper begins to show through the image, and to mingle its own colour with that of the ink. When ink is printed on a coloured background, whether it is another ink or a tinted paper, the image is the colour not of the ink alone but of image and background combined. Except by screen process, which can print a thicker film of a denser ink, ink cannot be legibly printed on a background as dark as or darker than its own colour.

In letterpress printing, and in the pulling of repro proofs, the pressure of the inking rollers has weight enough to convey ink evenly to the whole width or depth of the text pages on which they bear. As the rollers pass over any small isolated items, such as page numbers alone in a line, this pressure may be more than the small printing area requires, and the page number may be over-inked although the inking of the page is satisfactory. This is most likely to happen when octavo pages are printed on quad sheets, as the pages lie along the direction of the rollers, with their long edges parallel with the sides of the press. Page numbers may then be more evenly inked if they have an ornament or perhaps a bracket on each side, to share the pressure, or if they are set with other type in the headline.

The two main constituents of ink are *pigment*, which provides colour, and *vehicle*, into which the pigment is mixed and which conveys it to the printed image. The nature of the vehicle influences the appearance of the image. Gravure inks, for example, have to be thin enough to enter and depart from the tiny cells of the cylinder at speed, and have to dry rapidly throughout the thickness of the ink film in darker tones. Accordingly they are solvent-based; a high proportion of the vehicle evaporates, leaving behind an image of powdery pigment with a matt and almost dusty surface. In the same way, different presses, processes, and speeds of printing, with different thicknesses of ink and different methods of drying, produce images which differ in appearance. In offset printing, the film of ink may be only half as thick as that printed by letterpress, and may in addition be emulsified to some extent with damping water. Ink may also have to be run somewhat thinner on a perfector than on a single-cylinder press, because of ink drying problems.

Apart from this effect on the printed image, ink drying usually concerns the designer only when it goes wrong. If the ink does not dry quickly enough, successive sheets laid on each other at the delivery end of the press receive ink from the damp image immediately below them in the pile. This pale reflection of the image on the wrong side of the sheet is *printing set-off* — printed, as in offset, from something other than the printing surface, but unintentionally. If the ink dries very slowly, whether or not printing set-off has occurred there may be *binding set-off*. This is caused by pressure on the whole book after *casing* (§ 16-10), and takes the form of verso and recto images repeating themselves on the facing page. Thin fibrous paper is liable to *strike-through*; the vehicle penetrates so deeply into the paper as to be visible on the other side. This is likely to be the fault of the paper rather than of the ink, and can be distinguished from set-off which usually deposits some pigment, visible through a magnifying glass, in the set-off image. *Show-through*, the appearance of one or more printed images through the page, indicates lack of opacity in paper and is not an inking fault; all too often the next two pages can be seen through the recto.

The constituents of any ink are adjusted to suit the paper on which it is to be printed. The printer may have to obtain ink specially for any paper to which he is unaccustomed, and is most likely to use the most suitable when the paper is delivered well in advance of printing, for test, examination, and if necessary a special order for ink.

§ 12-4 Flat-bed presses

Proofing presses apart, only a letterpress printing surface lies flat in the bed of the press during presswork. From the invention of typographic printing to the present day, a forme of type has been essentially rectangular in section and in plan, and essentially heavy. The hand press, in which a flat sheet of paper is tightly shut between a horizontal platen and a horizontal forme, is inadequately rigid over more than a limited area; the maximum printing dimensions of such a press are indicated by the traditional sheet sizes of crown, demy, and royal. The nineteenth-century introduction of the cylinder press, in which the paper is wrapped round a revolving cylinder and rolled across the forme instead of pressed flat against it, reduced the area on which pressure was brought to bear at any one moment to a narrow strip, and enabled much larger sheets to be used. Until the middle years of this century, the quad, or double-quad sheet, four or eight times as great in area as the traditional sizes from the day of the hand press, printed on a cylinder press such as these, provided the foundation of industrial book production.

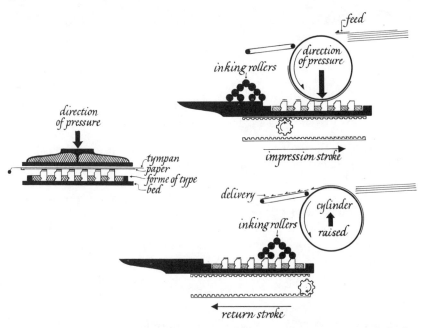

Figure 12-4a. *Left,* the hand-press principle. *Right,* a two-revolution single-cylinder flat-bed press.

In these large cylinder presses the heavy forme is shunted to and fro, under the control of massive buffer mechanisms. Make-ready keeps such a machine immobile for much of its working life. Printing speed is limited not only by the forme's weight but by that of the reciprocating bed, solidly built to maintain a level surface under pressure, in which the forme is laid.

Machinery of this kind has always been capable of excellent work, but is ponderous and occupies more floor-space and manpower than is economically compatible with its output. Quad and double-quad presses, and the compatible sizes of folding machines which helped to make them economic, are no longer being built. Double crown and double royal presses may still be competitive for editions smaller than the few thousand at which the economic claim of offset becomes compelling. Printing from type and blocks will not be obsolete while metal type is still commercially available; but the proportion of titles printed in this way shrinks year by year.

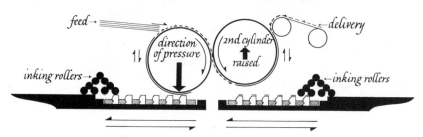

Figure 12-4*b*. Flat-bed perfector.

§ 12-5 Sheet-fed rotary presses

The principle of the flat-bed press is partly rotary and partly reciprocal; the impression cylinder, on which the paper is mounted, rotates while the bed moves to and fro beneath or beside it. In a rotary press the printing surface is also mounted on a cylinder, and the two or more main cylinders of the press rotate continually while the press is running. The cylinder or cylinders of a flat-bed two-revolution press also revolve continuously, but print only during alternate revolutions, since during one revolution the bed returns under the cylinder without contact. A rotary press prints at every revolution, and even at a similar speed of turning would therefore print twice as fast as a flat-bed. But having no heavy rectangular bed to halt and reverse at the end of each revolution, the rotary press is not limited in speed by the weight of its own construction, and normally turns faster than the flat-bed. The resulting difference in output, without necessarily more manpower, floor-space, motive force, or ink, combines with the rotary press's reliance on a curved printing surface to reduce industrial capacity for printing from type.

Since metal type is rectangular in section and in plan, it has to be moulded and stereotyped when it is to be printed on a rotary press. The same image can be converted for offset printing by means of repro proofs and film. In either instance additional processes intervene between image and reproduction, bringing with them the risk of deterioration and unevenness, together with an increase in cost which bears heavily on the economics of short-run publications. In terms of reproduction quality and cost, an image composed on film is better suited to offset printing than metal type.

The printing surface of an offset press is a cylinder covered with a rubber blanket; this receives from the offset plate an inked reproduction of the image, and transmits it to the paper. Two such cylinders can oppose each other on opposite sides of the sheet, and can print both sides at the same moment, because the blankets are resilient and form a plane surface across the cylinders. Half-tones backing half-tones on sheets so printed may cause problems of adhesion before delivery, and presswork quality is limited by the soft backing of the sheet at the moment of impression. Better presswork in general, and better half-tones in particular, tend to emerge from presses printing one side of the sheet at a time, even when a second impression cylinder mounted at a distance on a different kind of perfector enables both sides of the sheet to be printed at one pass through the press. Whatever their construction, perfectors generally restrain the thickness of the ink film, but offset perfectors are becoming the standard press of the book production industry.

Letterpress becomes competitive again when large editions are to be printed. A rotary letterpress machine, maintaining no delicate balance between inking and damping, runs faster than an offset press. Half-tone printing of good quality, however, seems incompatible with the resulting speeds and with a curved relief printing surface, and offset is to be preferred even for long runs of illustrated books. The speed of rotary letterpress imposes limits on the density of ink, and text printed on such presses tends to be adequately printed but hardly sharp or black. Letterpress rotary therefore tends to be found useful for paperback editions, printed in large numbers, in black only, on inexpensive and absorbent paper, either unillustrated or with line illustrations only.

The cost of preparing a printing surface which can be mounted on a cylinder compared with the cost of printing from a surface composed or engraved in metal, is partly absorbed by the lower make-ready time required by a rotary press. These additional costs are also incurred by the first printing rather than by reprints; flexible stereos last longer than type and may be economically stored in imposed form, and an image developed on good film may also be kept imposed and should be fit for use for years. Rotary reprints, even in short runs, tend therefore to cost less than reprints from type and blocks, which usually require laborious re-imposition and always require laborious make-ready.

§ 12-6 Web-fed presses

A web-fed press is rotary and prints by letterpress, offset, or gravure, as do the presses described in the previous section. Unlike them, it is fed from a continuous reel of paper, not a stack of separate sheets, and usually perfects and folds the paper. The word *web*, suggesting a woven material or some kind of membrane, could mislead; the paper is much the same as any other, but the continuous strip of it which enters the press is generally understood to be the web. The term *web offset* is more commonly used than *web letterpress*, so that *web* alone usually indicates offset.

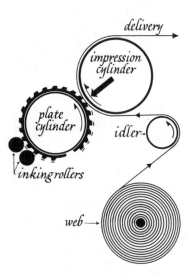

Figure 12-6. Web-fed letterpress rotary machine.

Whatever the process, the web itself enables presswork to reach its highest speed; the pace of a sheet-fed rotary press is limited by the rate at which separate sheets can be fed and gripped in exact position before impression. At the paper mill, the web can be delivered on to reels as fast as it can be made, and paper in this state costs less than similar material which has passed through the slower process of division into sheets. The inking of a web-fed press, on the other hand, is not fully effective until the press is filled with paper and running, and at this stage a considerable amount of paper spoilage occurs. This spoilage, a significant factor in web presswork, can be reduced by standardizing paper for different editions to be printed from the same make-ready. The speed of web printing can be prodigious; national newspapers in Britain have long been printed by web letterpress, and local newspapers have been converting their printing to web offset in recent years. Great speed is inimical to rich inking, and very fast presses, printing special heat-set inks, carry the sheet after printing through a momentary blaze of heat or ultra-violet radiation.

The width of the web, and hence one dimension of the page, can be varied when paper is made to order for the edition; but the page's other dimension is usually fixed by the circumference of the cylinders and by the cut-off mechanism of the folding attachment of the press. Web printing lends itself to very long runs in which quality is secondary to cost, to standardization of paper and page size, and to editions which can be batched at the presswork stage to exploit this standardization. Web presses, like several photo-composition systems, have been developed primarily for newspaper and magazine production; the book designer who wishes to make the most economical use of them will have to plan the production of his books as though in some ways they were newspapers or magazines.

§ 12-7 Reprints

A new edition of a book may be defined as a version which differs substantially in text or illustration or both from preceding editions. The scale of alteration need not be great; a single paragraph of conclusion may be enough to change the purpose of the author's message. A reissue is usually understood to mean a revived publication of an existing text in a physically different form, such as a different kind of binding or even a different type-face. If the text is re-set for reissue, it should follow the preceding edition page for page, as changes in the make-up, of a scholarly work at least, might justify publication as a new edition. A reprint is produced from typesetting already published, to which corrections and minor amendments may be applied but which in essence records the same message as before. The production of some new editions and reissues may provide much the same kind of task as that of a new publication; in production terms, other new editions and reissues will have more in common with reprints.

Alterations to the existing type-setting may affect the appearance and arrangement of the text, and will therefore require design attention. At any stage of book production, work done in one set of circumstances tends to differ in appearance, however slightly, from work done in other circumstances. Almost any textual amendment inserted into the existing text area is likely to show an unwelcome degree of contrast. The designer should take care to ensure that this contrast will not be so pronounced as to attract the reader's attention. If the reader notices a difference in amended lines, he may form the impression that the author has somehow changed his tone of voice, perhaps in mid-sentence. Alterations in reprints should whenever possible be so planned that no words are transferred from one page in the previous printing to another page in the new, in case the index has to be re-set and the reprint re-published as a new edition.

The number of copies reprinted may differ radically from that of the first printing on which the method of production was based. Since almost any industrially produced book may have to be reprinted, perhaps in small numbers, the design of the first printing should be economically suited to such reprints. An unusual size of paper, for example, may be made to order for the first printing, but a reprint may not require enough paper for a *making* (§ 15-3), any remaining stock of such a making might prove useful for no other title, and the cost of using larger sheets from some other source might prove too high for a short run. Variety of page size may support the presentation of a book, but may also tip the economic balance against a repetition of short reprints.

Other matters concerning reprints have been touched on in other sections, and one of them is worth repeating in this one. The designer's task is to plan not a single volume, nor even an edition, but an edition and a series of reprints. A reprint may never be ordered; but if it is, its quality should not be permitted by inadequate planning to fall short of that of the first printing.

Colour and printing

COLOUR is a sensation of the eye, stimulated by light. Everything visible reflects some light, or it would not be visible. Its colour is determined by the wave-lengths it reflects under colourless light. In the red light of a dark-room, white paper appears to be red, but its own colour reappears in daylight. In this sense, black and white are colours, since black ink and white paper are both visible. In some printing contexts, *colour* is intended to mean any colour or colours other than black and white. In this book, the meaning of each reference to colour is intended to be clear from the context, since the inconsistencies of printing terminology do not lend themselves to definition.

Different colours are normally printed from different printing surfaces, either by repeated passes through a single-colour press, or by the use of a multi-colour press. This drastically increases the cost of production beyond that of monochrome printing. For any edition, the choice between monochrome and colour is more likely to be the publisher's than the designer's. But the designer is entitled to propose colour printing when he sees an opportunity for its effective use, and he has a duty to make the most effective use of it when it can be afforded.

Colour-printing is gaining ground in book production. The reader's eye welcomes colour — as decoration, as a form of emphasis, as a means of clarifying information, and as a representation of visible reality more convincing than black and white. One of the purposes of book design is to deserve this welcome. The market-place of communication is flooded with colour, in the cinema and on the television set, in domestic photography, in magazines, and even, after a fashion, in newspapers. To compete for attention, books are likely to become more colourful as time goes on.

In more ways than one, the printing of additional colour is more than a multiplication of processes; it is, rather, a process distinct from others at almost every stage. The planning of colour-printing is likely to call for care and skill, for an eye capable of sensitive assessment and comparison, and for much of the technical knowledge outlined in other chapters. Even a minor irregularity at certain points in production, which would hardly be noticed in monochrome printing, may exert a disproportionate effect on colour quality.

§ 13-1 Black, white, metal, and varnish

For reasons of economy, most pages of most books are printed in one colour, on paper which is of another colour throughout the book. Traditionally, the paper is more or less white, and the ink is more or less black. The novelty of any obvious breach of this tradition might either reduce the visual appeal of the book, by appearing to intervene between reader and message, or enhance its attractions as artefact rather than communication. The contrasts between image and background, and between the different intensities in any monochrome half-tone, are diminished by any combination of colours other than black and white. Other colours might be found equally legible by readers accustomed to them, but no other combination has been generally introduced.

Black ink, already discussed in § 12-3, usually contains some pigment of another colour, such as blue or brown. The effect of this is more likely to be observed in half-tones and solids than in text. Blue intensifies contrast, but may cause drying problems; brown adds a touch of warmth, to relieve the chill whiteness of coated paper.

Large areas of solid black are now used as a design feature in books, usually as a frame or contrast for illustration. Such solids are difficult to print well by letterpress on uncoated paper, so they have been little used in the past, and are still unfamiliar enough to attract attention. On uncoated paper, an offset solid is less dense but more even than a letterpress solid, but the effect is all too likely to be ruined by *hickeys* — spots of foreign matter, usually from the paper edge or surface, which seal themselves to plate or blanket, and appear in the printed image as black dots with white haloes. Hickeys can appear in half-tones as well as solids, and in colours other than black, but are most obvious and unwelcome in a black solid. *Coated papers* (§ 15-4) tend to develop fewer hickeys, but the defect is not always due to the paper.

White as a colour is used in printing, but is not usually printed at all. The white image reveals the unprinted paper surface, surrounded and defined by black or some other strong colour. If the surrounding colour is weak, the white image will be inadequately visible. Printing ink is more transparent than it appears to be, and a light colour printed over a dark one, except by the *screen* process (§ 11-8), produces a third colour but not a clearly visible image. Type of text size seems to lose accuracy of outline from reversal; it looks legible enough for short passages such as captions, but with diminished clarity and grace. In larger sizes, white letters on a dark background can be conspicuous and attractive. White lettering on a four-colour background is expensive because the photographic reverse is applied to all four colours. When type is reversed into half-tone or a similarly interrupted background, the whites of the background serrate the outlines of the letters, and distort their shapes unless they are large and bold.

The translation of a coloured picture into monochrome half-tone is not always accurate. The camera tends to assign comparative intensities to red and blue, for example, which differ from those observed by the eye. The tonal balance of the half-tone may be distorted by this effect. A designer who wishes to preserve that balance will have to compare the coloured original with its monochrome photograph to make sure a half-tone reproduction from the photograph will adequately represent the various colours.

Four-colour printing reproduces the colours of reflective metal surfaces, but not their metallic sheen. A convincing imitation of gold or silver, for example, is possible only by means of another unbroken metallic surface, produced on books by the binder's *blocking* process (§ 16-8). Intermediate in similitude between colour-printing and blocking lies the use of metallic ink, in which a transparent *vehicle* (§ 12-3) conveys a pigment of metallic powder. The effect is decorative rather than accurate, and tends to add little of value to the illustration of metalwork. Blocking can be used on paper in place of printing, but the heated blocking-surface has to be pressed hard into the paper; a thick cover or jacket is better for this purpose than a book page.

Varnish is not a colour in itself, but rather a kind of emphasis which can be added to or laid over any printed colour. Varnish in ink brightens its appearance, but tends to sink into uncoated paper, which conceals its effect and may postpone its drying. Printed over colour, varnish brightens and intensifies all colours. Printed varnish may lend to coloured illustration an element of the spectacular, but may also dazzle the reader with brilliant reflections of white light.

§ 13-2 Colour in typography

Whatever its purpose, colour is likely to succeed when its use is planned from the beginning, and is likely to fail when an arrangement or an illustration designed for black has an additional colour introduced. This is true in particular of typographic colour printing; a black title-page, for instance, may be spoilt by the arbitrary addition of colour, which can alter the balance of the composition.

Legibility depends to some extent on the contrast in colour between type and its background. If the contrast is inadequate, as when a grey paper is used with black type, the letters cannot be seen clearly enough; if the contrast is too emphatic, as when extremely black ink is printed on extremely white paper, reading may become a strain. Legibility may also be impaired if areas of colour spread from an illustration into the text area; this may not only obscure the letters but may interrupt the rhythm of reading by differentiating one part of the text from another.

The rubric in church services is an example of colour as a means of clarifying information. The words printed in red are liturgical instructions, rather than part of the service to be spoken or sung, or they are the priest's part of the service as distinct from that of the congregation. The use of two colours in the same line entails the risk that the slightest error in register will spoil the alignment of the two parts of the line. But if in drama, for instance, the name of the speaker is to be printed in colour and the speech in black, the risk of inaccuracy can be diminished by placing the name in a line by itself.

Colour used for decoration on a page otherwise printed in black should be placed with care because of its different emphasis. The custom of using red for the first letters of displayed words has fortunately died out; it was bad not only because it divided the word into two differently coloured parts, but because it scattered spots of red in random positions on the page. Colour in typography is better used purposefully, to strike either a subtle or an emphatic note rather than any in between, and in a manner which sets up or continues a pattern applicable to other pages.

Colour is often assumed to be more emphatic than black, but this is a matter of opinion and of the colour to be used. If on a title-page the smaller type is all in colour, and the title itself in larger type is black, there will be little question of where the emphasis lies. Colour benefits from contrast with black; a display line such as the title of the book may, if printed in colour, look particularly well with a line of black type above it and another below.

Type printed in a comparatively pale colour tends to look thinner than when printed in black or in some strong colour. Display lines to be printed in light colours may therefore look better if set in a comparatively bold type.

Trichromatic inks (§ 13-6) do not serve well for display lines in colour when used alone; yellow is illegible, and cyan and magenta unattractive. The register required in four-colour printing is so close that display type — preferably large and strongly drawn — can be effectively printed in several colours, combining to present the appearance of one. In such a combination colour variations can be produced by printing some of the colours in line and others in half-tone.

§ 13-3 Colour selection and use

When colour selection is not determined by the illustrator or by the nature of the illustration, the designer is likely to find he has an opportunity for colour selection, particularly in the printing of covers and jackets. Selection cannot be regulated by principles, because it is a matter of taste and fashion, both of which are continually changing. The range of colours which can be printed is wide, since inks can be mixed to represent almost any colour. Very light shades are only occasionally useful, because of their inadequate contrast with white paper, and because they are too pale to satisfy the eye in the comparatively small area on which they appear. A pastel colour which would decorate the expanse of a wall may seem watery within the narrow confines of a page. Very dark colours may be too nearly black to be readily distinguished from it. Between these extremes there lies a range of useful colours, many of which possess the appeal of novelty. Some ink-makers provide colour catalogues to aid selection; good examples of these demonstrate not only the effect of mixing different colours but also the effect of intensity variations derived from variations in the printing surface. A single colour of ink can be printed not only as a flat area, but also textured with a drawn or mechanical grain or stipple, or with a plain half-tone screen, so that varying intensities of the same colour can be reproduced from a single printing. The dimensions of the area on which a colour is printed also seem to affect its intensity. A few small spots of red, for instance, printed on white paper, mingle with their background and take on a pink appearance; an ample panel of the same colour will look perceptibly redder. Colour is most intensely reflected from type or illustration which is bold and massive rather than subtle and detailed.

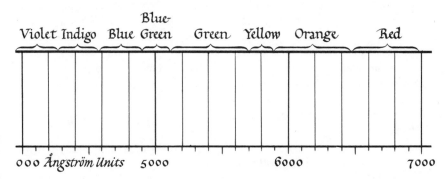

FIGURE I 3-3. The spectrum of white light, showing the *cold* colours on the left and the *warm* on the right.

Light is a form of radiation, and its colour is determined by its wave-length (figure I 3-3). Any light may consist of radiations of different wave-lengths — the effect of a prism on a beam of sunlight reveals the presence, in white or colourless light, of all the colours of the rainbow. All these radiations are reflected from a white surface such as that of paper. The shorter wave-lengths convey the violet, blue, and green group of colours, sometimes referred to as *cold*; yellow, orange, and red, projected by the longer group of wave-lengths, are known as *warm*. Cold and warm colours seem to induce differing emotional responses — funeral at one end of the spectrum, carnival at the other. The darkest colours in the warm group, including red and such mixtures as vermilion and brown, have long been more popular in typography than the cold colours. Red is the traditional colour for an extra printing on the title-page, and in various shades is still more widely used for that purpose than any other colour.

Since colour is a sensation, it cannot be precisely described to another person. If a designer specifies red ink to a printer, the result may be red but unexpected; the printer's favourite red may include a touch of blue, while the designer's might have been tinged with brown. Specification relies on comparison. Samples should be printed on the whitest of paper and compared in the whitest of light. Colour proofs should appear on paper which matches, at least in shade, the paper to be used for printing; and if colour original, proof, and print are not all compared in white light, some tinge of colour in the light-source may exaggerate a negligible disparity.

§ 13-4 Colour combination

The wave-lengths (and hence the colours) of light reflected from an opaque paper surface are those which have not been absorbed by the surface or by ink printed on it. In order to print a black image, an ink is used which absorbs nearly all the wave-lengths and reflects very few. The absorptive nature of black materials may also be noticed in black clothes, which absorb more heat radiations than white. The ink for a red image absorbs a majority of wave-lengths, which may be grouped as wave-length bands of violet, indigo, blue, blue-green, green, yellow, and orange. That leaves only the mainly red band, about one-sixth of the spectrum of colourless light, to be reflected.

Since printing inks are not usually opaque, one ink printed over another does not obscure the first printing. The pigment in each ink remains visible, each absorbing its own bands of wave-lengths — its own absorption range. The combined absorption ranges of the two inks leave fewer bands to be reflected than either ink would reflect alone. The result is reflection of a third colour.

When solid areas of two or more different colours are printed side by side, a fractional error of register will cause either an overlap and a joining line of a third and darker colour, or a divergence and a separating channel of white.

The shades of colours appear to be affected by those of the neighbouring colours, and are affected by the colour of the surface on which they are printed. A patch of blue surrounded by areas of dark grey and dark brown appears to differ in shade from a similar patch surrounded by pink and yellow. A creamy paper perceptibly darkens the tint of any coloured ink, in comparison with the same ink on a brilliantly white paper. When a transparent coloured ink is printed on an opaque coloured paper, the absorption ranges of the two materials combine, and release into reflection a complementary band of wave-lengths.

The colour of reflected or transmitted light is influenced by the colour of the light-source which enables it to be observed. Any tint in this source is likely to combine to some extent with colours submitted to the judgement of the eye. The best source provides colourless light, as does the midday sun on a clear day. A white fluorescent tube is better for this purpose than most light-bulbs. The colours of a transparency will be distorted if it is held up for viewing against a background of blue sky.

The intensity of colours printed together, and of the colours they produce when they are printed over each other, can be adjusted by the use of stipple or half-tone for one or more of the colours. Blue printed solidly over solid yellow, for instance, results in a strong green: a half-tone of blue over solid yellow, a yellow green: and a half-tone of blue and yellow together, light green.

§ I 3-5 Copy for colour printing

Separated copy for colour printing is prepared as though it were intended for separate black printings; for example, when a diagram is to be printed in red, blue, and black, a different black drawing is made for each colour. When in drama the speakers' names are to be red and the speeches black, names and speeches can be set together as though for a single printing, two sets of negatives made for offset reproduction, and on one set the names can be painted out with opaque paint and on the other the speeches can be painted out. Some computer-aided systems of photo-typesetting are capable of producing two separate and different sets of film for two colours from one set of data.

Separated copy for two-colour diagrams is often drawn on transparent material, to facilitate the artist's register between the separate drawings. The same method will do for pictures other than diagrams, for more than two colours, for half-tone as well as line, and for trichromatic effects. There is then no need for expensive colour separation and correction. But all depends on the artist's skill in work of this kind. Its effectiveness can be assessed, in advance of printing, only by means of expensive colour proofs, and the more imagination he concentrates on planning his colour separation the less he may have to spare for his picture. Publishers usually prefer to see illustration copy which in all but size resembles the printed picture. When copy for colour printing is drawn or painted in colour, publisher and printer can compare copy, proof, and print at each stage of production.

Comparison between colour print and original subject has limited value. The purpose of much illustration is to introduce the reader to landscapes he has not visited and to paintings he will not see. In most work, when the printed colours are consistent enough with each other, and the effect is convincing, he is bound to be satisfied, even if not all the colours are true. If the publisher were better placed for such comparison, he would have a duty to see to it; since the opportunity is exceptional, he is entitled to neglect it. Colour is now within reach of the least skilled of photographers, and prints or transparencies are all too likely to prove the difficulty of colour balance. They may look unlike their original subjects, but skilled colour separation can usually restore the balance and print a more faithful record of the colours as they seem likely to have appeared. Local alterations to colour, in part of an illustration only, is usually possible. Some cameras can alter the angles of a subject, for instance to rectify the effect of parallax when vertical walls photographed from ground level appear to diverge. Scanning equipment may even be able to distort the proportions of part of the picture, for instance to lengthen a girl's legs but not her torso, but such entertainments are more likely to interest advertising agents than publishers.

Although they share the advantage of comparability, photographic prints in colour are less effective for trichromatic reproduction than colour transparencies. Colours reflected from paper are less intense than those transmitted through a colour film, and trichromatic separation needs all the intensity the copy can provide. A painting fresh from the studio, particularly from that of an artist experienced in the effects of colour printing, is likely to be more brightly coloured than a colour print of the same painting.

For trichromatic reproduction, colour transparencies are the most effective copy. Some techniques of *colour scanning* (§ 13-6) require transparent and flexible copy, which will be mounted on a drum and illuminated by light passing through it. Most transparencies can be enlarged to the size of the biggest page, but enlargement is not unlimited. Some kinds of film, capable of dark-room development, when enlarged to more than five times their size, begin to reveal a grainy effect in the photographic emulsion; some others can be enlarged to as much as ten times their size without ill effect. Few book pages are ten times as big as a colour transparency, but such enlargements may be useful when only part of the transparency is to be reproduced.

§ 13-6 Trichromatic printing

Colour copy which contains more than three colours is usually reproduced by trichromatic processes. *Trichromatic* means three-coloured, but in this context the term implies the mixing of three colours to produce more colours than three. Inks of three specific colours are optically capable of adequate reproduction of most colour copy. More colours would be better for quality, and can in fact be used, but for the standard process the number of printings must be kept to the minimum for economic reasons. The printing of each colour is based on an almost complete duplication of the methods and equipment utilized in monochrome printing. Techniques of colour separation and correction are elaborate. The cost of trichromatic printing in any edition deserves to be approached with caution and handled with care.

The trichromatic ink colours are yellow, cyan (blue-green), and magenta (blue-red). The inks are transparent enough for their absorption ranges to mingle. When the inks are printed solidly over each other, they combine to absorb nearly all wave-lengths of light, and so to produce a tolerable black. Whether printed by letterpress or offset half-tone, by collotype, or by photogravure, each ink is variable in intensity, so that the three absorption ranges leave a variety of wave-lengths to be reflected. The screen angles in processes other than collotype are different for each printing, to prevent the dots from clustering into moiré patterns which distract the eye from the intended image. The positions of the yellow, cyan, and magenta dots in relation to each other — some under, some over, and some beside their neighbours — contribute to the brightness of the result.

Each impression provides a different image. A red area in the copy, for example, will be reproduced by large magenta dots, medium-sized yellow dots, and small cyan dots or none at all. A fourth printing is usual, to intensify the darker tones with some black; even if it is not entirely necessary for one illustration, black is likely to be useful for others on the sheet, and for captions.

Trichromatic printing is an imperfect medium. The inks are more light-absorbent than they should be (figure 13-6); better reflection and brighter colours would necessitate pigments with a tendency to fade. One ink printed over another partly conceals the first, and the lightest colour, yellow, is printed first because its pigment is the most opaque. These imperfections are rectified only by skilled colour re-touching by hand or by electronic scanning equipment.

The inks cannot do justice to any colour with a reflection range narrower than that of one of the inks. Hardly any such colours are seen in nature, though the violet-purple of the cineraria is an example. If two such flowers of this difficult but attractive colour, one slightly more violet than the other, were to be trichromatically reproduced, the result would probably be pleasing enough, but both the printed flowers would appear to be the same colour, and the colour would not be that of the natural flowers. Most colours are a mixture of wave-lengths, spreading across more than a third of the spectrum; exceptions are likely to be found only in the arts.

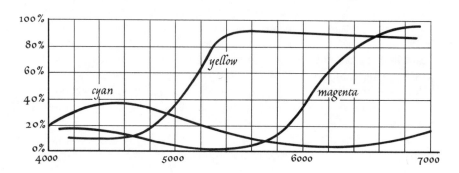

FIGURE 13-6. The reflection range of a set of trichromatic inks, showing a general weakness in cyan and an inadequate violet-blue reflection in magenta.

Conditions for good work are exacting. A four-colour press is best for register, since the dimensional instability of the paper has no time to demonstrate its effect between the first impression and the fourth. The completion of the four-colour image in a single pass through the press enables the operator to adjust all the colours to each other during make-ready, and to maintain the balance between them during printing. But good work can be done on two-colour and even single-colour presses, sometimes at a lower price.

Colour separation – analysis of the colour copy into the trichromatic colours, usually combined with black in the four-colour process – can still be carried out by a process camera equipped with colour filters, and corrected by masking or hand re-touching or both. Masks are positive, transparent, continuous-tone images, photographed from the copy through an appropriate filter, and laid in register over the negative before exposure. This locally reduces the effect of exposure, and minimizes the darkened areas in the negative; this increases dot-size and the intensity of the selected colour. Re-touching may be a matter of painting or drawing on the copy or on a continuous-tone negative or positive before screening, and may also include dot-etching (locally to reduce the semi-opaque vignette round the dot) and dot-intensification (locally to strengthen opacity and perhaps increase dot size). The method therefore includes four separate exposures in the camera, perhaps one or more additional exposures for masking, and extensive handwork. It is capable of the highest quality when the costly element of handwork is fully utilized.

The use of *electronic scanners* for colour separation is growing. The wavelengths reflected from or transmitted through the copy are measured, corrected, translated into impulses, and applied in dot form to film. The information offered to the eye by half-tone dots is conditioned mainly by dot size but also by dot shape and by the position of dots in relation to each other. Scanners offer the possibility of planned variations in shape and position, and hence an enhanced resolution in the image. Four separated positives are usually made at the same time from each item of copy. Positive film is normally used for colour work, to facilitate close register; positive register marks and dots on a transparent background are easier to place in position than when they are negative on an opaque background.

Colour proofs are no less necessary than text proofs. They are expensive, but they are the only means by which publisher and printer can agree exactly about what is to be printed. The proofs are *progressive*; they show each of the four printings first alone, then combined with one or more of the other printings in the intended order of colours, and finally combined with all the others. Designer and publisher do well to concern themselves only with the completed illustration; the rest of the progressive proof is a guide for the printer at each stage.

Perfect faithfulness of colour to all areas of the original may have to be insisted on for special reasons, but to insist unnecessarily increases the general cost of colour separation. That cost is high, and includes an allowance for revision anticipated for such work in general. The less revision there is, the more accessible the expense of colour, and the wider its use. To propose alteration to one colour alone might have unexpected results, as the illustration is brought into being by a balance of mingled colours. If alterations in proof are unavoidable, they are better indicated in layman's terms than as technical instructions, and by leaving the method of correction to the expert.

The printer's task is to match the edition to the colour proof as closely as possible. Here again, inessential precision on the part of publisher or designer can be disproportionately expensive. Ideally, colour proofs should be pulled on the paper to be used by the printer, but in practice this is usually difficult to arrange. Colour printing is a small proportion of the total volume of book production, and few book printers do their own colour separation. When colour separation and offset plates are prepared under the same roof, there may be some gain in quality.

§ 13-7 Paper and printing

When ink is transparent, much of its colour is reflected from the paper surface through the ink, rather than from the surface of the ink itself. The brightest reflections are therefore radiated from glossy white paper; a colour printed on matt or tinted paper is dull by comparison.

When a transparent reflective surface, such as that of a glossy *laminate* (§ 17-4), is laid over a coloured image, reflections of colourless light and hence additional brightness are added to the coloured reflections from below. The colours themselves gain in intensity. This is part of the value of lamination and varnish on jackets and covers. Such materials enhance the brightness of the exterior image of the book, and attract the eye.

In letterpress in particular, evenness of paper surface lends itself to regularity of dot outline, in the printing of colour by half-tone, and to regularity of dot size. For the brightest reproduction by colour offset, glossy coated papers are essential, but the reader's preference for matt papers, which reflect less glare from the light, has caused publishers to demonstrate how well coloured illustrations can be printed on matt coated papers. Coating also provides a level base for adhesion by laminates.

The impermeability of coated paper tends to keep ink on its surface, where it dries by oxidation after printing unless artificially dried on the press, and leaves a higher proportion of pigment on view than when some of it has penetrated into an uncoated paper. The same impermeability also suits varnish, which can remain sticky for years if it has penetrated. Varnish and laminates are commonly used only on jackets and covers, but varnish has also been used to brighten colour illustration in the book.

The characteristics of printing processes are much the same for colour as for monochrome; the individuality of colour printing as a process lies in its multiplicity. Sets of letterpress half-tone blocks for colour printing are expensive to engrave, mount, and make ready: small blocks cost more per square inch than medium-sized blocks: and the total cost of large blocks, charged for by area because of the high proportion of materials and etching time, is almost prohibitive. On a glossy white paper, letterpress is capable of more contrast and sharpness in coloured images than other processes, and indeed may rise at times to an almost harsh blaze of colour. Local adjustments of colour intensity may be effected after colour separation during make-ready, so it is sometimes possible to improve on the colour balance of the engraver's proof. Four-colour letterpress half-tone wanes in effect on a matt paper.

Photogravure is also expensive in the preparatory stages, because of the cost of etching cylinders. The process is out of favour with book printers, because it is not well suited to the reproduction of type or of line illustration.

But high initial costs can be afforded for long runs, and photogravure is capable of excellent colour work. The deeper tones are rich and lighter tones delicate. Variation of screen angle, which tends to generate dot-patterns, is not always necessary, and screens may be extremely fine. When the cell walls are overlapped by ink, they become all but invisible, so that the print gives an illusion of continuous tone. Matt paper does not detract from the printing quality.

Photolithography, now predominant for black printing in most forms of book production, is the most widely used process for colour printing in books. The number of copies in the average edition is only a few thousand, and offset plates cost less to make than letterpress blocks or photogravure cylinders. The area of an illustration has a minimal effect on its cost. Owing to the thinness of the ink film, colours are soft rather than intense. Very light tones are sometimes a little uneven, because of the difficulty of maintaining the smallest dots against the effects of the damping system.

Screen printing, briefly described in § 11-8, is particularly apt for opaque and even luminous colour.

CHAPTER 14

Illustration

Iɴ much of his typographic work, a book designer can rely on traditions evolved in centuries of printing history, and on well established customs of his own time. History and fashion provide less support in the planning and presentation of book illustration. Book production of a hundred years ago, before the general use of cameras in printing, was capable of much that was memorable, but the purposes and methods of illustration were more limited and formal in those days than they are now. Today's pictures look different, describe different subjects, and appear in greater variety than in the past. Plans for the arrangement of a few incidental illustrations may still be based on familiar conventions; but precedent is not the best guide when pictures are many and may be the dominant theme of the book.

As a conspicuous element of the book's appearance, illustration deserves the designer's skill. Knowledge of reproduction methods, and ability to adapt and arrange pictures for presentation in book form, are essential to success. But much illustration is textual in purpose rather than decorative, and such decisions as the choice of an illustrator or of illustrations may be reserved by author or editor. Those initial plans are still likely to benefit from any influence the designer is permitted to exert over them, and from his technical knowledge and visual imagination. His own part in the handling of illustrations is likely to provide scope for his ability.

§ 14-1 Commissioned illustrations

Every item in the book gains in appeal to the reader's eye from its relationship with all the other items. Something of a family resemblance, an appearance of being a set of pictures rather than a collection from disparate sets, may confer this advantage on the illustrations of any edition. In diagrams an even size of lettering and thickness of line after reduction is to be preferred for this reason. Similarities of visual texture, of dimensions, and of proportions may demonstrate the connection between text pages and illustrations; contrast of texture and size between illustration and type may establish a different kind of relationship, and emphasize the link between the pictures. When illustrations are to be commissioned for an edition, these possibilities are more easily exploited than when existing work is to be selected from a variety of sources.

Drawing and painting for reproduction in books is a special skill, not always possessed even by experienced artists. The upright or *portrait* shape of the usual page is awkward for *landscape* subjects. The pictures after reproduction are smaller than they would be if planned for other forms of presentation. A set of illustrations should be drawn for one scale of reduction only, however much detail there may be in one and however little in another. The effect on the picture of reduction and perhaps of a half-tone screen is not easily visualized while the artist is bringing his picture into existence. To an artist familiar with the appearance of his work in print, the special requirements of book illustration may present a challenge, but a talent new to book production is likely to need development.

An *autographic* picture is one that is printed from itself; the picture drawn or cut by the artist forms the printing surface. Wood-cuts and wood-engravings are one example, and may be printed with the type, at least in a limited edition; but wood wears out before type metal. An artist may be able to draw with special ink on special paper for pressure transfer on to an offset plate, or even draw on the offset plate itself. Autographic reproduction has advantages over other methods, in its fidelity to the intention of the artist, and in its distinctive appearance caused by the nature of the tools and surface of the medium. The weakness of most autographic methods is the short life of the printing surface, and therefore the limited usefulness of the artist's creation; any material hard enough to withstand a long run is likely to be too hard for the artist to work on.

Commissioned photographs can be made to order to much the same extent as other kinds of commissioned illustration. The photographer must go about his work in his own way, but once aware of the requirements of his set of illustrations he can adapt his methods to them. While he works on each subject, he will then be influenced by any specific plans for angle, distance, depth of focus, and lighting. If he develops his own monochrome film, he may be able to adjust tones locally within the picture, for instance to increase contrast and to regulate emphasis. In any circumstances, he should be able to provide a set of prints well suited to reproduction, even in size, and upright in shape if required, with a glossy surface and a wide range of contrast.

Diagrams usually add little to the visual appeal of a book, but much to its clarity. Editors and authors tend to think in terms of words rather than of pictures, and are slow to notice such possibilities as the expression of a statistical paragraph, encrusted with figures, in the form of a table or graph. If a qualified artist can be found, sectional drawings of machinery and other subjects, exposing the viscera, provide an example of diagrammatic illustration capable of a depth of explanation almost beyond the written word.

Authors of explanatory texts usually provide copy for their own diagrams, if only in order to explain what they need. When the author's contract stipulates that he is to supply illustration copy, he may choose to draw it himself or to get it drawn by somebody else whose main qualification for the task is that he will make no charge for it, or next to none. The resulting material may be clear enough to explain its meaning but incapable of adequate reproduction or too irregular in drawing to appear in a well-produced book. The only remedy is to have the illustrations redrawn by an expert, but whether this can be afforded is likely to be for the publisher to decide; prudent publishers know they cannot afford to make a habit of publishing conspicuously inept illustrations for all to see. Some diagrams may need nothing more than local improvement or repair, which should not require the attention of a specialist; a book designer will gain from developing his own ability to handle pen, brush, and other instruments with confidence and accuracy.

In spite of photographic developments in printing, line illustration is still popular. It is inexpensive, similar in texture to the typographic page, and usually easy to print. But often it is so little more than a stark assembly of outlines that its acceptability to the public is surprising. Illustrators of the nineteenth century, whose work once drawn had to be cut in wood by journeymen engravers, filled their scenes with hatched and shaded tones to suggest shape, movement, and depth. But those book illustrators have been overtaken by the camera, and today there are hardly enough commissions to provide a career even for the most talented. Pencil drawings, black crayon on a rough-surfaced paper, pen-and-wash can all be reproduced faithfully enough by today's process cameras, but the versatility of the available reproduction techniques is all too rarely tested or exploited.

Continuous tones which are to be reproduced by offset or letterpress have to be analysed by a half-tone screen; pencil and wash drawings are examples. The effect on such work of an exposure suitable for photographs may be to reduce contrast between shadow and highlight by laying a dot pattern over the whole area. A *drop-out exposure* is to be preferred, in which the dot pattern is dropped out of the white background of the artist's image. But if the image is not very black or the paper not very white, this kind of exposure may weaken the image itself.

Line diagrams and other commissioned illustrations are sometimes drawn very large, for drastic reduction, perhaps to make detail easier to draw, or because the drawings will be saleable after printing. An artist's work reduced to less than half its size presents an appearance he may not have been able to imagine and intend at the time of drawing; lettering for instance may become illegible. Whenever possible illustrations should be drawn for reduction to something more like 75 per cent of the linear dimensions, and if a standard reduction can be applied to a whole set of illustrations they will look like a set when they are printed.

§ 14-2 Prints and photographs

Illustrations selected from existing material are usually reproduced either from photographs or from prints of some other kind. A photograph ready for the camera is a *bromide print*; a *print* is an illustration already printed by any means.

Printed line drawings can usually be reproduced accurately enough for practical use, though any thickening or other imperfection in the original printing will be exaggerated in its reproduction. Printed half-tones must either pass through a half-tone screen a second time, or must be reproduced dot-for-dot. A second screening reduces contrast and obscures detail, and the two screens, whatever their angles, combine to interpose an obtrusive pattern of dots between image and reader. *Dot-for-dot* is simply line reproduction of an image already analysed into half-tone; the method is ineffective when the original screen was a fine one, and even at its best has a tendency to alter the tonal balance.

Prints by older processes usually simulate tone in a way that deserves careful and appropriate reproduction. Steel engravings are an example; if the ink is still black and the paper white, and if a slight reduction will be enough, they can be faithfully reproduced without a screen, though the most delicate lines may break or disappear. Otherwise, drop-out half-tone will be better, if enough contrast survives in the faded ink and discoloured paper of some old engravings.

Most illustration copy today is provided in the form of bromide prints, which sometimes have been used for reproduction before, and are not in the best of condition. When such inferior material has to be used, making the best use of it is part of the designer's skill, and the worst blemishes will have to be repaired with an air-brush. A good bromide print is newly made from a negative with strong contrasts, and developed on the whitest of paper with a glossy reflective surface; a paler image with a matt surface, watered-down by the reproduction process, is likely to look washed-out when printed.

Camera and studio instructions are still marked on the back of photographs, which is just about the worst place for them. Almost any mark on the back has some effect on the front, visible when the surface is turned at an angle to the light, and capable of being recorded by the process camera. Now that so many illustrations are to be printed by offset, most pictures can be shown in proof with at least a narrow extra margin surrounding them, and with picture identification and instructions in that margin. They are better not marked on the photograph itself, but should appear on a mount of waste card to which the photograph is lightly fixed and from which it can be detached. The advantage in working on a set of illustration proofs with the designer's instructions, identification numbers, and trim marks surrounding each is well worth the trouble of mounting. Aqueous adhesives should not be used, as they cause the photograph to cockle. If areas within the image need instructions or marks, they can be written or drawn on a transparent overlay, stuck to the back of the mount, folded over the photograph, and positioned with register marks at the corners. If the overlay has to be marked when in position over the photograph, a felt pen lightly used is unlikely to indent the vulnerable surface which is to be reproduced; but pen and overlay must first be tested to make sure no ink passes through the overlay on to the copy. The same treatment is likely to be convenient for other kinds of illustration, but is most often needed for photographs because of their narrow glossy margins.

The prime requirement of illustration is that it should illustrate; other qualities are secondary. Some of the pictures for an edition, or indeed all, may be the wrong size or shape or style for visually appealing illustration; photographs in particular may be lacking in contrast, inaccurately aimed at the subject, distorted by focus, movement, slant, and parallax, and ill-matched as a set. Whatever their faults, the illustrations are likely to be the best that can be found, and the designer must do his utmost for them, as does an editor for an ill-constructed sentence in the text.

Many illustrations are informative rather than decorative; photographs taken for scientific publications tend to offer little to the uninformed eye, but few illustrations deserve more care in arrangement and reproduction, and indeed electron photomicrographs, enlarging a cell to the scale of a landscape, have an eery beauty of their own. There will be other editions in which the illustrations will be expected not only to inform the reader but to attract his eye, satisfy his perception, and lodge in his memory. For such editions, the designer will do well to have ready in his mind the foundations of good quality in book illustration, in order to convince illustrators who may be able to provide the best, once they know how.

§ 14-3 Size and shape

If designer and illustrator can agree on the dimensions of the pictures before work begins, both the typography of the book and its illustrations are the more likely to succeed. Unevenly-sized chapter-head drawings, for instance, may cause the chapter titles to be uneven in height on the page; originals of the wrong proportions may not fit comfortably into the book. Small tail-pieces are best not drawn until the book is in page proof; only then can the designer be sure how much room remains at which chapter-end.

For the presentation of a set of illustrations, the pages of an edition provide a series of mounts of restricted size and uniform proportions. Within these limits, the size and shape of pictures can be adapted to the subject of each. The width of a battle-field, for instance, may be stretched across the upper part of two facing pages above a residue of text; a single soldier, standing with his rifle on his shoulder, fits into a single upright page.

The maximum area for any illustration in a book, larger than the page itself, is offered by a *folding plate* (§ 14-6). This is printed separately from the rest of the book, because of its finished size: folded into a size a little smaller than that of the cut pages, because unlike them it is not to be cut at the folded edges: and preferably *hooked round* the spine of a section (§ 16-3), because sewing it into the book secures it against detachment when unfolded and re-folded in use.

The paper for such a plate needs to be tough enough to withstand such handling, but as thin as possible; when the book is pressed during binding, the folded edges of the plate, however thin, indent the neighbouring pages. The plate is placed in the binding by hand, and this like all handwork increases the cost of binding. Height and width should be such that when the plate is folded, all its edges lie within the text margin; the folded edges of a plate smaller than this would indent the text area. If the unfolded plate is too big, detail at the far edges will be uncomfortably distant from the reader. Twice the page height and three times the page width should usually be enough.

When some subject such as a map is referred to on several pages of text, the folding plate should either precede or follow those pages, and lie face up when it and they are open, to be seen together. That part of the open plate which lies within the closed book is best left blank, as it cannot be seen when the text pages are turned. The plate unfolds upwards and outwards past the head and fore-edge, away from the reader and from the spine of the book.

Bigger illustrations still may have to be folded and slipped into an *endpaper pocket* (§ 16-3), but they are certain to be awkward in use, and likely to be mislaid because they are not fixed to the book. The *endpapers* of the book (§ 16-3) are sometimes used for maps and other illustrations, but nothing essential should be printed there as the endpapers are removed when librarians have the book re-bound. A folding plate is a more convenient presentation for an illustration larger than the text page, but it may be prohibitively expensive.

The next biggest area for a single illustration is a pair of facing pages, known as an *opening*. Only the centre pages in any *section* (§ 16-2) are printed on the flat sheet side by side with verso left and recto right, and these two pages combine to make a *spread*. In all other openings within a section, facing pages are folded into position opposite each other, and the two outer pages of each section are *gathered* into position opposite the outer pages of the neighbouring sections (§ 16-4). Register in both folding and gathering is imprecise, and edges and other obvious horizontal features in any illustration are likely to be slightly out of position after sewing, when they run across the spine of any opening other than a spread. The pages of a spread, on the other hand, are printed beside each other, and while part of a continuous two-page illustration printed on them may disappear into the *backing* (§ 16-6), the horizontal alignment of the illustration cannot be interrupted.

After a two-page illustration across a spread, the biggest available area is a single page, and this is more often used than spreads or folding plates. Given the maximum space available for an illustration, the designer has to decide how much of that space to use. An illustration which occupies the entire area of a page is said to *bleed all round*; its printed area extends beyond the cut edges at head, tail, and fore-edge, and up to the folded edge in the spine. Photographers and other artists, authors, and editors often prefer this maximum size for full-page illustrations, probably because most pages are by no means ample for pictures. Bleeds discolour the cut edges of the book, and the mottled effect of this may encourage the buyer with its promise of illustrations within. The bleeding edge of the illustration has to extend beyond the cut edge of the page, to be cut away by the binder, in case it falls short of the edge after cutting. When the dimensions of a sheet with bleeding illustrations approaches the maximum capacity of the press, the illustrations are likely to encroach beyond the press's maximum printed area, and to leave too little room for the grippers which hold the sheet in position round the cylinder. This space is otherwise provided by conventional margins, which have other uses during presswork; the channels of unprinted paper, between the pages printed on the sheet, may come into contact with controlling wheels or straps which hold the newly printed sheet steady during delivery. If these touch the fresh ink of bleeds instead of the clean paper of margins, the presswork may be marred.

Bleeding all round may deprive the page of space for caption, illustration number, and page number. If two such pages face each other, there may be no room for captions to appear in the same opening as their illustrations, and then picture and caption cannot be seen at the same time. When a succession of illustrations is to be presented with bleeds all round, the captions can be printed as a folding plate, to lie open beside the book while the illustrations are turned over. Otherwise the reader will have to turn the pages to and fro, in search of captions hidden in the text pages.

Some of the disadvantages of bleeding all round are avoided by bleeding at one or two edges of the page instead of three, and by arranging for the inner edge of the illustration to fall short of the spine fold. If at the foot of the page, for example, the edge of an illustration aligns with the last line of text on a full text page, there will be room in the tail margin for the caption. If illustrations facing each other across an opening align with the inner text margins, they will not appear to mingle with each other and so to form a single illustration.

Since a bleed means extension of the illustration beyond the cut edge, part of the printed illustration will not appear in the book. If the book is re-bound for library or other purposes, the edges of the illustration will be further cropped. The area of illustration remaining after such cutting needs to be considered before bleeds are planned, to make sure no detail of importance will fail to appear in all copies of the edition. If the illustration includes a straight line very close to and parallel with a cut edge, there is a risk that in some copies of the edition at least the line will either appear to be slightly out of parallel, or disappear altogether.

Whether illustrations should be presented within a frame of white paper, or in the extra size provided by bleeding, may be decided edition by edition, even within a series, and not necessarily by the designer alone. If there are to be margins at all, the narrowest should not be less than 6 millimetres or $\frac{1}{4}$ inch wide. This is partly to withdraw the straight edges of any illustration from any parallel cut edge, in order to disguise any slight deviation from the parallel between edge and picture, and partly to preserve the whole picture even after re-binding. The use of standard margins throughout an edition emphasizes the relationship between illustrations as separate items in a single set. When in an edition full-page illustrations vary in depth, a standard position for the last line of the caption, perhaps aligned with the last line of text on a full page, may remind the reader of the standard margins, even at the expense of varying distances between caption and picture. On the whole, varying spaces within a standard area of illustration are less disconcerting than varying margins, and variations of space at the head of the page are more conspicuous than variations at the foot. In the same way, extra space at the head of a text page usually means a new chapter, and the first text line of succeeding pages should be occupied by something, whether text or illustration. Whenever possible the last line too should be occupied, in case a short recto page appears to signal the end of a chapter. When picture and a long caption are to share the area allocated to illustration, the dimensions of the picture may depend on those of the caption, and for some editions the easiest way to determine illustration size will be to have such captions set and shown in galley proof first.

The shape of a well-presented illustration is influenced by its subject and by its medium. A man standing in open fields is a natural subject for a portrait shape of illustration which excludes most of his surroundings unless they have some significance and should be seen in the book. On an upright page, the obvious size for an upright illustration is full-page; a smaller size will fit well on to the page if two or more such pictures can be arranged into a group. A line of seated people tends to form a landscape picture, and to be printed as a half-page illustration on the same page as another illustration of much the same shape or over part of a page of text. The group may be editorially more important than an individual shown on a full page, but conventions of neatness and regularity in book production permit the designer to make a minor illustration from a major subject. Observance of a higher convention, allocating size to illustrations in proportion to editorial importance, may have to be preferred even when the result does not fit neatly into a pre-conceived illustration plan.

For the pages of books, oblong shapes have been preferred to squares for centuries, since before the European invention of typography. In illustration, the eye from habit prefers oblongs, whether portrait or landscape, to squares, which look squat. The inelegance of square illustrations is disguised to some extent when they are grouped on a page, with other squares or oblongs or both, but an inelegant shape has to be tolerated when it fits neatly into the available space.

Regularity of shape is emphatic when some kind of rectangular frame is drawn round the picture, or in a squared-up half-tone when the dot pattern, continuous along the edges, defines the illustration's area. Printers favour the rectangular because it lends itself to make-up in a rectangular page. Any illustration which diverges slightly from the rectangular, as many photographs do, is likely to be squared-up in the course of reproduction, unless the designer gives specific instructions to follow the outline as it is.

A succession of squared-up illustrations may be pleasantly interrupted by an occasional *cut-out half-tone*, from which the background has been removed. The original picture may lend itself to this treatment, when the background is white or nearly so. Otherwise handwork and some extra expense will be involved. The uneven shape of cut-out half-tones lends them emphasis, and unless they share some similarities of scale and outline they seem to compete with each other instead of resting together on the page.

Drawings seem to need no particular outline or shape, and unless they present the appearance of rectangles their exact size and their alignments may be handled more freely than those of squared-up illustrations. But the vertical axis of the drawing on the page may not always be clear from the original drawing, and even when it is, it may need adjustment in proof. Sets of squared-up illustrations may be placed in position by the printer in accordance with the designer's plan; any drawing with an irregular outline needs to be placed exactly by the designer, as a visually satisfactory position can be identified only by eye.

§ 14-4 Scale and trim

The *scale* of an illustration may be defined as the ratio between the linear dimensions of the original after trimming, and those of the printed illustration. The *trim* is that part of the original which is selected for omission from reproduction. When an illustration is to be trimmed, scale and trim depend upon each other, and are normally planned together.

Any illustration is reproduced either *same size* (the same size as the original, before trimming), or in enlargement, or in reduction. Scale is best calculated and expressed in percentages of linear dimensions, with 100 representing same size; the percentage marked on the copy is that to which it is to be reduced or enlarged. An electronic calculator is useful for millimetre scaling and sizing, and there are also special slide-rules with millimetre or inch measurements.

Unless different illustrations for one edition have been drawn for different scales, the same scale should whenever possible apply throughout the book to the commissioned drawings of one artist. Uneven scale causes uneven thickness of line from page to page, and the illustrations may then appear to be an ill-matched set. Not all artists plan differential reductions within a set of drawings, or make allowances for scale while they draw.

The original scales of scientific and archaeological illustrations and of maps are often stated on the illustrations or in their captions. These scales can be falsified by enlargement or reduction, and may have to be amended. If the designer does not see to this, he must refer the matter to editor or author, with a reminder of each original scale and a note of each of his own scales. Otherwise the reader may be mis-led.

Enlargement is usually necessary for colour transparencies; Kodachrome, for example, is made in 35-millimetre size only (just over $1\frac{3}{8}$ inch). In monochrome work, the comparatively small size of most book pages and the larger sizes of most bromide prints indicate reduction more often than enlargement. Enlargement tends to emphasize any faults in the original, and to deprive line drawings of some of their sharpness of line and detail.

Severe reduction can transform line drawings into delicate miniatures, and any reduction tends to conceal unevenness of line in the original. But the effect of reduction on detail may also impose a kind of illegibility on the illustration. Lettering on diagrams may become too small to read without a magnifying glass: fine lines in the original may be too thin to reduce more than slightly, or may be impossible to reproduce continuously: and the diminished interstices in cross-hatching may darken shaded areas. When a map or other diagram includes a dry-transfer tint, the pattern may become too fine for clear printing after reduction, and will tend to close up. White lines on a black background may be reduced to such straits that they fill with ink. Half-tone screens blur fine detail, and when the detail has also been reduced this effect is proportionately increased.

When illustrations exist, or have been planned, before the design of the book is begun, the proportions of the illustrations and their capacity for scale may well influence the size and proportions of the pages on which they are to be printed. When page size and margins have been planned, even provisionally, before the illustrations, the size and proportions of the latter can be designed to suit the former. When text illustrations are many, some cost can be saved by commissioning line drawings in such a size that they can be pasted up for exposure with the text in which they are to appear, if the text is to go before the camera. Otherwise the drawings are separately exposed and assembled in film with the text.

Many photographs benefit visually from being trimmed for use as book illustration, but some cannot properly be trimmed at all without the agreement of author and editor. Without understanding a medical text, for example, no designer should confidently trim X-rays or photomicrographs. Vital details may appear inconspicuously near the edge of any illustration. A photograph of a work of art should usually be trimmed only to the inner edge of its frame and should never bleed; some galleries and museums which hold copyright in paintings insist on this as a condition of reproduction, and anyway to amputate part of a carefully planned background is an injustice to the artist. With the permission of the copyright holder, and probably after his approval of the proposed reproduction in proof, part of a work of art may be reproduced, so long as the caption indicates that the illustration is 'detail' only and not the whole picture. When other kinds of photograph are to be reproduced, and the photographer is named anywhere in the edition, permission to trim should be requested.

Photographs are trimmed for one or more of a variety of reasons. The photograph is not always rectangular, and if the designer does not make it so the printer probably will. The subject of the picture is not always square on the photograph; naturally, it is not always intended to be, but an unintentionally tilted picture may have to be straightened up, by trimming to new edges. When camera parallax causes vertical features to diverge or converge in the photograph, the designer may have to choose a datum vertical, preferably near the centre of the scene, with which he can align the sides of the illustration. The effect of parallax can be corrected by a printer equipped with an enlarger in which the original can be exposed on the slant, and this is certainly worth doing with colour transparencies.

More often, a photograph is trimmed in order to adjust its proportions or to concentrate its subject. By trimming away the upper and lower parts of a portrait-shaped illustration, the designer may be able to reproduce it in the proportions of a landscape. Often there is no need to reproduce the whole of a photograph; the irrelevant parts can be trimmed away from around the essentials, so that the space available for illustration can be filled with whatever the reader needs to see, in the largest possible size.

§ 14-5 Position

The position of illustrations — in relation to each other, to the text, to the page, and to the rest of the book — is regulated by the designer in every illustrated opening of a meticulously designed edition. Illustrations are a conspicuous part of the book, and their arrangement is a conspicuous part of design. The position of each, in relation to any passage in the text which refers to it, is a sub-editorial matter which is best decided in terms of book design.

The opening is the standard visual unit of the book. The appearance of an illustrated opening benefits from as much vertical and horizontal alignment as possible between the various illustrations and other elements printed on it. Illustrations can be sized to align at one edge at least with each other and with the text area, in an arrangement reflecting an appearance of order and purpose. At the other extreme, drawings can be scattered across page or opening, preferably in a pattern that indicates their relationship with each other. Illustrations of irregular outline need to be in balance rather than in alignment.

A conspicuously rectangular illustration, such as a squared-up half-tone, is usually placed with one or more of its edges in exact alignment with other elements of the opening. The hard straight edges then seem to have been arranged with intent, parallel with or in extension of similar edges of illustration or text. For this reason, squared-up part-page pictures are often reproduced in a width which matches the text measure, so that they can be neatly fitted into it.

If half-tones, or indeed any other kind of illustration, are too close to each other or to text, the reader may have difficulty in identifying and appreciating them as separate items. They may even appear to blend into a single picture, or distract the reader from the text. Every picture is improved by a frame of white space, even a narrow one; and when text and illustration approach within a pica of each other, technical problems of page assembly may begin to appear.

When half-tones or solids are placed back-to-back on the same leaf, there is a risk that each will show through the paper to confuse the other. The printer may also be cautious with his inking, in case too thick a film of ink causes set-off on the printed sheet. Whatever the process, the best way to print illustrations will often be separately from the text and unbacked; but

by Brendan Behan

Borstal Boy
Brendan Behan's Island –
with Paul Hogarth
Hold Your Hour and Have Another –
with decorations by Beatrice Behan
The Scarperer

PLAYS
The Quare Fellow
The Hostage

by Paul Hogarth

Defiant People: Drawings of Greece Today
Looking at China
The Face of Europe
People Like Us
Brendan Behan's Island –
with Brendan Behan
Creative Pencil Drawing

BRENDAN BEHAN'S NEW YORK

by
BRENDAN BEHAN

with drawings by
PAUL HOGARTH

PUBLISHED BY
BERNARD GEIS ASSOCIATES
DISTRIBUTED BY RANDOM HOUSE

Figure 14-5. Illustration and display typography have been arranged to balance a

TO AMERICA
MY NEW-FOUND LAND.
THE MAN THAT HATES YOU
HATES THE HUMAN RACE.

The F.B.I.
comes to Town

I am the child of the King of Greece's son
    who married the King of Ireland's daughter,
    and travelled west.
The Red Sea opened for us. We were quakers before God,
    but quaked before no earthly King.
We are people of the Bible,
    Old Testament,
    New Testament,
    or no Testament.
We carried Los Angeles with us,
    and Luther and Francis of Assisi
    and Robert Ingersoll and Tom Paine.
We are white, coffee coloured, black and beautiful bronze.
We are as painful as human life and as exciting.
We are men.
Birds, we are told at school, where the lessons are
    obviously slanted by crypto-bird teachers,
    are very clever and build nests.
But how many birds would it take
    to build the Empire State Building?

'New York City is hell,' said an old Midwest lady who was living
with me at the Algonquin Hotel. She had a room on the same floor
and used the lift with us, that is.

times to reflect each other. Reduced to 45% from a large octavo format.

this separates the illustrations from those passages of the text which refer to them, introduces blank pages into the book, and increases the cost of illustration. When illustrations are printed as sections of *plates* (§ 14-6), they are usually backed, and when they are printed in the text, a textually appropriate position is likely to be preferred, whether illustrations back each other or not.

When an illustration shares a page with text, the reader may prefer it to appear above or below the printed lines, which are then uninterrupted. If text is to be divided by an illustration in mid-page, the act of reading will be favoured by placing the illustration between two paragraphs, rather than in mid-sentence within a paragraph. This may seem a little untidy when the picture has straight horizontal edges, which look neater next to full-measure lines. The division of a paragraph into two parts on one page, above and below an illustration, is now sanctioned by custom, whether the reader likes it or not, but to divide a word immediately before and after an illustration is clumsy.

Each text illustration should appear whenever possible in the same opening as the first reference in the text which identifies it. If the editor has not already traced and marked such references in the typescript, the designer should do so; if the printer is left to see to it, he should not be compelled to amend his make-up without charge, as a penalty for having accepted editorial responsibility. Any planning decision of this kind, which depends on reading the whole typescript, should always be assigned to author and editor; the designer who reads a typescript thoroughly is mis-using his time, unless he is also acting as editor. Such decisions cannot however be final, because the make-up of typographic pages is almost certain to vary from that of typescript folios. When a typographic reference to a half-page illustration appears in the lower half of a recto page, the illustration will have to be placed in the next opening, unless by means of re-composition earlier in the chapter the reference can be raised to the upper part of the recto or transferred to the verso. Text references to illustration as appearing *above* or *below* are likely to need amendment after make-up; references to illustration numbers are not. *Above* and *below* in books have come to mean *further back* and *further on*, but such conventions become unfamiliar with decreasing use.

Pictures should never be allowed to interfere with legibility, by coming too close to the text or by encroaching upon it, even in a different colour, however pale. *Run-round* illustrations, printed beside the text and within or partly within the text area, require a local abbreviation of the measure. This interruption of reading rhythm has become unpopular, perhaps for economic reasons, while in metal type it requires local re-setting during make-up; even when in film the text can be rearranged locally without significant expense, running round will be better avoided. Marginal illustration, usually placed in the outer margin, and sometimes extending into space left open within the text area, tends to reduce the measure throughout the book, and drastically to increase the unprinted area; a computer could be programmed to make up different pages or openings in different measures, to adapt them to the presence or absence of marginal illustrations, but readers who prefer the main text of any book to be set in a single measure might not care for the result. Placing illustrations in the margins, however, has the advantage that throughout the book a number of small or at least narrow pictures can be placed beside the actual reference to each, without extending the text or interrupting the reader.

Some of the traditions of book production are deeply rooted and widely spread; they are too well-established to ignore, and they are conspicuous enough to be familiar to anybody concerned with books. One example is the preference for placing certain features on a recto page; recto is major, verso is minor — the emphasis starts with a major recto announcement on almost any title-page, and continues with the minor information on its verso. Accordingly recto pages are often preferred for full-page illustrations when they are few; and if they are also unbacked, the blank verso which follows each is accepted as an inconspicuous result. Even one blank recto within the printed part of the book may appear clumsy, as though designer or printer had miscalculated. The first blank verso, on the other hand, is likely to be seen among the preliminary pages, and subsequent blanks in the text cause no surprise. The editor of a symposium may reward his contributors by providing *offprints* (§ 16-10), the production of which is facilitated by starting each new contribution on a recto page, so that some end with a blank verso. Otherwise, chapter headings which start a new page can be either recto or verso, since unnecessary blanks in the text might appear wasteful. In the industrially produced book, a reasonably compact and economical arrangement throughout is generally preferred by readers, and blank pages as well as unusually wide margins may be supposed to contribute nothing except some addition to the price.

When a table or an illustration cannot be fitted into the upright proportions of the page, it may have to be turned on its side. This uncomfortable treatment is better avoided whenever possible, particularly if the book is to be large or heavy. When it cannot be avoided, turned illustrations and tables should have their feet to the right, and should if possible be printed on the recto. The reader then will need to turn the book in one direction only to view all the turned items, and after turning the book will have the recto closer to him than the verso. All the illustrations on one page should invariably be the same way up; and to turn one illustration in an opening but not another looks awkward, though it cannot always be avoided.

The convention that turned illustrations should have their feet to the right is not always observed when diagrams and other lettered illustrations are drawn. As a rule, lettering on any illustration should start from the top and read downwards, unless it is horizontal. Any lettering which reads upwards on the illustration will be printed upside down if the illustration is turned. Graph lettering conventionally reads outwards from the illustrated axis, and may have to read upwards in the drawing even though it may be upside down in reproduction.

Most illustrations provide internal evidence to show which way up they are to appear, and there is no need to emphasize the obvious. But photographs taken from aircraft or otherwise from above, X-rays, photo-micrographs, and some paintings of the twentieth century, may look much the same to a printer whichever way up they are. The designer should unambiguously mark these, on original and on all proofs, 'THIS way up'; there may still be room for a mistake between proof and printing, without such a reminder. Air photographs taken in the northern hemisphere should be printed with the north at the foot, so that shadows fall towards the reader. If the shadows fall away from the reader, the relief may appear to be reversed, so that valleys look like hills.

Illustration, as distinct from decoration, should not as a rule be printed on endpapers. Without significant extra cost, endpapers can be made of tinted paper and printed in a colour other than black, and they are always easy for the reader to find. They offer a tempting position for a decorative map, where it will undoubtedly improve the appearance of the book. But they may have to be removed during re-binding, and then the library reader will not be able to find them at all. If the design printed on them is textually inessential, they are not strictly illustrations.

§ 14-6 Plates and frontispiece

Illustrations printed or gathered separately from the text are known as *plates*. A folding plate could be printed on part of one of the text sheets, but it would still have to be separately gathered. More often, plates are printed separately because they differ from the text pages in printing process, in colour, in paper, in inking, or in size. The *frontispiece* is not necessarily a plate. By definition, it appears in the preliminary pages; by custom, it usually faces the title-page.

The separate printing of plates is a matter partly of economy and partly of quality. Whatever the process, illustrations of the highest quality must often be printed separately from the text. Paper and presswork which best suit the text may be unsuitable for illustrations which require a more extensive film of ink on each page, and at the same time a more precise outline for an image composed of much smaller elements such as half-tone dots. Paper colour may justify separate printing; creamy paper was at one time almost obligatory for text, and may become equally popular again, but white paper will always be best for half-tone illustration. The economic advantage of plates is most obvious when the illustrations require methods or materials or both which would be too expensive for the text. For example, eight pages of four-colour illustrations would usually be printed separately from a monochrome text of average length, unless all could be printed on one side of one text sheet, in which event they would be rather awkwardly combined with the text of a single section.

The disadvantage of plates is a matter of economy and quality of a different kind, and also a matter of position. Even text illustrations cannot always be placed in positions most convenient to the reader; plates can hardly ever be placed in those positions, for reasons discussed in § 16-3. Readers have come to accept the textually random position of plates, but it is still a fault of industrial book production, now often avoided at the expense of various qualities of both illustration and text. A long text, for instance, is now sometimes printed on a stiff, opaque, slightly reflective, very white paper, in order to extract the best result from text illustrations in textually apt positions, although the book may be uncomfortable to hold and read. The cost of binding plates into the book depends on their arrangement; the most expensive binding method is to place single plates opposite or very near the appropriate passage of text, and the least costly method is to present groups of plates as separate sections for binding like the text sections. These two methods are extremes in terms of handwork; the former requires a maximum, the latter a minimum, and other methods are outlined in § 16-3. The quality factor in these methods is examined there, and is mainly a matter of the durability of the plate's attachment to the rest of the book.

A frontispiece should be chosen purposefully; if it is to face the title-page, it should be a picture suitable in every way to be seen beside the title of the book. The object of a frontispiece which faces the title is to enhance the title-opening, and the picture is therefore best arranged as the left-hand part of a single design. If the picture stands upright on the page, it will be more closely related to the title than if it lies on its side. The title refers to the whole text of the book, and the subject of a suitable frontispiece may well do the same, unless its intention is simply decorative. A really striking frontispiece may appear to unbalance the title-opening, and to deprive the title itself of its proper emphasis; colour pictures are however often used, and indeed when there is only one colour plate in the book this is the most conspicuous place for it. A line drawing harmonizes better than a half-tone with the title-page, and provides an apt frontispiece for a book which contains line and half-tone illustration. Half-tones are however more often used for this purpose, as though to advertise the presence of other half-tones later in the book.

The reader will usually see frontispiece and title as a single opening, and the two should be designed together as such. When the rest of the book obeys the centred convention of bookwork typography, illustration in this opening is normally restricted to the left page and type to the right, apart from the frontispiece caption. When this convention is abandoned, and when the frontispiece is not a plate, illustration and type may effectively share the opening, one or both spreading from its own to the opposite page, preferably in the general style of illustration arrangement later in the book. Folding accuracy must not, however, be relied on to maintain conspicuous horizontal alignments between the two pages; the title-opening may not be a spread unless the preliminary pages are specially imposed to make it so, for instance as the fourth and fifth pages of an eight-page section. A frontispiece plate is usually verso and unbacked, so that it offers the reader a blank recto, and perhaps a blank opening, almost as soon as he opens the book.

Unbacked plates in any book tend to be comparatively few, and will be seen at their best if each appears on a recto page, in front of a blank verso. When the plates are separate from each other, *tipped-in* as single leaves (§ 16-3), such plates can be printed on one side of the sheet; when they are to be *wrapped round* sections (§ 16-3), they will have to be printed on both sides of the sheet in order to appear on two recto pages out of four.

Perhaps illustrations printed separately as plates will always provide the best available quality in terms of reproduction. In terms of economy and of the reader's convenience, illustrations printed with the text are preferable, and now tend to be generally preferred for monochrome work.

§ 14-7 Typography, lettering, handwriting

Explanatory letters, figures, signs, and words which form part of maps, diagrams, and other illustrations deserve no less care and skill than the illustrations themselves. Whatever the method, allowance must be made for the probable scale of reduction from camera copy to printed image; drastic reduction may leave drawn lines clearly visible while rendering explanatory words illegible. Any contact, or even too close a proximity, between explanatory characters and any part of the illustration tends to induce a degree of illegibility; a local white background for wording, or bold type reversed white on a dark background, may be inelegant but will certainly be clear.

Diagram artists need to plan drawing and wording together. Dry transfer characters, which are convenient for studio use, may be typographic in form but typographically incompatible with the rest of the book. If the wording is to be set in the same type-face as the text, repro proofs of it are best pulled on paper with an adhesive backing and sent to the artist before he begins drawing. But the artist's own handwriting is likely to provide a greater variety of size, weight, design, and spacing than stencils or even dry transfer. The writing needs only to be clear and compatible in style with the rest of the illustration.

Apart from the wording embodied in some illustrations, handwriting and lettering are little used within books, or for decoration rather than for illustration, except of course in books about lettering and handwriting. *Lettering* may be defined as the construction of letters by a multiplicity of separate strokes applied to each letter, as when for instance a large letter is drawn in outline and then filled-in between the outlines. In even the most formal *handwriting*, a single stroke is enough for each part of the letter. The main use of lettering at present is on covers and jackets, where the biggest possible letters can be designed to fit exactly into a restricted width. But lettering and handwriting have made distinguished appearances in the preliminary and text pages of books in the past, and might still supplement the dwindling repertoire of decorative display types with something of a flourish.

Artists should not sign book illustrations; the repeated name is likely to annoy the reader as though it were an advertisement. Every contributor to a book's illustration is entitled to acknowledgement in the preliminary pages, and that should be enough. When illustrations drawn for the book by one artist are likely to be considered by the public to be a feature of the edition, the artist's name should appear on the title-page.

The right to reproduce copyright illustrations is sometimes granted on condition that acknowledgement to the copyright holder appears next to the illustration itself. The reader is better left free to concentrate on the picture and perhaps on its caption. All sources of illustration should be acknowledged, as a matter of courtesy, and also because the source may one day have to be identified for another edition or even another title. But the lists of illustrations or acknowledgements in the preliminary pages are available for this. If acknowledgement next to an illustration is obligatory, it is best included in the caption, and for the sake of consistency other acknowledgements throughout the book should be similarly placed. Few books these days continue the distracting habit of setting acknowledgements separately and off-centre in the smallest available type over a centred caption.

In the numbering of illustrations, the reader's convenience is commonly subordinated to production economy. When an illustration is referred to in several chapters, the reader's search from reference to illustration is not often made easy. The number of the page on which the illustration is to appear cannot be known when the type is being set or until the pages are made up. To identify and insert page numbers at that stage can be laborious and will cost as much as any other alteration. Illustration numbers are therefore usually preferred for all references in the text. A list of illustrations in the preliminary pages, with page numbers, is then essential rather than merely helpful.

Each plate page is usually numbered in a series and style distinct from any other series of illustration or page numbers in the edition. Roman numerals are still used for this purpose, though when the number is always qualified by the term *Plate* it may as well be arabic. Whatever the style, a separate series of numbers is essential, as the location of plates interspersed throughout the book cannot be determined until the text pages are made up and their allocation to sections decided. When in the same edition text illustrations are serially numbered, the position of plates within that series is indeterminate at the time the series is compiled. For the same reason, colour plates and folding plates are often numbered separately from ordinary monochrome plates. The location in the book of each kind of plate can then be selected without reference to the location of illustrations numbered in other series, except from a visual aspect, as a plate should not face a text illustration by chance but only by intention. Each series of illustrations is usually listed separately from the others in the preliminary pages.

Text illustrations are usually entitled *figures*, and textual references to them are all too often contracted to *fig*. Like contractions or abbreviations for *part*, *chapter*, *plate*, *volume*, and *page*, this has nothing to commend it, unless references occur in such numbers that to set the words in full would appear uneconomical or unsightly or both. Nor is there any apparent need for an initial capital, as in *Figure*, within a sentence of text. Capitals are better reserved for specific purposes, and contractions and abbreviations, with their marks of punctuation, tend to interrupt the act of reading.

Captions are usually set in sizes one or more points smaller than the text, to differentiate one from the other. Captions often look neatest when they are no wider than the illustration under which they appear. When they are to be narrower than their measure for this reason, they should be marked for indention left and right by a calculated number of ems of set. This is the way the printer will set captions to a reduced width when they make more than one line, and the best instructions to send him are those he can follow exactly.

When several illustrations are printed on one page, the captions are sometimes grouped together, linked with the pictures not by position but by illustration numbers. When a caption appears between two illustrations, one above and one below, it should be nearer to the upper picture, to which it refers, than to the lower. The caption of a turned illustration should itself be turned, so that both can be considered at the same time.

When there is no room for captions on an illustrated page, they are often placed at the foot of a facing page of text, in the tail margin if the illustrations are plates (since the appropriate page for such captions may be identifiable only after make-up), and often ranged inwards so that they do not resemble a footnote. When several illustration pages of this kind follow each other, the captions may be printed with page numbers in the preliminary pages, as a list of illustrations. Strictly they are not then captions, but they may be easier to find there than elsewhere.

Captions printed on the same page as illustrations deserve a neat and modern arrangement, to avoid visual competition with the picture. When text typography is mainly centred, captions are usually centred on the illustration to which they refer. When such captions make more than one line, the lines are sometimes indented to match the width of the picture. When the last line of a caption is centred, it looks like an awkward compromise between text and display typography. The caption is better set like the first paragraph of a chapter, with no indent in the first line; or displayed like a chapter heading, divided into lines of approximately equal length, all centred.

The illustration number, acknowledgement, and other typographic items referring to the illustration may be run on with the caption as one paragraph, to simplify the illustrated page. In works of information, the number will be easiest for the reader to find if its position on the page is standard throughout the book, perhaps at the left-hand margin. Captions need consistent punctuation, just as the text does, and throughout the book should end consistently, either with or without a full-point.

Headlines, and page numbers in the headline, should not appear immediately over an illustration, since they do not refer to it but might appear to do so. Both should be omitted from a page showing a turned illustration, as they would be at right-angles to the picture.

CHAPTER 15

Paper

Paper is the only material from which all industrially produced books can at present be made. No practicable alternative, such as flexible plastic, is plentiful and economical enough. The variety of techniques in industrial book production is more than matched by the variety of paper characteristics.

Effective presswork and binding, and the visual, tactile, structural, and mechanical qualities of the book depend to a large extent upon the materials from which it is made, and books are made mainly of paper. Ordering or selecting paper is an aspect of book design which requires technical knowledge and an ability to balance contending factors. The requirements of the publisher with regard to price have to be met, together with those of the reader with regard to the appearance of the page, those of the librarian with regard to durability, those of the printer with regard to presswork of appropriate quality, and those of the binder with regard to the effective construction of the volume.

The purpose of this chapter is to describe in outline the materials and processes now general in the making of text and illustration paper, and the characteristics of the most widely used grades of such paper made in this way. There are other processes and materials, and many other kinds of paper, towards which a general understanding of paper can be extended.

§ 15-1 Fibres

Paper has ingredients, a structure, physical functions, and a lifetime; and it is not by nature inert. Its main ingredients are fibres, *loading* (§ 15-2), and water; others are described in each section of this chapter except the last.

The structure of paper is that of a felted mat of vegetable fibres, strengthened for printing by adhesive additives. Paper's functions are to receive and retain printing ink, to withstand the strains of presswork and folding, to bend easily enough for pages to be turned and to lie flat when open, and to endure this latter movement as long as the text printed on it is needed, which may be a matter of a day (in newspaper form), a year, or a century. And if libraries of the twenty-fifth century are to possess books of the twentieth, as today's libraries possess books of the fifteenth, we must seek out and use papers which will survive in store for hundreds of years.

The fibres of which paper is made are caused to swell or shrink by fluctuations of humidity, so that the whole sheet or part of it expands and contracts, and may become locally contorted. Heat and pressure generate in paper enough static electricity for it to become magnetic so that it clings to a machine instead of passing through it. Different fibres and other ingredients (the *furnish*), and different methods of paper-making, have different effects on the images printed by different processes.

Paper can be made of almost any vegetable fibres, but only the most suitable and effective of these provide a basis for printing paper. All are cellular in structure, and are known as *cellulose*. Different kinds of fibre are often mixed in paper, to obtain the desired combination of characteristics.

Simple cellulose fibres are obtained from cotton, linen, and hemp. These strong fibres are rarely used alone in any *making* (§ 15-3), and are now restricted to paper of the best quality — often hand-made, and rarely used in book production because it is costly and difficult to print and bind well by industrial means. The fibres are long, consist of a high proportion of cellulose, and resist the drastic processes of purification better than other materials. Cotton is characteristically white, linen is stiff, and hemp is particularly strong. Papers made from simple cellulose resist tearing and the effects of folding longer than other papers, mainly due to the length of the fibres. Resistance to tearing and endurance of folding are two of the three main kinds of strength needed in book papers. Simple comparative tests can be applied anywhere, but measurement is based on laboratory equipment. The third kind of strength is the tendency of a paper to resist the effect of age: this can be assessed from its furnish, and scientifically measured by the use of artificial ageing techniques.

Various kinds of grass, including esparto and straw, provide *compound cellulose*. The fibres are shorter than in the simple cellulose group, and also resist purification less effectively, so that paper made from compound cellulose is less strong and less durable. Straw is less soft and bulky than esparto; *bulk*, or thickness of fibre, is one of the factors examined in § 15-5. Esparto is opaque, bulky, capable of fluffiness, dimensionally stable, and not particularly strong; its small but even expansion and contraction when damped and dried out make it ideal for offset. Esparto papers are pleasantly creamy, with a clean, soft surface which is receptive to ink. Esparto was for years the most common material for the better qualities of book paper made in Britain, but has now lost much ground to wood fibre. A proportion of wood fibre is added to esparto, for easier running on the paper-making machine.

During the nineteenth century, wood became the world's main source of paper-making fibre, and is more plentiful and economical than any other

[279]

material. This is another form of compound cellulose. Wood fibres are obtained from soft-wood coniferous trees such as spruce and pine, or from such hard-wood deciduous trees as eucalyptus, poplar, and chestnut. The variety of trees and of pulping processes is exploited to produce *pulps* (§ 15-2) with properties which differ according to the kind of paper which is to be made. The cellulose content of wood is low, less than half that of cotton, and the fibres are shorter than those of grass.

§ 15-2 Pulp

Pulp is a solution in water of the other ingredients of paper. By the time pulp is ready for paper-making, water is 99 per cent by weight of the solution. Much of the paper-making process is a matter of water-extraction, and the finished paper contains only about 5 per cent of water. Water is itself one of the ingredients, and some of the quality of papers made in Scotland has been ascribed to the purity of water there.

Mechanical wood pulp takes its name from its preparatory processes, which are mechanical only and include no chemical treatment. Logs are stripped of their bark and ground to pulp, which is washed and strained to remove chips and lumps. The method is the cheapest possible, and the paper is fit only for ephemeral printing. Ligneous incrustations and other impurities are left in the pulp, and in the paper these deteriorate in colour and structure more rapidly than the fibre itself, particularly on exposure to light. Mechanical pulp is too weak to support its own weight when wet during making, or indeed to be used for publications when dry, unless reinforced with about 15 per cent of stronger fibres. Mechanical papers, with this reinforcement, can be absorbent, bulky, and particularly opaque, and the poor colour of the pulp can be brightened by bleaching. They are substantially less expensive than *wood-free papers* (made of wood fibre, oddly enough, but containing no mechanical wood), but no paper which contains any proportion of mechanical pulp can be expected to survive long in use.

Pulp intended for any edition of enduring value is more expensively prepared by chemical means, and may be described as *chemical pulp*. When it contains no mechanical fibre, paper made from it is more likely to be called *wood-free*.

All fibres other than mechanical wood are subjected to chemical processes after the raw material has been sorted, dusted, or chopped up, according to its nature. Boiling or *digesting*, in an acid or alkaline solution, removes ligneous and other unstable matter, and dissolves the lignin which held the fibres together in life. *Sulphite* pulp is made by boiling the pulp in an acid liquour, calcium bisulphite. Different methods of digestion produce different grades of sulphite pulp: *strong bleachable* pulp (which resists the weakening effect of bleaching) is used for *cartridge* papers, in which extra strength is needed: *easy bleaching* pulp can be mixed with esparto to make good quality book papers: and *bleached* pulp is used to make softer papers for which cleanliness, opacity, bulk, and whiteness are more important than strength. After digesting, the pulp is bleached for whiteness, unless this process is to follow at a later stage, and the chemical residue is washed away. The pulp is next strained through a screen to remove knots. The digestion of *sulphate* pulp is similar to that of sulphite, except that caustic soda is used; this is more suitable for esparto and deciduous woods.

The preparatory processes which the pulp has then to undergo, in the course of its transformation into paper for printing, depend partly upon its nature and partly upon that of the paper required. The processes briefly described here are the main processes usual at this stage, and each may take various forms. Refining or *beating* is the central process of this preparation, doing more than any other to determine the nature of the paper: from the same fibre, different kinds of beating can produce anything from blotting-paper to grease-proof. The beating-engine (figure 15-2) forces the fibres between a barred roller and a serrated bed-plate. This first splits open the fibres, and apparently exposes the adhesive material inside which makes them cling together when the paper is made; then it frays out the ends or fibrillates them, increasing the strength and compactness of the paper; and finally it cuts the fibres to different lengths, according to the kind of paper required. After a short period of beating, the paper will be bulky and opaque, as in *antique* (§ 15-3); beating of medium duration produces a thinner and tougher paper with a good surface; and long beating, a strong thin paper with reduced absorbency and opacity.

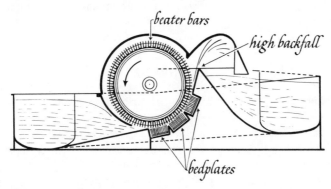

beater bars

high backfall

bedplates

Figure 15-2. Beating-engine.

The pulp may be bleached at this stage; separate bleaching is a more costly process, but produces a purer and stronger fibre.

Loading or *filling* is the addition of mineral to the pulp in the beater. A purely fibrous paper would lack opacity, smoothness, and resistance to ink, and most printing papers are loaded with about 15 per cent of mineral. The most common form of loading is china clay in various qualities. Titanium dioxide is expensive and is normally used with china clay; it is extremely white and opaque. Sulphur dioxide in the atmosphere of industrially developed countries is inimical to paper, and is one of the causes of its deterioration. One defence against this acid in the air is the use of an alkaline loading such as calcium carbonate. Since mineral, unlike fibre, is inert, loading reduces the expansion and contraction of the paper in changing humidity. Too much loading reduces the paper's strength, bulk, and durability, and diminishes the effect of sizing; highly finished papers which contain too much mineral also tend to deteriorate in damp conditions.

Engine-sizing is the addition of size — alum and rosin, casein, starch, or waterglass — to the pulp in the beater. This addition reduces the oil-absorbency of the paper, so that printing ink will not sink into it and pass through, and size also replaces the lignin by holding the fibres together. Alum and rosin have been used in paper-making since the seventeenth century, but they combine in an acidic compound to release sulphuric acid. No paper is likely to survive in use for many decades in the internal and external presence of acid. Neutral forms of size go far to ensure the durability of the edition.

Colouring material may also be added to the pulp in the beating-engine. Precise colour-matching in paper-making is difficult; the nature of the materials and of previous and subsequent processes all have their effect on colour. Separate makings of the same kind of paper sometimes differ slightly in colour, and whenever possible each edition should be printed on paper from one making.

§ 15-3 Paper-making

The paper-making machine on which printing paper is made is known as the Fourdrinier, from the name of the first holders of the patent in Britain. Fourdriniers are very big machines indeed; to make less than a tonne of any kind of paper at one time with such machinery would be uneconomic. The paper is made in a continuous web and reeled up when completed, even when it has later to be cut into sheets. A *making* is a batch of paper made with a specific combination of furnish and process at one time; the making is completed before a new combination is utilized on the same machine.

The pulp is poured out from the delivery apron of the machine on to the wire at a controlled rate of flow which determines the *substance*, the weight of a given quantity of finished paper (§ 15-5). Finished paper emerging from the other end of the machine is weighed, and the flow adjusted if necessary; the weight of two single sheets from different parts of the making may therefore differ slightly.

The *wire* is a continuous, travelling wire mesh which vibrates from side to side, shaking a proportion of the fibres out of their natural tendency to settle along the direction of travel (the *machine direction* or *grain* of the paper). The paper surface tends to be marked by the wire, and to be slightly rougher on the *wire side*. The pulp is prevented from flowing off the sides of the wire by the *deckle*, a strap which moves along each side of the wire and which can, if necessary, be moved inwards to produce a narrower web of paper. The maximum making width of the machine is also known as the *deckle*. In this latter sense, the deckle may be uneconomic for paper of a specific size with a specific grain, when a multiple of the required sheet dimension across the grain will not fit between the minimum and maximum deckle.

Figure 15-3*a*. A mould for making paper by hand.

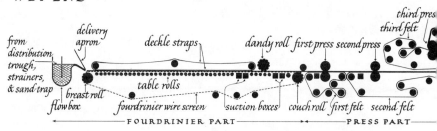

Figure 15-3b. Above and opposite, the Fourdrinier paper-making machine. The thick line running continuously from wet end to dry end represents the paper; the thinner lines between couch roll and calenders represent the felt webs which support the paper in its wetter and therefore weaker stages.

The wire, evenly spread with pulp, moves away from the delivery apron, the water drains out of the pulp through the wire, and fibres are felted together; draining is accelerated by suction-boxes under the wire. *Twin-wire* papers are made on a twin-wire machine; two separate wires make two separate webs of paper which are pressed together while still wet, with the wire side of each inwards, so that the paper has no wire side on its surface.

When the paper has begun to form, it passes under a hollow wire-covered roller called a *dandy-roll*. This squeezes some more water out of the web and closes it up, and can be used to impart a *water-mark*. At the point where the water-mark design of wire, in relief on the dandy-roll surface, touches the web, the fibres are slightly pressed down, so that the paper is thinner and more transparent at that point than elsewhere. The kind of dandy-roll most commonly used has a diagonally woven cover of wire, which leaves a fine and sometimes all but imperceptible pattern in the paper, similar to that of the main wire, and the paper is then termed *wove*. Another kind of dandy-roll has a grid pattern of wires running one way (usually across the grain), and thicker, more widely-spaced *chain wires* running the other; paper made in this style is termed *laid*. Laid paper is an imitation, so far as water-marking goes, of the old hand-made papers, which until the eighteenth century were made with a laid water-mark derived not from a roll but from the wire of the *mould* (figure 15-3a) on which the paper was made. Such paper is an anachronism, and is sometimes considered inappropriate for modern use, but the book trade is not generally intolerant of anachronisms. Any water-mark is a surface irregularity which can cause presswork difficulties when it appears within the printed area of the page.

The paper, formed but still wet and unable to support its own weight, is carried by endless reels of felt, first between pressure rollers, then round steam-heated cylinders, to dry most of the water out of it. During this drying, one of two optional processes may be applied. The paper may be

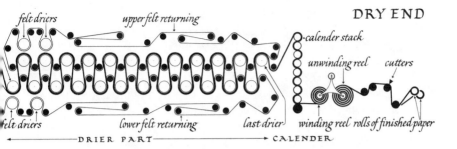

felt driers   *upper felt returning*   *calender stack*

unwinding reel   *cutters*

felt driers   *lower felt returning*   *last drier*   *winding reel*   *rolls of finished paper*

————DRIER PART———————→ CALENDER

machine-coated; a liquid coating of china clay or some similar mineral is caused to flow on to its surface and is regulated in thickness by a knife or blade. An uncoated paper may be *surface-sized* on machine, with a thin lick of starch. This does not replace engine-sizing, and may indeed supplement it; the sized surface is sealed to some extent against the emergence of loose fibres in the form of fluff, which causes trouble and mess during printing, especially by offset.

After more drying cylinders, the web reaches the *machine calenders*. This stack of polished steel rollers, smoothing the paper with adjustable pressure and friction, produces the kind of surface required. A paper which by-passes these calenders, or is only lightly *nipped* between them, is an *antique*, with a comparatively rough surface showing the impress of wire and felt. *Machine-finished* is a comprehensive classification including most bleached papers which receive some finish from the machine calenders but which are not further treated. Much the same kind of calender finish, combined with machine-coating, produces a *matt* or dull coated surface. Heavy calendering results in a smooth machine finish or a glossy coating.

§ 15-4 Off-machine processes

Further treatment, after paper-making on the Fourdrinier, is expensive but may be necessary. Off-machine processes are separate from those of the Fourdrinier, and are applied to paper which is otherwise complete.

Uncoated paper may be passed through a vat or tub of size. This may be defined as off-machine surface-sizing, but is called *tub-sizing*; it is more thorough than the on-machine process, improving strength and durability as well as resistance to ink.

Coated or *art* papers are those which consist of a substrate, core, or base of paper, coated on one side or both with a mineral material such as china clay, and an adhesive such as casein, which together may make up from 5 to 40 per cent of the paper's total substance. This kind of paper presents to the printing surface and the ink a finish which is even, flat, hard,

[285]

non-fibrous, and non-absorbent. The quality of the body paper is important, since any surface faults may be exaggerated by the coating. The technique of machine-coating is effective, but the best papers of this kind are coated off-machine.

Coated or uncoated papers may then pass through a separate stack of polishing rollers known as *super-calenders*. This may impart to uncoated paper a smooth and almost glossy finish. The glazing effect of super-calenders on coated papers can be regulated by differential pressures. There are four main degrees of finish — *bright* (glazed), *bloom* (semi-glazed), *semi-matt* (lightly glazed), and *matt* (unglazed). Matt art paper may approximate in surface to an uncoated paper.

The completed paper is re-wound on to reels, and is usually matured or conditioned by the addition of moisture, to bring its moisture content into balance with that of the prevailing atmosphere. Paper absorbs moisture from a humid atmosphere and releases it into a dry one, changing size slightly and sometimes undergoing some local distortion as it does so. The damping system of an offset press has a more drastic effect, and the speed and quality of offset work depends particularly on the condition of the paper. Changes of this kind cause trouble on any printing press, and interrupt register. A pause of a few days at the mill allows time for conditioning before dispatch; as at other stages in book production, haste may prove itself an enemy of good material and good work.

§ 15-5 Selection factors

The selection of paper for a specific edition, or the drafting of a specification for paper made to order for one or more editions, is complicated by the variety of factors to be considered. These factors, moreover, cannot be isolated from each other; all apply to the same material, and some affect each other. One example is the selection of paper for an octavo book in which a long text is to be illustrated by half-tones appearing on textually appropriate pages. To do justice to the half-tones without using glossy paper, which would be an injustice to the text, offset is likely to be preferred to letterpress.

The illustrations will probably look best on a white paper, but for the text most readers would prefer a more merciful shade of cream. Opacity will be important, because of the pictures; but a paper thick enough to be adequately opaque may prove to be a stiff one; when 32 pages are to be printed on each side of the long-grain sheet which the printer will prefer, the grain will run across the spine fold; and a book bound across the grain of a stiff paper will never lie flat when it is open. Emerging into daylight with his solution to these contradictions, the book designer may find that the paper he has specified cannot be economically made in the size he has in mind, or that its bulk will not suit the edition, or that it cannot be delivered in time, or that it will cost more than he can afford. A properly critical attitude to paper has to be tempered with compromise and supported with persistence.

Availability is alphabetically the first factor in selection, but all factors have to be reconciled in the right paper. If paper is made for an edition, perhaps as a special feature of the production, or if the sheet size is unusual, the same kind or size of paper may not be available for a reprint which does not justify a making order. If an edition is reprinted on a paper visibly inferior to that of the first printing, the physical quality of the publication will be diminished. A change in bulk may also necessitate new *brasses* (§ 16-8) and a new jacket design.

Bulk in paper is the ratio between weight and thickness. Sheet thickness is known as *caliper*, from the name of the measuring instrument, and in Britain caliper is expressed in multiples of the micron (μ), one-hundredth of a millimetre. The ratio between weight and thickness is called *volume*, the number of millimetres of combined thickness in 100 leaves (200 pages) of the paper if made in a weight or *substance* of 100 *grammes per square metre* (gm^2 or gsm). *Featherweight* is the bulkiest paper likely to be wanted in book production (but unlikely to be wanted at all by many book designers), and its volume is about 22; a smooth machine-finished or supercalendered paper may have a volume in the region of 10. Caliper tends to be proportionate to *substance* (defined later in this section), and 200 pages of a volume 10 paper, if made to a substance of 70 gsm, would be 7 millimetres thick.

The *caliper* of the paper limits to some extent the number of pages in the signatures into which the sheet is to be folded, since a sheet folded too many times may throw out diagonal creases or even split along the fold. Sheets over about 80 gsm are better not folded into signatures of 32 pages. Signatures of 16 or 24 pages are more common in good book production; signatures of 8 pages are rare for the main body of a book, probably advisable only when the paper exceeds 120 gsm.

The caliper multiplied by the number of leaves (half the number of pages) determines the thickness of the book. The public still has a tendency to judge the value of a book by its thickness, but too thick a book may intimidate the reader. Some idea of this dimension should be present in the designer's mind throughout his planning, particularly if the thickness is likely to be unusual in proportion to the leaf size.

Bulky fibres, treated to a short period of beating, are used for antique papers. These are subjected to a minimum of calendering, and whether wove or laid, are more *two-sided* than other grades of paper — the wire side is rougher than the upper side. Featherweight is a form of antique; papers of a volume between 20 and 22 are usually described as *bulky*; and other antiques are unlikely to extend below volume 16. Antiques make up in opacity what they lack in smoothness and solidity, and they continue to be used in book production of good quality.

When the caliper is not known, the thickness of the book can be calculated in millimetres by multiplying the number of leaves by the grammes per square metre and the volume, and dividing the product by 10,000.

In the United States, volume in text papers is expressed in terms of the number of pages which make a thickness of one inch. To convert the American bulk factor in pages to volume, multiply the number of pages in the bulk factor by the weight in pounds and divide the product into 200 (pages) × 25.4 (millimetres) × 67.5 (pounds, equivalent to 100 gsm) = 342,900. Volume may be converted for application to an American paper by multiplying the volume by the pound weight and dividing the product into 342,900. Conversions between American pound weights and British gsm are considered below under *substance*. Marshall Lee (1979) provides a table of *finishes* (defined later in this section) and bulks from which it is possible to deduce that a $67\frac{1}{2}$-pound paper would make an inch with about 290 pages of antique finish, and a volume of about 18; about 325 pages of eggshell finish, volume about 16; about 350 pages of vellum finish, volume about 14; and about 390 pages of machine finish, volume about 13. The weight or *substance* of paper is analysed later in this section.

The relationship between the *dimensions* of the paper and those of the leaves has been described in § 3-1; the dimensions of either may determine those of the other. An edition requires paper of certain dimensions and of a certain grain, but these two factors may be incompatible with the minimum and maximum deckle of some paper-making machines. Some papers are only available in *long-grain* sheets (with the grain parallel with the longer side) up to a certain size, and in *short-grain* beyond it. The size and squareness of sheets sometimes vary within a making, and for close register in printing and folding the extra cost of guillotine cutting may prove worth while.

The *finish* of a paper, its smoothness of surface, depends partly upon the treatment of the fibre, partly upon the proportion of loading or the method of coating, and largely on the pressure applied in calendering. Bulky papers are rarely smooth, and opacity is increased by such rough surfaces. A heavily loaded paper which has been strenuously calendered may be glossier than some coated paper. Two-sidedness is reduced by calendering, and eliminated by twin-wire paper-making. A laid dandy-roll affects not only the *look-through* of the paper — its appearance in transmitted light — but its surface for printing. The absorbency of the fibre tends to cause some outwards spread of any printed image, and a rough surface induces some irregularity of outline. This affects the clarity of small typographic detail and the sharpness of half-tone dots, to an extent which is visible to the reader. The design of most metal type-faces allows for a certain degree of thickening on the press, and such series as Monotype Centaur, Caslon, and Garamond look spindly and spiky without it. When printed by letterpress, antique papers are subjected to firm pressure from the printing surface, to ensure a solid impression over the whole of the printed image. This pressure accentuates ink *squash* (§ 11-1), and the ink forced outwards around the image tends to spread on an absorbent surface.

The mineral coating of art paper does not absorb ink to the same extent as does cellulose fibre, and the printed image has little tendency to spread across such surfaces. The outline of the image is less regular on a matt than on a glossy coated paper, and any irregularity thickens the image to some extent. The sharpest outline, closest to an exact representation of the printing surface, can be printed only on such glossy papers, and they are essential to the highest quality of half-tone illustration. They absorb little ink, retaining most of the pigment on the surface, and offering to the reader not only an extremely white background, but an extremely black image. They provide the most reflective of paper surfaces, so that monochrome and colour illustrations are equally brightened by reflected and transmitted light. But the reader will not welcome so much reflection from the text area, and may prefer a fibrous surface to a mineral one. Such public preferences in book production may be difficult to satisfy, but when they exist the book designer should be sensitive to their presence, even if he does not share them. The public may prefer paper that looks like paper, not like a flexible form of ceramic tile. To eye and to fingers, a mineral surface appears to be what it is, an artificial solution to printing problems. When the text extends to more than a few pages, the glossier art papers at least are best reserved for illustrations printed as plates. The popularity of a sharp, intense image printed on a glossy paper should not be identified with a preference for the paper itself.

Furnish — the ingredients from which paper is made — provides an indication of the general nature of the paper, and of characteristics which may have to be deduced but cannot be seen or measured. For example, the character and quality of paper containing any proportion of mechanical fibre, or made from too acid a furnish, are likely to deteriorate in all too short a time. Some of the visible and invisible properties of any paper tend to be those of its furnish, and those of the fibre above all. Most properties are derived from furnish, or from finish, or from furnish and finish combined; a high proportion of loading, for example, lends itself to a particularly smooth finish after supercalendering. Any paper's affinity for presswork depends to some extent on its furnish; the coating of offset papers has to adhere to its substrate strongly enough to resist separation from it by a flexible blanket covered with particularly tacky ink, which would otherwise pull away particles of coating to mar the next sheets printed.

Whatever the printing process, the *grain* or machine direction of paper is a bookbinding factor, and the spine should whenever possible be parallel with the grain. Paper is stiffer across the grain than along it, and unless the pages are very wide or the paper very limp, cross-grain pages will not lie flat when the book is open. The grain should be the same in all the paper used throughout the book, including plates and endpapers, as the cohesion of the spine may otherwise be impaired.

Changing humidity causes paper to expand more across the grain than along it. Most of the fibres lie along the machine direction, in spite of the lateral vibration of the wire, and in the presence of water become thicker but not much longer. The offset printer, who dampens the paper on the press, finds the resulting expansion is best contained when it affects the shorter dimension of the sheet rather than the longer, and when it extends round the cylinder rather than along it. For this reason, offset paper should usually be provided *long-grain*, the longer edge of the sheet being parallel with the Fourdrinier machine direction. The direction of the grain is sometimes represented by (M) in paper specifications after the appropriate dimension, as in '960 × 1272(M) mm'.

The *opacity* of paper is vital to the legibility and to the good appearance of every book, and this characteristic is usually examined in terms of the extent to which a printed image under the leaf shows through it. The type on the verso is usually visible through the recto, but it should not be permitted to intrude. Type on the next recto may be readable with effort through the leaf, even a leaf of decent paper. But if type on the verso after that is more than very faintly visible, through two leaves, the paper is uncomfortably transparent.

Opacity can be assessed comparatively by laying two samples of different paper, one of which is known to be satisfactory, side by side over a printed image of appropriate blackness. More precisely, opacity can be measured; the reflectiveness of a sheet of paper, printed in black on the reverse side, can be expressed as a percentage of the same sheet's brighter reflectiveness when unprinted and backed with a number of similar unprinted sheets. In these terms, bookwork papers usually measure from 92 to 96 per cent.

Strike-through is the extent to which ink printed on one side penetrates the paper towards the other side. This is not strictly a matter of opacity, since an absorbent but opaque paper such as featherweight may appear to be transparent where the ink strikes through.

Shade is perhaps a term more apt than *colour* for text and illustration papers, which are usually white or cream, to clarify the printed image with contrast. Whiteness in paper accentuates the high-lights in monochrome half-tones, brightens printed colours, and elucidates small type and illustration detail. Any other shade, including cream, has by comparison an opposite tendency, however slight, but increases opacity and is more welcome to a reader's eye.

The unity of the book's appearance is enhanced when the whole book is printed on paper of the same shade. If two papers of different shades have to be used for one edition, the designer may need to remind the printer that the papers should not be mixed in any copy of the book, since each shade of paper will have to be allocated to part of the printing of each forme, and printed sheets of each shade will have to be separated until *gathering* (§ 6-4) is complete. When illustrations are to be printed as plates, on a paper different from that of the text, the two papers may be chosen to match each other in shade. A shade between white and cream may be best for this purpose. Alternatively, a text paper of deep cream could be selected to accompany a brilliantly white paper for the plates. An obvious contrast in shade has the appearance of intention; slightly differing shades may look like an unsuccessful attempt at a match.

The main elements of *strength*, in paper for book production, are longevity, stiffness, and resistance to tearing and the effect of folding. In newly made paper, stiffness and resistance are not evidence of longevity. Stiffness is a product partly of substance and partly of furnish. It is not itself a welcome quality, since the page may refuse to lie flat in the open book, even when the spine fold is parallel with the grain. If a stiff paper is bound across the grain, the book may be almost unmanageable in use.

Resistance to tearing and folding need hardly be tested in most established book papers, since for most editions they will be adequate when they are new. For an edition designed for rough and frequent handling, simple comparative tests on a few apparently suitable papers may reveal where most strength is to be found.

Surface strength, which retains fibres, loading, and coating in the sheet against the adhesive pull of the ink, is particularly valued by the printer, whose opinion on this and other characteristics is worth seeking when a paper new to him is about to be selected.

Time and use deprive any paper of some of its quality. One long-life paper has an initial folding resistance of 1,200 folds and a tearing resistance of 73 grammes, and tests indicate it is likely to retain more than half this strength after two centuries. The same tests showed that some ordinary book papers were likely to lose more than half their initially inferior strength within ten years. Mechanical fibre also tends to crumble and to lose its whiteness after a few years, even with little use. The effects of age cannot be measured except in a laboratory, but the designer should always have their probabilities in mind. Long-life papers should certainly be selected for books of permanent value, but nobody can say for certain which editions of today will be valued a hundred years or more from now. Until books are replaced by some other form of communication between author and public, it is the survival of most of our editions which matters most, not that of a few unused copies in private hands. Any book may be valued many years after publication, even if a reprint of it may appear uneconomic, and a reprint would be impossible if the publisher could not find one or two copies which could be broken up for the camera. The best course for a book designer to follow, in the service of books and authors and readers, is to seek a long life for every edition by examining the longevity of any paper he selects.

Cartridge papers, whose fibres have not been substantially weakened by bleaching, are initially stronger than most other book papers, but their longevity too depends on their furnish, and on loading and sizing in particular. Cartridges may be coated like any other paper. Most twin-wire papers are cartridges, and these are usually made in substances of 105 gsm or more.

Substance is the weight of paper, measured in Britain by grammes per square metre of a single thickness, and in the United States by pounds weight of 500 sheets (a *ream*) of such a paper as though made in sheets 25 × 38 inches. To convert an American pound weight to grammes per square metre, multiply the pound weight of the ream by 1.48; to convert gsm to ream weight in pounds, multiply gsm by 0.675. Caliper, opacity, and strength tend to be proportionate to substance. Machine-finished paper, perhaps the most widely used in book production, is usually available in substances from 50 to 100 gsm, and in a limited number of brands down to 30 and up to 140 gsm. Under 70 gsm, paper may generally be classified as thin; owing to the increased cost of making such paper the price per tonne rises with the decrease in substance, and owing to the difficulty of handling and printing such paper, the printer may be able to justify a surcharge on his usual price for presswork.

Edition binding

Binding is a multiple operation, a series of processes converting printed sheets into one of various forms convenient for reader and shelf. Most of the preceding pages have dwelt on visual aspects of design. In the planning of binding, the designer concerns himself also with structural and mechanical qualities. The prime virtues of a good binding are mechanical — the ability to hold the pages together and to protect them, to open flat, to endure in use and store, and to withstand both climate and time.

The various processes and materials should be selected to combine effectively together in each volume, and to do so in terms of relative strength, mechanical function, and general appearance. If a book has to be printed on paper containing mechanical fibre, buckram over stout boards will not be the best binding for it, and will not look like the best binding. Disproportion in a book is a form of weakness, and disproportionate strength at one point may generate mechanical strain at another.

The visual and tactile qualities of a binding also deserve meticulous design. Decoration in industrial binding is still in transition, partly because in Britain it is so seldom and so sparingly used; its possibilities have not been generally explored. The design of lettering, for instance, and the choice of materials are all too often half-hearted; covering materials are often either drab or gaudy, and lack those qualities of surface that might please the hand or eye. Yet the outer appearance of the binding is that of the book itself, as its owner may see it for many years, and part of the evidence, visible in his own house, of his own taste. The development of package design has had little general influence on the design of bindings, as distinct from that of the jacket. Perhaps it is just as well, since a book should not compete too insistently for attention with its neighbours on a private shelf; in the bookshop, the jacket competes on behalf of the book it encloses. But almost any row of unjacketed spines could be more comely without being spectacular. Binding design is a conspicuously visible element in the appearance of any volume, and deserves the best attention of the book designer.

§ 16-1 Bookbinding by hand

If a book is intended to survive in use for a century or more, it will have to be bound by hand; no mechanical operation provides such endurance. *Folding* (§ 16-2) and *gathering* (§ 16-4) could be done mechanically, but subsequent processes must for the most part be manual. The sections are sewn to each other by stitches which pass through the spine fold, and which on emerging from inside the section are carried round bands or cords of strong material, which lie across and extend beyond the spine of the gathered section (figure 16-1*a*). Slips of vellum or equally strong material are laid across the spine, one at the head and one at the tail, and are sewn to the sections, usually with coloured silk thread and in a decorative style (figure 16-6*c*). These *headbands* and *tailbands* will take the strain of the finger which pulls the book from the shelf, and hide the joint between cover and spine. The *endpapers* (§ 16-3) may be sewn in the same way as the rest of the book, and are so *made* that they are not *tipped on* (§ 16-3) to the next leaf (figures 8-1 and 16-3*a*). The extended ends of the cords are laced into the boards (figure 16-1*b*) or, in library binding, laid between the two parts of a two-ply board which are closed over them and glued together (figure 16-1*c*); whatever its dilapidation, the binding will then enclose the book as long as thread, cords, and boards hold together, and that is likely to be very much longer than in any other form of binding. The covering material, at least over the spine and part of the boards adjacent to it, is usually some form of hide, commensurate in strength with that of the binding's construction and with the cost of the handwork by which it is assembled. This cover is pasted over spine and boards while they enclose the sewn sections, and the endpapers are pasted down inside the boards (figure 16-1*d*).

This kind of work is not part of industrial book production. Each binding is a separate task, and most orders for so costly an operation are for a single copy only. There are however times when an industrial book designer is called on to produce a presentation copy in this way. His first duty is to ensure, by his selection of the bookbinder, that the traditional method of construction and the traditional standard of materials will be used if they are likely to be expected, and that the sheets or sections supplied to the bookbinder are in good condition. His second duty is to achieve an acceptable design. He has no duty to design the binding himself, except in collaboration with the bookbinder, who is entitled to provide the most suitable materials available to him and the most appropriate lettering and decoration he knows how to apply. The assembly of any such binding is usually the work of a craftsman likely to hold valid opinions about its appearance.

The hand process may reasonably be called *bookbinding*, since it is applied

FIGURE 16-1

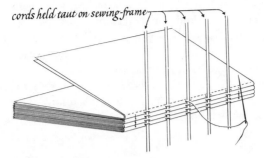

a. Sewing in hand bookbinding.

b. Lacing cords into boards.

c. Library binding.

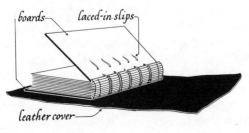

d. Cover and book.

to a single book, but the distinction between this and other forms of binding in book production is not always indicated by this use of the term. The essential difference is the method of assembly and construction; if this is not borne in mind when such work is ordered, the individual responsible for the order may one day hear from a disappointed potentate to whom an inadequately bound copy has been presented.

Bookbinding of this kind should never be hurried. If sheets are delivered to the bookbinder some two months before publication, the presentation copy is likely to be ready by publication day.

§ 16-2 Folding

Edition binding comprises a variety of different binding methods, all of which are capable of mechanization and are therefore suitable for the rapid production of editions and unsuitable in every way for that of single copies. No form of edition binding approaches the strength, durability, or magnificence of bookbinding by hand at its best, but edition binding by definition is economically expedient for industrial book production.

Folding is the first process in any form of binding. When a book is printed on a web of paper instead of on sheets, it is usually folded on the printing press. In a strict sense, folding is then part of the printing process; newspaper presses deliver complete and finished newspapers, which pass through no separate binding process. In book production, folding on the press may also be termed a bindery operation which is carried out in the press-room by certain machinery. Hand folding continues in industrial binderies, but is normally restricted to *oddments* (§ 16-3). Most sheet-printed books consist of a number of sheets of one size, sometimes slit on the press into smaller sheets, and these are always folded by machine.

The imposition of any printed sheet is determined by the designated folding method, since the purpose of any imposition scheme is to arrange the pages in such a way that after folding they appear in the correct order. When folded, a sheet, or a separately folded part of a printed sheet, becomes a *signature*, so called because it is identified by a letter or letters — also known as a signature — at the foot of its first page, and by the *collating mark* (§ 16-4) on its spine. Sometimes the collating mark takes the form of a signature and replaces it, which is all to the good as the signature contributes nothing to the printed page in a bound book. One or more signatures make up a *section* of the book, with a spine in common which can be penetrated by a single stitch if the book is sewn. Folded edges of the section other than the spine fold are known as *bolts*, and are often perforated during folding, to allow air to escape and so to minimize the risk of creasing.

The locations in the book, in terms of page numbers, of each *spread* (§ 12-1) and of bolts are likely to concern the designer when the spreads are utilized as such or when plates are to be *tipped in* (§ 16-3). When the location of any page in its section concerns the designer in this way, he needs either to instruct the printer about imposition in some detail, or learn the printer's intentions and make sure they are maintained.

Folding machines place their folds a fixed distance from the *lay-edge* of the sheet — the edge laid against a guide on the machine at the moment of folding. The position of each fold depends on the positions of the pages printed on the sheet; whether or not the forme is printed centrally on the sheet, and whether or not the edges of the section align after folding, the pages must align with each other. If one or more sheets were off-square when they entered the printing press, the pages on them may be out of parallel with the folding lay-edge, and will therefore be askew on the folded page. If the sheets before printing are square but vary in size, the pages are likely to be parallel with the edges of the book, but laterally or vertically out of position. Even when the paper is perfectly square and even in size, the accuracy of folding throughout the edition should not be relied on to deliver a perfect alignment between two facing pages except in a spread. The usual limit of tolerance is about 6 points.

The number of pages in the section is a decision for the designer. The thickness of the paper is influential, since thick or brittle material such as art paper does not take well to repeated folding. Some papers over 90 gsm are better folded in 16s than in 32s, and when such papers are over 130 gsm signatures of 8 pages may be better. One signature can be automatically *inset* in the centre of another to make smaller signatures for folding and larger ones for sewing. Too much thickness in the folded and sewn spine strains the paper, may stiffen the signature so that it does not lie flat enough, and can lead to creasing. If the book is to contain a great number of pages, the number of sections should be minimized, or the combined thickness of the sewn folds will become excessive. Sections of 16 pages are usual in well-produced editions, and sections of 32 are more economical.

The maximum and minimum capacity of the binder's machines for folding and for other processes may impose limits on the designer's plan. Folding machines are likely to be economical enough between crown octavo and demy quarto, but larger or smaller books may give rise to extra expense.

§ 16-3 Plates, oddments, and endpapers

After folding, the edition consists of stacks of sections, each stack comprising part only of the finished book; all copies of pages 1 to 64, for example, may be in a single stack, all of pages 65 to 128 in another.

Illustrations printed separately as *plates* have been described in § 14-6. They can be fixed into the book in various ways, none of them entirely satisfactory. Ideally, each should face that page in the text on which the first main reference to it appears. In practice, this is rarely possible by means compatible with their security of tenure and also with the economics of industrial production.

Except by chance, single plates can be placed in textually appropriate positions only by passing one or more of them through a bolt. In order to do this, the binder has first to slit the bolt at head and fore-edge, and insert the plate between the apposite pages. When a book is to be sewn, such a plate may be fixed into the section by *guarding* or *tipping in* (figure 16-3*a*). Guarding is a secure method; the guarded plate is 3 to 5 millimetres wider at its inner edge than a tipped plate, and is creased along this edge to make a spine before insertion. On insertion, the guard is passed round the spine of part of the section, and eventually the guard is sewn in. The guard is visible, though hardly conspicuous, where it emerges on the other side of the section. Too many guards in one volume tend to thicken the spine unduly. A plate which is to be tipped in is much the same size as the uncut pages, and is first pasted along about 3 millimetres of one side of its spine edge. Having been placed in its position in the section, the plate is fixed there by the paste in contact with the spine edge of the adjacent page. If the plate is printed on one side only, the paste must be on the other side, or the illustration will not lie flat when facing upwards in the open book; a backed plate should not be tipped in. Any frontispiece plate is likely to be tipped in after passing through a bolt in this way. Tipping into the centre of a section is less expensive than passing the plate through a bolt. Tipping is insecure; some reader, irritated by the plate's tendency to rise up when opened away from its adhesive side, is sure to pull it flat and loosen the adhesion. For a frontispiece plate, the method is all but unavoidable; for more than a few other plates, their position in the text is usually considered to be secondary to the use of economical and structurally effective methods. But without tipping, the position of plates in relation to specific passages of text is random.

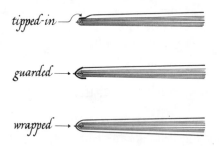

Figure 16-3*a*. Three methods of fixing plates into the book.

To *tip on* (as distinct from tipping in) is to tip on to the outside of a section. This can be done by machine, as for endpapers. In terms of security and of textual reference, tipping on is not preferable to tipping in, but the process is less expensive.

When plates are to be dispersed through the book, and textually random positions are acceptable (as they usually have to be), plates may be made up as signatures of four or eight pages, each *wrapped round* the outside of a section or *inset* into its centre (figure 16-3a). Both methods are secure, since the plate signatures will be sewn into the book, and both are more seemly than guarding, as no guard appears as an intrusive strip of paper at the spine edge of an unrelated opening. Eight pages may be too many for wrapping or insetting if the plate paper is not distinctly thinner than that of the text sections, since adjacent sections of substantially different thicknesses may weaken the spine of the book. The position in the book of one half of a wrap determines that of the other; the designer should check both positions, to make sure they are suitable for a plate or plates. They should not as a rule be allowed to intervene between a part half-title and the first chapter of a new part, or face a full-page line illustration printed with the text, nor if upright should they face a turned plate wrapped round the next section.

Plates may be grouped as separate sections, and placed arbitrarily in the book. Sometimes they appear at the very end, even after the index, where they look like an afterthought; elsewhere they are likely to interrupt a chapter or other item, and appear to intrude. The best place is between the end of one chapter and the beginning of the next, but to arrange for any chapter to end on the last page of a text section is all too rarely possible.

A binding *oddment* is a signature which differs in the number of pages from the sections in the book. The term usually refers only to pages of the main text; a text section with a signature of plates wrapped round it is not called an oddment. More often, for instance in a book sewn in sections of 16 pages, an oddment will be sewn as a section of 8, or wrapped round or inset into a 16-page signature to make a section of 24 pages, which then becomes a binding oddment itself. Such oddments are usually placed before the last standard section, since the two outermost sections of any book are subjected to more strain than others when the book is opened and closed, and oddments present a point of weakness. The position of plates in the book can be checked only when the position of oddments is known.

In edition binding, an *endpaper* is an opening of two leaves, plain or printed, tipped on to the spine edges of the outermost sections (figure 16-3b). It is not usually sewn through, and is therefore not considered to be a signature or section. Tipping on is a mechanized process because most cased editions have endpapers. The function of endpapers is to conceal the inner sides of the cover boards and the flexible joints between boards and

Figure 16-3*b*. Endpapers in a hand-bound book (left) and in a cased book.

outermost sections, to help in holding boards and sections together, and to share the strain of opening and closing.

The material used for endpapers should be stronger than the text paper, because they have a more exacting mechanical purpose. Uncoated cartridge paper of about 100 gsm is commonly used, since this offers a sound combination of strength, enduring adhesion, and opacity. A book cased without endpapers is said to have *self ends*; the outer leaves of the outer sections are pasted down to the boards, which are likely to show through such paper and before long to break it along the joint.

Books tend to break up after a certain amount of use, and the joint between board and *fly-leaf* — the inner or free leaf of the endpaper, as distinct from the leaf pasted down to the board — is one of the most vulnerable points in the volume's construction. In a heavy book, the joint is worth strengthening with a strip of linen pasted along it, through which the endpapers can be sewn to the sections between them. In any book, the grain of the endpaper should run along the joint, since paper withstands repeated folding better along the grain than across it.

Printed endpapers can be attractive and even useful, but nothing essential should be printed on them. Libraries often paste withdrawal cards on to the front endpaper, and destroy both endpapers when they re-bind. But when the book is to contain one or more discs or folded maps or other such items, these are usually slipped into a paper pocket or sleeve, pasted to the pasted-down endpaper inside the boards.

Coloured endpapers contribute to the decorative effect of any book, and since the extra cost is slight they could be used more often than they are. Darker colours are preferable to light, since these conceal the finger-marks that may otherwise mar the book in the bookshop or afterwards. Plain endpapers usually have the creamy tint of cartridge, likely to differ in shade from a white text paper, but there is no harm in this when the cover of the case is in one of the warm colours such as red or brown. If the text paper has an icy whiteness and the cover is of a chilly blue, creamy endpapers may look like a mistake. Coloured endpapers assert their membership of the book as a unit when their colour is at least compatible with that of the covering material and of any non-metallic *blocking* (§ 16-8). A colour match between paper and cloth — or indeed between any two different materials — is difficult to achieve and better not attempted without satisfactory samples of both materials.

§ 16-4 Gathering, collating, and sewing

The sections, with their endpapers, and plates if there are any, are next *gathered* into the order in which they will appear in the book, and are *collated* or checked. The purpose of collation is to ensure that all sections are present in the right order and the right way up. Evidence of this is provided by a *collating-mark* printed on the spine of each section towards the head (figure 16-4a). The mark on each section is an even distance below that on the section before, so that when the book is gathered the marks form a straight diagonal line across the spine. The designer is not responsible for the form of the collating mark, since it is intended to be hidden in the spine after binding, and it is not usually shown in proof. But a very slight inaccuracy in folding may reveal the mark or part of it on the face of a page, and the best form of mark is probably a small light letter, not over 6 points wide, which replaces the signature mark and if revealed does not advertise its presence.

For a very heavy book, the outer sections may have been *side-sewn* before being gathered, for reinforcement. This is done, as on a domestic sewing-machine, with small stitches from the first page of the section to the last, alongside the folded spine. The section does not thereafter open as widely as others, but resists the pull of the endpapers to better effect.

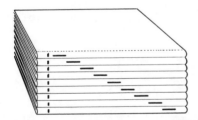

Figure 16-4a. Gathered sections with collating marks.

The double threads of the main sewing pass in and out through the spine, making four or five long stitches between head and tail. On emerging at the end of each spine, the threads are knotted as they continue to the next section, so that a single break in the thread will not allow the sections to fall apart. The back of a heavy book may be reinforced by two or three linen tapes laid across the spine before sewing; threads emerging from inside the section pass over or through the tapes, to hold them tightly to the spine.

Unsewn binding is an economy method which has not replaced sewing in well-made cased editions intended for long use. The spine edges of the gathered sections are cut off, to a width of 3–5 millimetres, and rasped to allow glue to grip the cut edges. A thick film of flexible glue is laid on the spine, to hold the separate edges together, in place of thread. Any failure of the glue to adhere or to set properly may cause the book to fall apart, perhaps after some use which may deprive the owner of any right to replacement. *Rounding* and *backing* (§ 16-6) may be possible, but unsewn books in general are unlikely to retain their shape as well as if they were sewn. The dimensions of the uncut leaf need an extra 3 to 5 millimetres of width in the spine margin, to allow for spine cutting. The technique is widely used for paperbacks and inexpensively cased editions.

Stitching — the sewing or stapling of a whole publication as a single section — is the cheapest replacement for sewing, but is unsuitable for any work of lasting value. *Saddle-stitching* passes through the spine in the same way as sewing, and may consist of thread or wire staples. The name originates in the saddle of the machine, across which each section is laid open before stitching. Wire should be rust-proof, as it will otherwise stain and corrode the pages. The sections are inset into each other, a paper cover is usually laid on top, and two or three stitches pass through all the spines at once. Too many pages prevent the book from lying flat even when shut; this shows at 64 pages even on quite limp paper, but 96 or more pages can be bound in this way, and the cheapness of the booklet is likely to forestall complaint. *Side-stitching* can be carried out with wire staples driven through the sections beside the spine fold; this prevents the leaves from opening fully and from lying flat when open, necessitates extra space in the spine margin, and is better avoided at all costs (figure 16-4*b*).

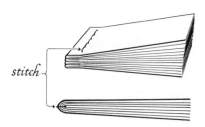

Figure 16-4*b*. Side-stitching (above) and saddle-stitching.

§ 16-5 Limp binding

Limp binding encloses a square-backed book in a flexible cover, and enables the edges of cover and text to be cut at the same time. This is an economical technique, but the comparatively low published price of many paperbacks is due only partly to binding economy.

Endpapers apart, the processes described in §§ 16-2 to 16-4 precede either limp or cased binding. A limp book may be endpapered, but this is a disproportionately expensive application of handwork in an otherwise mechanical sequence of processes. The cover, glued to the spine of the sections, strengthens the spine, but once it is in place the endpapers have to be pasted down by hand, adding little or nothing to the strength of the binding.

The spine, sewn or unsewn, is covered with flexible glue, and may be *lined* (§ 16-6) at this stage if extra strength is needed. While the glue is still molten, the limp cover is placed in position across the spine, pressed down, and folded round the book. Pressure on the spine forces glue outwards from the spine a little way between the cover and the outer pages, and within the overlapping cover at head and tail. Unless at this stage the cover overlaps a little at head and tail, glue tends to be squeezed beyond its edges and to foul the outside of the book. Along the spine between cover and page, however, the thin strip of adhesion between the two helps to hold the book together. Thick or glossy covers fold best alongside the edges of the spine when they have first been creased into hinges. A cover laid on the spine in this way, stuck to it, and folded round the book is said to be *drawn on*. The head, tail, and fore-edge of the whole book, text and cover, are cut together, and the paperback is complete.

Such light, flexible bindings are portable but far from durable. The cover, fixed to the spine, bends outwards when the book is fully opened, and creases and cracks; but their adhesion together strengthens them both. Cover and text tend to part company after a certain amount of use. The pages are unprotected at the edges; a limp cover which projects beyond them bends and breaks at the edges as soon as it is placed standing on the shelf. The spine is stiff, and likely to break if repeatedly opened flat. But a limp binding is itself an advertisement for a comparatively low-priced book, and paperbacks have vastly extended the scope of popular literature. In private and infrequent use, the flimsy construction may survive for many years.

§ 16-6 Edges and spine

Since the beginning of edition binding, the head of the book has usually been cut to a smooth solid edge, which prevents dust from penetrating on to the surface of the pages when the book stands closed on the shelf. At one time, fore-edge and tail were quite often left uncut, and the reader was left to slit the bolts with a paper-knife. Other editions were uncut at the tail only;

the tail bolts, imposed so as to project beyond the folded edges of the sheet, were trimmed off with a rotary knife, and the tail edges showed a pleasing irregularity of depth (figure 16-6a).

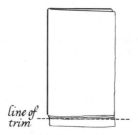

line of trim

Figure 16-6a. Trimming off projecting bolts.

Industrially produced books today are normally cut at head, tail, and fore-edge. All the pages are then separate from each other; all are the same size, easy to turn over, and neat in appearance. The edges do not retain dust or permit it to enter the book between the pages.

The corners, at the head and tail of the fore-edge, can be rounded, to prevent damage to an otherwise vulnerable part of the book. Rounded corners look incompatible between a rectangular type area and a rectangular case, but the corners of a flexible case in particular can also be rounded. *Round-cornering* is now mainly reserved for small books intended to be carried in a pocket or hand-bag.

Since the cut edges resist penetration by any substance when the book is tightly shut, they can be treated as a flat surface. *Gilding* is the most spectacular treatment; the edge of the closed book appears to be a single sheet of gold. The physical properties of gold, even in leaf form, enable each page to be separated from its neighbours without dislodging the thread of gold on the edge of any leaf. Gilt edges exclude everything which might harm the book, even atmospheric acid, but are too costly a feature for most edition binding.

The practice of *edge-colouring* is derived from that of gilding, but as colour does not exclude anything it is hardly a substitute for metal. To some extent a dark colour conceals dust on an edge, but on the whole colouring is no more than a form of decoration. Edges may be painted in solid colour or sprinkled; an absorbent or coated paper, or a loosely clamped book, sometimes admits a border of colour to the face of the page along the coloured edge, which though narrow may be conspicuous and irregular enough to spoil the book. Since the colour will otherwise be inconspicuous when the book is open, it should be congruous with the colours of binding and jacket rather than with those of the text.

Lying closed on a flat surface, the book is now thicker at the sewn edge than at the others, owing to the accumulated thickness of the thread inside the sections. The application to the spine of a coating of flexible glue forms the spine into a single unit, in place of a stack of folded edges, and by penetrating a little way between the sections the glue also contributes to the additional thickness of the spine. The thickness of the bound book must be even, since the front and back would not otherwise be parallel, and a wedge-shaped book would fit awkwardly into the shelf. The extra thickness at the spine is therefore disposed of by *rounding* the spine into its familiar convex shape; the fore-edge at the same time becomes concave in parallel (figure 16-6*b*). This is done while the glue is still malleable and before it hardens into its final dimensions.

The tendency of the backs of the sections to spread outwards during rounding is encouraged by *backing*, so that the spines of the outermost section form a hinge for the boards (figure 16-6*b*). This hinge helps to hold the sections in place between the boards when the book stands shut, and shares the strain of opening and closing.

before rounding   *after rounding*   *after rounding and backing*

Figure 16-6*b*. Rounding and backing.

A rounded and backed book tends to keep its shape, with the fore-edge withdrawn between the boards. When rounding and backing are omitted, and the book is bound with a square back, the straight fore-edge tends to swell outwards after the book has been used for some time. In a comparatively thin book this may be inconspicuous, but a stout one takes on a pot-bellied appearance. A square back is not the most secure construction for any cased book, but it has a certain elegance in the bookshop, and is often seen in books which are thin in proportion to page size. Its advantage is that the opening lies flat, with the surface of each page visible right into the spine fold. Rounding and backing, in contrast, bend the inner edge of each page into a channel at the spine, absorbing part of the spine margin and concealing part of any illustration which runs up to the fold.

The spine is re-coated with glue, and a kind of coarse muslin (*mull*) is laid on it, falling just short of head and tail and overlapping along each side. The open weave of this material allows adhesive to pass through, binding together the stronger materials on either side of the mull. The overlapping parts of the mull, perhaps reinforced by tapes already in position after sewing, will eventually be fastened between the board and the pasted-down leaf of the endpaper (figure 16-7b). There it will help to hold text and case together, and to strengthen the hinge at the inner edge of the board. The overlap of mull beyond the hinge should for neatness be wider than that of the tapes, and roughly equal to the turn-in of cloth round the fore-edge of the board, but a heavy book may need a greater width of mull and tape.

Mull is the first *lining* by which the spine is strengthened to retain its shape. The *second lining* is a strip of stout paper, glued down on most of the area of the spine over the mull, to hold its threads in place, to add to the stiffness of the spine, and to share the strain when the book is opened and closed. A *third lining* of linen, also covering most of the spine but in addition overlapping like the mull, serves as reinforcement of spine and hinge in particularly heavy books. A single lining of linen, or of paper impregnated with latex, covering the whole spine from head to tail and overlapping across the hinges, sometimes replaces all other forms of lining. This provides enough flexibility for the book to withstand treatment that might break the back of another kind of binding, but does little to hold the spine and fore-edge in shape for long. This kind of single lining is well suited to an unsewn spine.

Bookmarks or *registers* are now rarely provided in cased editions, and have perhaps become no more than an expensive form of decoration. One end of a suitably coloured silk ribbon, however slender, is placed between two linings before they are glued together. The free end passes between the pages to extend a little way beyond the tail, so that both ends can be seen when the book is closed. A bookmark would be useful to the reader, but readers are now accustomed to doing without one, or to providing their own.

Headbands and *tailbands* are another form of decoration in binding. In traditional bookbinding the headband, sewn to the sections, takes the strain of the finger which pulls the book from the shelf, and conceals the narrow joint between cover and spine (figure 16-6c). In a cased binding the headband is an anachronism, since it is only glued to the spine, and neither strengthens the head nor conceals the *hollow* (§ 16-7). In unsewn cased bindings, headbands and tailbands sometimes seem to be used to conceal the fact that the sections are unsewn, though this should not need to be concealed.

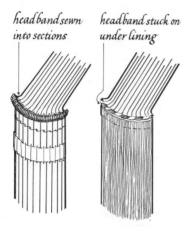

head band sewn into sections   head band stuck on under lining

Figure 16-6c. Headbands in binding by hand (left) and by machine.

The pages and cover of an unsewn binding can be punched and secured to each other by a spiral of wire or a plastic clip like a curved comb. This replaces glue and lining, and provides a flat opening interrupted at the spine by a wide gap and the apparatus of wire or plastic. This method of binding is best reserved for special use, since authors and their public may often agree that the product does not look like a book at all.

§ 16-7 The binding case

The *boards* which cover the front and back of a case-bound book are a thick and usually rigid form of paper. Nearly all case-bound books are intended to be kept on shelves, standing on one edge between other books. The weight of the whole book then rests on two edges of board, projecting beyond the edges of the pages between them, and so protecting them against pressure and abrasion. The strength of the boards should therefore be related to the weight of the book.

Made of fibre, sandwiched between two different kinds of fibrous material (cloth and endpaper), and coated with two kinds of adhesive (glue and paste), the binder's board has a strong tendency to warp, particularly but not exclusively in conditions of changing humidity. Even if the board itself resists warping, any shrinkage in cloth or endpaper will tend to pull it out of shape. The rigidity of the selected board — its ability to resist warping — should therefore be appropriate to the conditions in which the book is to be used and kept.

The thickness of the boards should be proportionate to their own dimensions and strength, and to the bulk and nature of the book they enclose. Particularly important books benefit from extra thickness in the board, providing a pleasant solidity as well as a durable cover. Very thick boards can be bevelled, to prevent each outer side of the board from offering two sharp edges to the effects of abrasion, but this is now rarely afforded.

Warping has to be minimized, and the first precaution is not to hurry the binder into using immature or *green* boards. If nothing else is available at the time it is needed, postponing publication will be better than producing books which may be flat enough in the publisher's warehouse but which may distort themselves into hideous outwards curves in the bookshop or on the shelf. No great harm is done to the book by a slight inwards warp, even across the board as well as along it; a second precaution, not usually necessary, is to make the case with the *wire side* outwards (§ 15-3), as boards tend to warp away from that side. But the essential precaution is to ensure that the grain of all the fibrous material in the book — text paper, endpapers, boards, and covering material — lies parallel with the spine. This not only minimizes warping but protects the adhesion between case and contents against the strain of differential expansion and contraction.

Until a few years ago, boards were specified in Britain in terms of the weight in ounces or pounds of a single board in size 25 × 30 inches. The caliper of a single board in fractions of an inch was also used until the metrication of British book production. Since then weight has been expressed as grammes per square metre and caliper as microns (μ), as for printing paper. The ratio between weight and caliper varies according to quality; the stronger the board, the less its caliper in proportion to its weight. A 32-ounce or 2-pound board of average quality is likely to have an inch caliper of about 0.090, a micron caliper of about 2286, and a weight of about 1876 gsm. Standard makings of board are graded in round numbers rather than in such precise equivalents, and a board of 2300 μ is the nearest available equivalent of a 32-ounce board. This caliper may still be considered appropriate for a demy octavo book; a thicker board might be lavish, and a thinner the economical custom in much of today's book production.

Chipboard, made from chips of recycled cardboard, is one of the least expensive forms of binders' board, but has a greater tendency to warp and crack than other boards, and is better not used for books of enduring value. *Strawboard*, once the standard material for the general run of books, is no longer made, and has been replaced by *greyboard*, which is somewhat less brittle. *Millboard* is made of longer fibres and is accordingly stronger. Over about 1175 μ these classes of board are commonly produced in *paste-board* form — two or more layers are pasted together, with the grain of each layer at right-angles to its neighbour or neighbours, as an additional precaution against warping. The hardest, stiffest, and most expensive boards, the least likely to warp, are *hot-rolled* or *hard-rolled* (as paper is calendered) *mould-made millboard*, which has a substantially higher weight for its caliper than less compact forms of board. *Flexible boards*, now usually plastic, are used for books intended to be flexible for the pocket rather than rigid for the shelf.

Many different kinds of covering material are used in cased binding. Cases can be hand-made with leather, vellum, or other kinds of hide which are normally used in bookbinding by hand, but which do not readily lend themselves to machine operations. The strength and durability of hide are incompatible with those of the cased binding itself, and the case is likely to outlast its adhesion to the rest of the book. Cloth was introduced with the binding case in the first half of the nineteenth century, and in the second half of the twentieth is giving way to other materials, though it is still preferred in production of good quality.

Any covering material must be impermeable by glue, which would otherwise emerge on the outer surface of the case. Resistance to abrasion, at least in cloth, is of greater value than tensile or folding strength. When a binding breaks down after long use, or because it is too weak for its own weight, the endpapers tend to split first, and the mull and tapes begin to give way at the hinge; the cloth begins to wear out at corners and edges first. Material weaker than cloth in its resistance to folding may, however, break at the hinge before the endpapers do. Covering material should be capable either of resisting dirt or of being cleaned; light colours are vulnerable unless they have enough water-resistance to withstand a wipe with a damp cloth. Most colours fade in light after a time; fadeless cloth may be necessary if a specific binding design is to retain its effect, but few readers seem particularly to regret the fading of an exposed spine in contrast with the unfaded sides of the same book enclosed among other books on the shelf. The smoother and more compact the material, the sharper and brighter the blocking; the surface irregularity of a hard cloth tends irrepressibly to survive the impact of blocking, and to impart a patterned effect to the wider areas of the blocked design.

The furnish of most bookcloth is based on cotton fibre; linen, which is considerably stronger, is used only in cloth of the highest grade. The *warp* of the cloth is the thread that runs along the loom; if the threads running in one direction are thinner than those running in the other, the thinner threads are usually the *weft* which runs across the loom. The cloth is usually more flexible along the warp, and expands and contracts more across it; the best direction for the warp is therefore up and down the book. The binder cannot always take this into account, as the cloth may have to be cut in the most economical manner possible. Most cloths are *backed* with size, to strengthen their resistance to glue, and some are backed with paper for the same reason, and also to stiffen and thicken the material. They may also be *filled* between (and sometimes over) the threads with casein, starch, or water-resistant plastic, to reinforce impermeability and to add thickness. Cloth may also be plastic-coated for resistance to humidity, but the surface of any covering material is exposed to abrasion, and the coating may have a short life. When cloth becomes worn, filling tends to dissolve, taking with it any pattern printed on the surface rather than blocked into it.

Covering material which is coloured only on the surface shows the effects of abrasion sooner than material which is the same colour all the way through. Cloth can be woven of threads mingling strands of differing colours, or of threads of two colours for warp and weft. The durability of cloth can be assessed from the weave; strength lies in a high frequency of thick threads in both directions, and perhaps because most people realize this they like the look of a high-grade cloth. Embossed patterns on bookcloth are no longer popular; a woven material provides its own natural pattern.

Buckram is a cloth of fine quality, the furnish of which contains linen. Ordinary buckram is woven of cotton and linen; the strongest and most costly buckram is made of linen only. Buckrams are usually calendered, and then have a smooth, solid surface which takes blocking with admirable sharpness; they contain little or no filling. Buckram is the best type of bookcloth, and should be used whenever possible on all heavy books, reference books, and books of permanent value. *Canvas* is a coarse unbleached cloth, the furnish of which often includes hemp or flax; it is strong, unpolished, and rough, and only slightly less expensive than buckram. Its surface does not as a rule lend itself to delicate blocking or to offset or letterpress printing; but the area to be blocked can first be flattened by *blind blocking* (§ 16-8), or by first blocking a coloured panel to enclose the lettering. *Crash* is a rarely used but attractive kind of cloth. It is distinguished by a marked contrast in thickness between warp and weft, and a wide space between the warp threads. The individual threads too are uneven in thickness and colour, and warp and weft sometimes differ in colour.

There are many other kinds of bookcloth, with varieties of furnish, thickness, weave, and surface, and whole ranges of colours. Among these the designer should have no difficulty in finding material to please the eye and hand; and for the larger edition or for a series he may be able to have cloth made to his own specification.

Imitation bookcloth is usually a stout form of paper or card embossed with a pattern imitating the weave of cloth. This is now widely used for novels and other comparatively inexpensive titles. Plastic covering material, usually patterned in imitation of some kind of hide, is smoother, and probably the most humidity-resistant of all covering materials. These imitation finishes can be justified on such materials, which would otherwise enclose the book in a featureless surface.

The covering material can be printed, even in half-tone and colour, before being used for case-making. The design should not be an essential part of the book, and should exclude the wording on the spine, as it is likely to fade and to be rubbed away in time. The essential parts of the design should be blocked after case-making.

The covering material is cut to size for case-making, and the corners are trimmed off (*mitred*) with a diagonal cut (figure 16-7a). The back of the cloth is glued, and the boards and a *hollow* of stout paper, also trimmed to size, are laid in position on the glued surface. The hollow will stiffen the cloth which covers the spine, keeping it in shape and helping it to accept and retain the impression of the blocking. The cloth is *turned in* round the head and tail of boards and hollow, and is glued down to the inner side of the case.

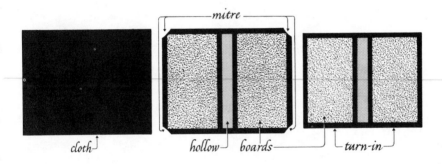

Figure 16-7a. Case-making. Left, the glued rectangle of cloth. Centre, boards and hollow in position, corners mitred. Right, cloth turned in. Below, case and contents, as separate elements; compare figure 16-1d, in which the whole book is assembled into a single unit piece by piece.

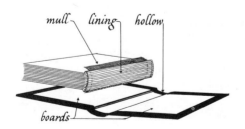

The turn-in of the cloth is usually 10 to 15 millimetres or $\frac{1}{4}$ to $\frac{1}{2}$ inch, and should be even at head, tail, and fore-edge, though a slight increase at the longer edge or edges may improve the appearance of evenness. When the weight of the book allows, the overlap on to the board of mull or linen from the spine should be limited to the width of the fore-edge turn-in, to give the inner side of the board an even panel of extra material all round. This inner side is a point in the book at which part of its structure cannot be concealed; since it will be visible, it should be presentable.

The boards are so trimmed that they will project outside the head, tail, and fore-edge of the text pages, and they are roughly equal to the uncut size of the book. That part of the board which projects beyond the pasted-down endpaper is the *square*. Its width is usually 3 to 5 millimetres ($\frac{1}{4}$ to $\frac{3}{16}$ of an inch), though in quarto and folio books it may be larger. Too wide a square is unsightly, and too narrow a square places the leaves vulnerably close to the edge of the board. The squares should be similar in width at all three edges, at the front of the book and at the back; but big books sometimes have visually even squares — the longest, at the fore-edge, is slightly wider.

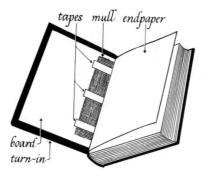

Figure 16-7b. A cased book before casing-in.

In traditional bookbinding, a book wholly covered in hide such as morocco leather is said to be *full-bound* or *full morocco*. If the spine and corners only are covered in hide, and the sides in some less costly material such as buckram, the book is said to be *half-bound*. A book of which the spine only is covered in hide is *quarter-bound* (figure 16-7c). These terms, and two of the styles, have been inherited by cased binding, and a book may be described as *full cloth* or *quarter cloth*. Cased binding in quarter cloth entails a *three-piece case*, consisting of one piece of cloth covering the spine and overlapping on to the boards, and two pieces of other material covering the rest of the front and back boards. The other material is usually more decorative and less expensive than cloth, but in Britain the cost of three-piece case-making usually deprives quarter-binding of economic advantage.

As a decorative form of edition binding, the three-piece case offers opportunities. The cover can be made of three pieces of material which differ from each other, but the front usually matches the back. The width of the spine piece can be varied at will, and could even cover a quarter of the front board and three-quarters of the back board, if so eccentric an effect would suit the book.

A case made with a cloth back and paper sides may be pleasant but is unlikely to have the endurance of full cloth. The wear on corners and edges is at least equal to that on the hinge between board and spine, and if the book remains long in use the spine will outlast the rest of the case. Where endurance is less important, however, the use of a reasonably stout decorative paper for the sides, and even for the whole case, has attractive possibilities.

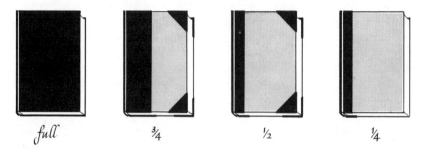

Figure 16-7c. Four styles of cover, three of them composite, in hand-binding.

§ 16-8 Blocking

Most books spend most of their useful lives among their neighbours on the shelf, each visible only as part of a row of spines. The lettering on each spine identifies the text within, and bears witness to the nature of its presentation; without this evidence of identity, the spine may be supposed to be that of a booklet or catalogue, but hardly that of a book in the full sense of the term. An edition has its best chance of being read when the information on the spine is clear, unequivocal, and durable.

The information on the spine will be useful only if it remains legible as long as the binding lasts. The wording printed on a paperback spine is likely to last as long as the cover adheres to the book within. The spine of a cased edition should be capable of surviving longer and more frequent use, and the lettering on it should be more durable. For this reason, lettering on cased editions is usually *blocked* rather than printed. A printed image lies on the surface; a blocked image is recessed perceptibly below it, protected by its depth from the abrasion of use.

Unless the printed material is *laminated* (§ 17-4), an image printed on the cover of a binding case is vulnerable. Even if the covering material is paper, by which the ink is absorbed to some extent, the pigment on the surface will be rubbed away in time. The filling of cloth is fugitive, and takes at least part of a printed image with it. Coloured inks in particular tend to fade after long exposure to light. For permanent decoration of the binding, the binder's methods are more durable than those of the printer, and may include a design blocked on the front board, as well as on the spine, perhaps in pigment as well as in metallic foil: a carefully selected covering material: headbands and tailbands: and a coloured top edge.

The blocking-press employs heavy pressure and heat to overcome the roughness of the covering material, and to force the image into the surface. Heat also fixes the blocked image firmly to its base in the cover material.

Various metals can be used for the application of pressure and heat, but few withstand the combination for long. Unless the relief is comparatively deep and almost vertical, the image cannot be struck deeply enough into the covering material; printing surfaces such as type, blocks, or stereos are not only liable to soften in continuing heat, but are too shallow for the best work. Brass, an alloy of copper and zinc, was for many years the best metal for this purpose, and *blocking dies* are still called *brasses* at times. New etching techniques now enable bronze to be used, and this alloy of copper and tin gives way to acid more readily than brass. The image is printed down photographically on to the metal, so that reproduction proofs of type, or drawn lettering, can be enlarged or reduced. The image is protected with acid-resist, and the non-blocking surface round it is etched away. The wider areas of the non-blocking surface are cut away to greater depth by a hand-guided drill known as a *router*.

Blind blocking is the application of heat and pressure from a blocking surface, without the interposition of any blocked material. It produces a recessed area or pattern in the cloth without any change of colour, and can be used to flatten the grain of the cloth locally in preparation for a subsequent impression with a blocking material. If a rough-surfaced canvas is used, for instance, the lettering may be blocked in metal foil on a blind panel, and even small lettering is then likely to be sharp and clear. Boards are still sometimes panelled with blind rules, and blind ornament or large lettering provides a discreet form of decoration for the front of the case.

Ink is an inexpensive blocking material, with the advantages of an unlimited range of colours and an ability to adhere after pressure without heat. Light colours on a darker covering material may have to be blocked more than once in order to achieve the necessary opacity. The comparative cheapness of ink is best utilized in large areas. Pigment foil is more opaque, but is also more expensive, and it is limited in the range of colours. Metallic foils are perhaps the most popular blocking material, at least for spine lettering in editions of the better kind. Imitation gold does not always resist tarnish as the true metal does, but imitation silver and gold foil of good quality are likely to remain bright for some years. Metallic foil is fed into the blocking-press from the reel; more than one reel can be used at a time, so that different parts of the case can be blocked in different foils in a single pass through the press. Inks and foils can be combined in the same area; when one is to be blocked over another, pigment foil should be over ink, and metallic foil over all. The area of blocking is limited by the width and number of foils; the most economical image is contained in a narrow oblong, across or along the board.

The best material, and the most costly, is gold leaf. Its high cost is due not only to the value of the material itself, but to the labour involved in blocking. The design is sometimes first blocked in blind, to smooth and prepare the surface, which is next rendered adhesive by the application of *glair* (the white of eggs). The area of the image is then covered with gold leaf, which has to be cut and laid on by hand, being far too delicate for mechanical handling. The blocking surface is applied again with heat, and the surplus gold is carefully rubbed off by hand, for return to the gold-beaters. This material and method of blocking gives the sharpest definition and the most lasting brightness, and in spite of cost is still sometimes used for books of permanent value.

Contrast between the blocked image and its background is necessary for legibility. Lettering in metal, for instance, does not show up well on light-coloured covering material, and benefits from the darker background provided by a panel of darker pigment or ink.

Blocking should be confined to the area of the boards and the spine hollow, which receive and retain the impression better than does the cloth alone, and should not spread into the hinge which consists only of covering material. Blocking over the area into which cloth is turned inwards at head and tail may also be awkward, owing to the difference in thickness.

§ 16-9 Identification and decoration

Bookshelves are likely to be scanned at arm's length at least, and the wording on the spine should be easy to read at that kind of distance, even if large lettering across the spine enforces awkward word-breaks. If the spine is decorated in any way, decoration should not be allowed to mingle with information to the point of confusion. The spine wording may be defined as an editorial matter, but not one which interests all editors, and the designer may have to take responsibility for it. The title alone may be enough, but if it would benefit from explanation, or if it is known to be similar to any other title ever published, the sub-title when there is one should be added. Title and sub-title should be exactly as shown on the title-page. The author's name may be abbreviated but only with care; any misunderstanding is likely to annoy the reader before he begins. Whether the publisher's name is to be included is a matter for the publisher. Certain imprints on certain spines are a kind of guarantee; most convey nothing to the general reader, and serve only as an unnecessary advertisement.

The arrangement of lettering on the spine of books periodically stirs up argument among publishers and designers. All agree that lettering should read across the spine if possible, because the usual position of a book is upright on a shelf, so that lettering which reads across the spine is horizontal. The argument concerns lettering which is to read along the spine, either because the wording is too wide or the spine too narrow for any other arrangement, or because the lettering is to be too large to read across. No general and international agreement has been achieved, and book-shelves present the curious spectacle of spine lettering running in three directions — upwards, downwards, and across.

Those who favour the upwards style maintain that the head is more easily turned to the left for vertical reading than to the right, and that a whole row of upwards titles on a shelf can be read in the natural direction, from left to right on each spine and then downwards from book to book. This argument would be more cogent if all spines read upwards. The advantages of the downwards style are less abstruse. If the lettering travels down the spine, the title can be read when the book is lying closed with its front board uppermost. The publisher's imprint, usually at the tail, follows the author's name and the title in much the same sequence as in textual references to book titles.

The advantage of lettering along the spine rather than across is that, unless the title is a long one, the thickness and length of the spine permit the use of large and spacious letter-forms, and even of decorative letters.

Lettering across the spine should not look cramped, or it will be difficult to read; although space is limited, words set in capitals or small capitals are still clearer when letterspaced. Unless the words are short and the spine wide, narrow letter-forms are preferable, and italic lower-case is particularly useful for this purpose. Display sizes of legible type-faces tend to occupy rather too much lateral space; the letters are comparatively wide and loosely fitted, with thick strokes. Type can, however, be quite easily linked in style to the typography of the book itself. A designer who draws his own spine lettering or has it drawn can resort to narrower and more closely fitted letters, constructed of thinner lines.

The front and back boards of most binders' cases seem a little drab to some eyes when they present a plain rectangle of cloth. An absence of inessential detail, however, is by no means unfashionable in industrial design. The boards are not seen until the jacket is taken off, when the book is on the shelf, or while it is being read. There is usually no particular need for a book to be identified on the front board, and decoration outside the area of the spine may be expensive out of all proportion to its value. But the case as a whole offers an opportunity for ornament which is still welcomed by creative people; apart from the endpapers, the rest of the book offers no such area, since whole pages of decoration would assert a distracting contrast with text and illustration. In style and subject, binding decoration should be more apposite to the book itself than the jacket needs to be, and less emphatic than a jacket design tends to be. The jacket is a detachable piece of advertising; the case is part of a possession. Lettering designed for the spine of the jacket may, however, be suitable for that of the case.

§ 16-10 Completing the book

The binder cuts the jacket at head and tail, either to match the height of the cases exactly, or to be fractionally smaller. If the jacket projects even slightly beyond the case, it is likely to be creased at the tail before the book is sold.

The outermost page of each endpaper is pasted, contents and case are fitted together in position (figure 16-7b), and the book is pressed until the paste is dry. The endpaper, pasted to the inner side of the board, fixes mull and tapes between itself and the board. If all goes well, the edge of the mull is parallel with the spine, the tapes are flat and straight at right-angles to it, and the *squares* are even.

Pressing is important; released from pressure while the paste is still damp, the boards would begin to warp. Some binders build the books into piles, with stout wooden boards over and under every book. The boards may have a brass flange along the outer edge; this fits into the hinge, forcing the cloth down into the spine and forming a *groove*, which helps a thick board in particular to open easily. After pressing, the books are jacketed, and are at last complete.

Some editions are enclosed in boxes or slip-cases. Boxes take various forms, for instance with separate lids or with hinged lids, but all completely enclose the book. Few books are convenient to keep in boxes, so the box has no continuing function and may be covered in paper. The lid and its sides may be printed in much the same way as the front and spine of a jacket. Slip-cases leave the spine of the book showing, and can protect the binding throughout the book's life without much inconvenience to the reader. They are occasionally useful to enclose a set of two or more volumes, whether cased or paperback. A good slip-case is covered in cloth for strength, perhaps matching that of the case. Withdrawal of book from slip-case is made easy either by thumb-holes in the open edges of the slip-case, or by projection from the slip-case of the book's grooved hinges. Booksellers prefer jackets even on books in slip-cases or boxes; and the slip-case itself may be identified by a label on its spine, which encloses the fore-edge of the book within.

Offprints are something of a rarity in book production, but when they are required of publishers and printers unused to them they can cause a disproportionate amount of confusion unless arrangements for them are made at the outset. When a *symposium* is to be published — a collection of articles by separate authors — and some or all of the authors are holders of academic appointments likely to be assessed for promotion in terms of their publications among other factors, the editor requires for each author a standard number of offprints of his own article together with the opportunity to order any number of extra offprints.

An offprint consists of the printed pages of an article in a symposium or a periodical, fixed together in one of a variety of ways which might be termed 'binding', or even sent as loose leaves to the author. Since the extent of any article is likely to include oddments of two pages, and since the centre pages of very few articles indeed will ever be the centre spread of a section, offprints are not sewn through the spine but are cut into separate leaves and sewn or stitched through the side near the spine edges, or glued at the cut spine as in unsewn binding. As offprints are sometimes the only payment offered to contributors to a publication, they should have a limp cover printed at least with the words 'offprint from' and the title of the publication.

They should also consist of the author's contribution only. If his article begins on a verso page and ends on a recto, the leaves of his offprint will include the last page of the preceding article and the first page of the succeeding one, unless those two leaves are reprinted for the offprint. Reprinting was all very well by letterpress but is distinctly expensive by offset, and is better avoided by starting each new article on a recto page, leaving the preceding verso blank if necessary. The advantage of such an arrangement is the only justification for mentioning offprints in a book about book design.

The printer's estimate should include a price for the standard number of offprints of all articles, together with a price for the same number of extra offprints including postage to the author. The extra offprints have to be ordered before the paper quantity can be known. In capable hands the administration of offprints is no more troublesome than any other detail of book production, but they do affect the design of the book.

CHAPTER 17

Jacket and cover

THE JACKET of a book is the loose paper wrapper placed round the binding case after the book is bound. The original purpose of the jacket was to protect the case before sale against handling and exposure to light and atmosphere, and the jacket is still sometimes referred to as the *dust-wrapper*. Since any unprinted jacket, unless it is transparent, conceals the identity of the book, the title and the author's name at least appear on its front and spine. The identity of the book can indeed be made more obvious by the use on the jacket of bigger lettering than might be seemly on the case. This emphasis on the book's identity has developed into the most lavish form of advertisement used by the book trade, and the jacket has been aptly described (Rosner, 1952) as a poster wrapped round a commodity.

The place for advertisement in book design is the jacket, and only the jacket. A well-designed and well-made book is its own advertisement; its binding and text do not need that touch of the remarkable — the bright colours, for instance, and large bold types in striking arrangements — which thrusts forward a message rather than offers it. Advertising devices, designed to catch attention rather than to hold it for long, become tiresome in a book which may be owned for years and read or referred to over and over again. Jacket design is part of book design only to the extent that a book designer should be capable of planning a presentable exterior for his product; the best jacket designs are likely to be planned and executed by illustrators and art editors, who might not claim to be book designers themselves. The best typography and illustration do not always make the best jacket. The best judge of an effective jacket may be a bookseller, a publisher's representative or art editor, or even the author — the book designer will do well to defer to their verdict. If he designs the jacket himself, he may justifiably use a more emphatic style than he allowed himself in planning the book. The jacket should however imply nothing that the book does not fulfil in some way; advertisement and commodity are close enough to each other for comparison.

The style of the jacket may be governed by the policy of the publisher, who may wish the jackets of his books to wear a family resemblance, reminding the public of his imprint. This limits the individuality of jacket design, which should grow out of the nature of the text and illustration, and use colours compatible with those of the binding. But publishers do from time to time standardize some idiosyncrasy of jacket design which has enabled an imprint to become a selling factor. The jacket is itself a sales device for the book trade; an edition which is to appear in bookshops should be jacketed, a private publication need not.

Covers for paperbacks serve much the same purpose, while remaining part of the book itself, and apart from the absence of flaps (§ 17-1) are designed in much the same way.

§ 17-1 Dimensions

The dimensions of the jacket depend upon those of the bound book. These are not easy to calculate exactly at the planning stage. A bound *dummy* (§ 19-5) will have to be made up, probably from a sample of paper. There may be some variation in bulk between sample and making, and a hand-made dummy is likely to differ in size from a mechanically bound book. After publication the book may be reprinted on paper of a slightly different caliper, and an unaltered reprint of the jacket may not fit well round the reprint. The best jacket designs, then, are those which allow some latitude for thickness. Spine lettering which occupies the whole width of the spine, and front lettering which begins too near the spine and ends too near the fore-edge, may be equally troublesome.

Once the possibility of some variation in the spine has been allowed for, the whole area of front, spine, and back is available to the designer. A series jacket, which with adaptations is to be used for a whole series of titles each of differing length, will probably need a separate spine for each title, to allow for different bulks, so the image should not continue across the spine from front or back. Apart from this, a picture may run fore-edge to fore-edge from front to back round the spine, or across front and spine, or only across the front. The picture may even extend to the flaps; as long as any text there remains legible, and the picture is worth its extension, a lavish gesture has been made at little if any extra cost.

The height of the jacket will be that of the case. Jackets are delivered to the binder uncut at head and tail, and the binder cuts them to exact size. The image printed on the jacket may bleed at head and tail.

The jacket *flaps* are those extensions of the paper, beyond the fore-edge of each board, which are folded inwards round it. In the closed book, the flaps are held between board and fly-leaf, and so hold the jacket in position round the book. To do this effectively, the flap should be at least half the width of the board, and two-thirds as wide is more effective still. This kind of width leaves room for a reasonably wide measure of such text as the *blurb*, a short passage of advertising copy usually printed on the front flap. When a picture covers the front or back of the jacket, it is probably best if it turns right round the fore-edge fold and continues a little way on to the flap, so that in effect it bleeds round the fore-edge and comes to a concealed end.

The intended position of the image printed on the jacket in relation to the book is not always indicated by the design. A typographic arrangement, for instance, with ample margins at head and tail, printed on larger paper than necessary (as jackets are, from time to time), may leave the binder bereft of guidance to the intended position of the cut. If as a result the jackets are cut at the wrong points, the designer should blame himself, not his suppliers, and take care next time to mark the intended position of cuts on a jacket proof and send it to the binder. Cutting guides may also be printed in the form of fine lines at the edges of the jacket, outside the intended cuts.

§ 17-2 Text and display

For some editions, the jacket need do little more than identify and introduce the book. The title and the author's name usually appear on the jacket front, and also on the spine with the publisher's imprint. Type is accessible and inexpensive for the display lines; fleurons and rules can be arranged for decoration. To print one colour on paper of another colour, particularly when the colours are in harmony with those of the case and the endpapers, may provide enough individuality.

Display types of the more elaborate and emphatic kind are better suited to the jacket than to the book itself, but only a small number of display types have the boldness, the individuality or grace of form, and the economy of set which are the most useful qualities in letters for the jacket. Most alphabets of dry transfer lettering imitate the forms of printing type and tend to be used in its place. The advantages of type are precision of form and fit, and low cost: those of lettering, freedom of style from the established designs available to every other jacket designer, freedom of proportion, and a potent appeal to the eye. Whatever the extent of the displayed words, they can be printed in the maximum height and width when the letters are drawn; the set of type, fixed in proportion to its height, may prevent the use of a fount of apparently equal size. Dry transfer lettering is sometimes crowded too closely together, when width is reduced at the expense of legibility.

On the front of the jacket, the title of the book and the name of its author should be capable of being read at a distance of not less than eight or ten feet; this is the approximate distance from the back of a bookshop window to the eye of a passer-by. The spine of a jacket does not usually allow of such generous display, but here the lettering should be large enough to be read when the book is on the highest or lowest shelf in the shop. Langdon-Davies (1951) suggests that 'no customer will look at a book lower down than three feet or higher than about eight'.

Since the spine and front of the jacket are often visible at the same time, the display lines which appear on them may as well be aligned horizontally with each other as far as possible and particularly at the head.

The relative emphasis of the words to be displayed on the jacket is not always a matter for the designer to decide; the publisher or editor may have to be consulted. The name of an author of continuing popularity often has to be more conspicuous than the title. When a long title is to be emphasized, a few words may have to be selected, with editorial agreement, for display in larger letters than the rest.

The words selected for the main advertising impact on the front of the jacket are best placed near the head, where they can be seen in the bookshop even when part of the jacket is hidden behind other books. For the same reason, the principal words on the front should be horizontal or nearly so; but to arrange lettering at an angle from the horizontal seems to provide a kind of emphasis like that of the slant in italic type.

However fancifully drawn or set, the display lines have to be capable of being read without difficulty at some distance. The principles of legibility suggested in §§ 6-3 and 7-5 apply to jackets as to display composition in the text; thickness of stroke and size of counter, for instance, should be in proportion to each other, and the lines of type or lettering need some space between. So long as these requirements are met, the larger and bolder the letters, the more legible they will be at a distance.

Legibility in jackets is complicated by the use of colours and of illustrations. Letters printed in pale colours will be clear only if their background is dark. Since ink is usually not opaque, arrangements of this kind necessitate reversing from black to white. Conversely, letters in dark colours are clearest on light-coloured backgrounds. Where part of a display line appears on a light background and part on a dark one, part of the line may be printed in positive and the other in negative or reverse. A background broken up with different strokes and colours can make almost any lettering illegible. Combinations of picture and words in the same area need care; wording may have to appear in an area of plain light colour such as the sky, or in a plain panel fitted into, or part of, the picture.

White or light-coloured letters set in a darker background are more conspicuous than black or dark-coloured letters on a tinted background. When the front of the jacket is to be printed all over in four-colour half-tone, letters reversed in white are distinctly emphatic, but at the cost of a line reverse in each of the four-colour half-tone separations. The reverse may be partial, leaving for instance a solid yellow and a dot pattern of magenta to combine into orange letters. Variations of emphasis in display can be derived not only from differences of size, weight, and spacing in the letters, but from differences of colour between letter and background. Letters and decoration which are to be reversed should be neither small nor delicate;

text types do not look well in reverse, and hairlines and serifs in larger type either close up or appear to do so.

In a picture jacket, the spine is often treated as an extension of the front, and the spine wording may be handled in much the same way as that on the front. Size in the spine lettering has more value than on the spine of the case. Larger letters, even when turned sideways, are easier to read at a distance than smaller letters arranged horizontally. Since in the bookshop the book is likely to be either standing up or lying on its back, turned spine lettering on the jacket should invariably read downwards.

The back of the jacket has little advertising value, except when front, spine, and back form a single design so striking that the bookseller may choose to show the book open in order to display the whole design. More often, the back is spare space, used perhaps for an advertisement of related titles published over the same imprint, or for a picture of the author (if that seems likely to enhance the book's appeal to its public), or for excerpts from reviews of his other books. Even when the book designer has not himself dealt with the front and spine, he may have an opportunity to arrange the back in a style compatible with the rest of the jacket and perhaps also with the typography of the book. If no copy is provided for the back, the design on the front may be repeated there without much extra expense, to avoid a partly blank exterior at the point of sale.

When a potential buyer first opens the book, he is likely to look first at the front flap, the traditional position of the price. This flap is therefore the right place for the blurb. It may be headed by the book's title, to indicate the subject of the blurb; it should leave room for the price in the bottom right-hand corner; and it should be contained within the front flap. Not every reader's enthusiasm for blurbs will send him hunting for its completion elsewhere. A provocative or impressive blurb, when brief enough, may be placed on the front of the jacket as part of the design.

The price should be printed on one of the right-hand corners, in such a position that a single diagonal cut will remove it if the book is to be privately presented. When the jacket has been designed by a freelance whose services are available to other publishers, the name of the artist or designer should appear on the front flap. Standards of jacket design are likely to improve if all can see by whom the better kind of work is done. An artist's name is better placed on a flap than beside or within the picture on the front, and there is no need for it to appear in both positions.

Not all readers examine the back flap at all, and it should not be used for anything essential. If there is no room elsewhere for such items as excerpts from reviews or a short biography of the author, they can be printed on the back flap, but if they are to have a good chance of being seen before sale they will be more effective under the blurb. A note on the front flap, between blurb and price, can be used to draw attention to the back flap.

Text and display printed anywhere on the jacket should be far enough from intended cuts and folds to preserve the lettering from inaccuracy in cutting and jacketing, or from any discrepancy in dimensions between book and jacket image. Text on the jacket is likely enough to be read if it is designed with the same care as the text of the book itself, and placed within tolerable margins however narrow.

§ 17-3 Ornament and picture

Some kind of decoration, which conveys no message but attracts the eye, is commonly applied to the front of the jacket, which otherwise may look like (and even be) a version of the title-page. A picture may be used as a form of decoration, for instance when a small wood-engraving appears with type or lettering. Jacket pictures are not strictly illustrations, which are intended to illuminate the text; they are, rather, part of a pictorial advertisement, and they may symbolize the text or introduce it or merely be compatible with it, but they need not refer directly to it.

Until a few years ago, most composing rooms equipped with metal type could offer a selection of typographic fleurons, plain and ornamental rules, and border units. Compared with the immense variety available, the selection was usually quite a small one, but it offered some scope for ingenuity. The composition by hand of such material naturally accompanied the composition in metal of the wording on the jacket. But this kind of equipment was little used on the whole; composing rooms which hold it are becoming fewer, and those that survive are jettisoning under-utilized material. There are comparatively few dry-transfer ornaments to take its place. Photo-matrices of decorative units are available for use with display photo-typesetters and even with photo-typesetters, but the compostion of a rectangular border is likely to test the ingenuity of the operator as well as the versatility of the machine. Until a wider variety of decorative material is readily available for relatively easy composition, typographic ornament is likely to remain unfashionable, though all the more welcome on its occasional appearances.

In the small rectangular area of a jacket, decoration is usually applied in one of four main styles, which may be called *spot, line, panel,* and *background.* Each style may be used alone or in combination with one or more of the others, and more than one application of any style other than *background* is possible on any jacket. Spot decoration is based on a single item, of any shape and size, surrounded by undecorated space. A single small fleuron, a heraldic device, or a diamond-shaped pattern of fleurons, between title and author's name on the front of the jacket, are examples. Line decoration is arranged in a line, usually straight and often parallel with the lines of lettering. It is likely to consist of a rule or a row of border units or a combination of rules and units. A typical form of panel decoration is a rectangle, which either bleeds or is bounded by edges parallel with those of the jacket. This style may include frames, perhaps of border units and rules, usually enclosing lettering. Lines or panels are not essentially straight or rectangular.

Background decoration covers at least the front of the jacket, or it would be a panel. Unless the background is light-coloured, the lettering tends to appear within panels superimposed on the background. Panel and background decoration may include undecorated reverses of lettering, which are decorative simply by providing colour.

The four main styles are not often classified in these terms by jacket designers, but they represent an analysis of methods of book decoration which provide a starting-point for the planning of a jacket or a series of jackets.

Freehand drawing in lettering and decoration attracts the eye by its freshness — every stroke without an exact precedent anywhere — and by the evidence of a human hand at work. Lettering in jacket design need not necessarily be beautiful, formal, or even precise, when every character is drawn for a specific jacket by a competent lettering artist. Such letters can themselves provide all the decoration there needs to be, but some artists can also add anything from a simple *paraph* or calligraphic flourish to a picture composed of freehand flourishes and loops. This kind of decoration usually accompanies italic or at least formal and conventional lettering; adaptations of typographic letter-forms and odd innovations are equally valid in the field of jacket design.

Pictures are widely used on jackets; for books which rely mainly on casual sales they are almost obligatory. The jacket design of an illustrated book is often based on one of the text illustrations or on similar work, specially drawn or photographed for the jacket by the illustrator. Perhaps the best jacket for an illustrated book is one that serves as an introduction to, or sample of, the illustrations within — a photograph is best used for the jacket of a book illustrated with photographs. The jacket of an unillustrated book should be compatible with the text; the artist will need to see the typescript or a proof, and should be required to conform with the text as far as possible in letter and spirit.

Line drawing in a single colour seems to lack impact as advertisement, as though it were too diagrammatic and inadequately representational for the public taste. Those drawings which approach the appearance of a photograph, with an emphatic statement of solidity and perspective, are exceptions. Even line drawings with colour washes are no longer much used for the jackets of popular books. Photographs or paintings reproduced in colour half-tone seem more likely to appeal to the public eye.

Selection from existing photographs may be easy enough. If the final decision is to be that of the editor or the publisher's sales manager, a designer's selection can take place in competent company. Commissioning photographs is far from simple. The range of subjects is limited to those which can be prepared for and reached by the camera without excessive cost. Photographic studios and their lighting, stage properties and costumes, human models, and seasonally appropriate scenery are all difficult to find for the right money or at the right time. The art editor's imagination is better concentrated on the composition of everyday scenes, people who are not models of people, still life, and evocative detail.

Drawing and painting in colour offer wider possibilities. The artist can use photographs to take the place of human models and scenes, and much of his work, unobtainable from any other source, may spring from his experience, knowledge, and imagination. Colour, shape, proportion, perspective, and contrast are all at the service of advertisement rather than of reality. But the best kind of work in this field, as in that of photography, is expensive, as most of it is done by artists active in the lucrative advertising market.

The colours of such striking and expensive jackets are under the artist's control; those of simpler jackets remain for the designer to decide. Coloured paper and coloured ink can be combined even in the cheapest jacket. The colours of paper and print should suit not only each other but the subject and appearance of the book within. This is partly because the head and tail edges of the case can be seen even when the jacket is on, and partly because the discriminating buyer of books, removing the jacket to look at the binding, sees the two side by side. Any discord of colour between jacket and case will be obvious at that moment. An unsuccessful attempt at a close colour match is likely to be equally disappointing. Matching the reflection ranges (and therefore the colours) of physically dissimilar surfaces such as cloth and paper, whether plain or printed, is always difficult, and different but compatible colours are more likely to be successful.

Since the jacket is to be seen from some distance, and is to appear in the bookshop among many other jackets, the best jackets (at least for books intended to be popular) are rather strong and bright. Combinations of colours gain from pronounced contrast. Delicate tints are best used in large areas, as background for lettering and ornament in stronger colours.

§ 17-4 Paper, print, and surface

The thickness of jacket paper depends on the size, weight, cost, and importance of the book. A large and heavy book needs a fairly stout jacket, since it may tend to fall out of a flimsy one. A thin jacket is apt to seem shoddy round an expensive or particularly important book. All jacket papers need to resist tearing more effectively than text papers, and thickness contributes to strength.

Bulk without proportionate weight contributes to weakness, and bulky papers should not be used for jackets. The best kind of paper is the strongest, and cartridge papers are to be preferred. A suitable coated or uncoated cartridge can be found for almost any kind of jacket. The grain of such comparatively stiff papers should lie parallel with the fore-edges, round which the jacket will be folded, to help the jacket to lie close to the case.

Unusual texture and colour in jacket paper have an advertising value of their own. Papers made from separately coloured fibres are an example; a thin mixture of darker fibres on a lighter background give the paper an individuality hardly to be matched by any other means. Laid paper should be printed on the dandy side, to reveal the surface texture. Imitation cloth, made of embossed paper, provides an almost luxurious jacket. Rough surfaces, however, tend to collect dust in the bookshop.

Whatever the outer surface of the paper, it serves best when fairly rough on the inner side. Smooth papers tend to slide off the case, and paper coated on the outer side only is better than an art paper suitable for text use.

Coated paper is not ideal for jackets, because of its sensitivity to damp, and its tendency to show marks after being handled and to crack along a fold. Coated papers on the whole are better suited to varnishing or laminating than uncoated papers, which absorb some of the varnish; this may be slow to dry, and reduces the thickness of the surface varnish and hence its glossiness. A hard smooth art paper offers a receptive surface to the adhesion of a laminate. The reflective surface of a glossy art paper adds to any jacket that brightness of colour which is more of an asset in advertising than in illustration.

If letterpress printing fails to hold its own among jackets, much of the quality of the plainest and most economical kind of jacket will have to be relinquished. Very few type-designs look well in sizes above 18-point when they are merely enlargements of a set of all-purpose type drawings; but in sizes above 18-point the excellence of letter forms skilfully designed for reproduction in large size is clearly appreciable, and the wide diversity of display founts in metal is not yet obsolete.

Paper can be blocked in metallic foil in much the same way as cloth and imitation cloth. On a jacket or paperback cover, the effect can be spectacular enough to be worth reserving for special use. When lettering is blocked in gold or silver foil, it looks pale or dark according to the angle of the light; and at one angle, reflecting a bright light directly to the eye, it shines brilliantly out of its background.

An experienced jacket printer chooses inks which will not fade on exposure to light; an experienced jacket designer keeps a few recent jackets exposed to light for a while, to make sure the right choice has been made.

Any printer will need to know in advance that a jacket is to be varnished or laminated. Either of these processes may cause the colours of certain inks to *bleed* or run slightly, and may also induce some separation of pigments, so that a black ink, for example, may bleed a shade of blue. Unless varnished or laminated, dark ink printed on the front or back of the jacket may set off on adjacent jackets in the bookseller's shelf, and may show scratches and rub-marks after being handled and moved.

Varnishing and lamination protect the ink, and brighten its colours by reflecting colourless light with the coloured light reflected by the ink. *Varnish* can be printed over the whole jacket by the printer, in much the same way as ink; it should be colourless, but sometimes shows a slightly yellow tinge. *Lamination* is a separate film, fixed to the jacket by an adhesive; it provides a more reflective surface, more pleasant to the touch than art paper or ink. It resists dirt from the atmosphere and from handling, can easily be cleaned, and adds strength and stiffness to the jacket. When the whole jacket is laminated, the flaps may curl when released, and the film tends to peel away from corners after cutting. The laminate should preferably extend round the fore-edge fold but not far across the flap, so that the rougher side of the flap can be held firmly between the pasted-down endpaper and the fly-leaf.

CHAPTER 18

Extent, price, and cost

THE EXTENT of a book is the number of pages of any kind, including blank ones. Endpapers are not counted as pages, except *self ends* which consist of the outermost leaves of the text, which passed through the press with the rest and are pasted down to the boards. Usually plates, and sometimes preliminary pages, are included as separate totals in any statement of the extent.

The *price* of a book is usually understood to be the retail price at which it is sold to the public.

Production cost, from the publisher's point of view, is the amount of money he pays to his suppliers for the manufacture of the edition. The publisher may choose to include such items as editorial fees in this category; the designer may not be free to decide about the inclusion or omission of these charges, but he will need to be aware of them. The publisher's estimate is a forecast of production cost; the supplier's estimate is a forecast of the price he expects to charge.

Estimates, from the principal suppliers of printing and other services and materials, are based on an estimated extent or *cast-off*. These estimates are combined in the publisher's estimate, together with items the publisher may rightly expect without having been warned to do so; a notional sum, for example, may be allocated for alterations to type, although the cost cannot be forecast before the alterations have been marked on the proof.

If the designer's production plans for an edition are not economically valid, they may have to be changed. To make sure they are valid before they are complete is laborious. As a rule they should be assembled as part of an economically rational plan, and economically tested by the publisher's estimate. Alternatively an experienced designer, given the intended published price, may be able to work out for himself what production expenditure can be afforded. Whether he can then formulate production plans to fit within that budget is quite another matter. In order to produce an economically sound plan, he will have to understand the economics of book production and published price, and he may have to work out for himself the approximate extent of the book.

Good quality in book production tends to cost more than inferior quality. Anybody can quantify the difference in cost. Nobody finds it easy to prove the commercial advantage of good quality. Unless the management of an organization positively intends to maintain a specific standard of quality, the designer may have to assume responsibility for the difference in cost between first-class and third-class production. The publisher's production bill in any year is likely to be the major part of his total expenditure, and therefore represents an area in which marginal cost reductions appear to be worth most.

A commercial case can however be made for affording production of good quality. A generally low standard of production is likely to restrain the publisher from fixing prices the public might well pay for better workmanship and materials. Some authors observe production standards with critical interest, wishing to have a well-printed book to show for all their work; and if they do not notice much at first, they may soon be enlightened by reviewers. Inferior printing may convey an impression of financial insecurity. Almost any publishing organization is in competition for the good opinion of authors, agents, reviewers, and the public; books are the most conspicuous evidence of the publisher's taste and skill; and the evidence should not be allowed to favour the competitors. A high standard of production may or may not sell more copies of a specific title, but it is likely to bring in more authors and more book sales to a specific imprint, even at higher prices. A low standard, in an area of competition, is a risk unidentified and ignored. The only standard worth striving towards is that of the best in the field.

Working within or for any organization, the designer who wishes to move towards the best will have to earn the confidence of the management by thrifty technical planning and by a vigilant sense of proportion. There will be editions which benefit from that appearance of authority which springs to the eye from good materials and good workmanship; and there will be others, long established and in demand perhaps, for which a low price or rapid replacement or both may be more important. True economy is not niggardly, it is a purposeful disposition of expenditure, but an excess of expenditure at the wrong point is extravagance.

§ 18-1 Casting off

To *cast off* any kind of copy is to calculate the space it will occupy in a specific style. A single word may be worth casting off, for instance to determine the width on the title-page of the title itself, when set in a provisionally selected fount and alphabet. Illustrations are cast off, when the size of each has been planned. The printer has to cast off the whole book as accurately as he can, unless he is to proceed without estimating or with a provisional estimate only. His estimate for composition will be based partly on the estimated number of characters in the whole typescript, and partly on the number of characters he expects to fit into the average typographic line and page.

All design plans for an edition should be based on some kind of cast-off, however approximate. Merely to recognize a typescript as unusually long or unusually short may at times be enough, but there will also be editions which at the planning stage should be fitted into a specific extent. Headlines consisting of several words, to be set in letterspaced small capitals, should be cast off in that style to make sure they will not have to be either abbreviated or set in italic upper and lower-case, if they are to fit into the measure. Casting off is the method by which the designer fits copy of irregular extent into specific areas.

The most effective casting off is based on an assessment of the number of characters in the copy. The number of words is an unreliable base for calculation, as the average number of characters per word varies from book to book. Measurements of a variety of texts in English have indicated an average word-length of 5.24 characters, including the space after the word; the figure for this paragraph is 5.65.

When dealing with a few words only, the designer counts the characters, but when he has to do with more than one folio of typescript, he may not have time to do more than assess the character-content by means of averages. A character, for this purpose, is anything which causes the carriage of the typewriter to move on one space. This may include a letter, a punctuation mark, a figure, a sign, and the space between words.

Accuracy in casting off is a matter of time. Working on evenly typed copy for not more than 30 minutes, a designer may be able to assess extent within 10 per cent. A printer is likely to spend several hours, and would usually expect to estimate within 1 per cent of the actual extent.

The process of casting off the whole book falls into three parts. The first is an estimate of the number of characters in the copy. The second is an estimate of the average number of characters on the printed page. The third part is a comparison between the character-content of the copy and that of the printed page, resulting in an estimate of the total extent.

Most typewriters used for publication copy have a fixed set — the carriage moves on the same distance for every character, whatever its visible width. When typescript is regular, in the use of one typewriter throughout, in an even average length of line, and in an even number of lines per folio, unevenness in the length of line can be measured in terms of characters, and the content of the folio calculated by multiplication. Irregularities in the typescript, such as disproportionately wide indents in the first line of each paragraph, too much space between sentences, and extra space between paragraphs, have to be assessed and compensated for. Anything to be set in a style different from that of the main text may have to be separately assessed; extracts and footnotes in smaller types are examples. Space in the last line of each paragraph in the copy may be assumed to balance with similar space on the printed page. The character-content of the whole text in copy form is the average number of characters per folio multiplied by the number of folios.

Having planned the style of composition, at least provisionally, and assessed the character-content of the copy, the designer is in a position to calculate the typographic area likely to be occupied by any number of characters.

To cast off an item of less than one line, measurement of individual letters is best. A display line may contain a proportion of capitals or other wide letters which would distort an average. Type specimens, showing all or most of the founts in a series, and all the standard characters in each fount, are essential for this measurement.

The average number of characters in the text line is more often necessary, more important, and more exacting. The surest guide to this essential information is a specimen page, or several pages, set from the actual typescript by the actual printer in the actual style selected. Such a specimen may well guide the printer in his own casting off, but is unlikely to be ready while the designer's plans are still provisional. Earlier editions, in proof or after publication, specimen pages in type specimen books, and casting-off tables may have to serve instead of a bespoke specimen.

Given such a specimen, the designer selects five or ten lines he considers to be typical in character-content, and counts the characters in each to find the average. Allowances may have to be made for variations in the width of word-spaces, caused by unusually long words (as in a post-graduate work on chemistry) or by a narrow measure. The fewer the word-spaces in the line, the greater the variation in their width.

Verse is an example of copy which is cast off by lines. However short the line of copy, it will make one printed line. A line of copy which has too many characters to fit into the measure will make two lines. Long passages of curt exchanges of conversation may need the same treatment, which may have to be applied to any passage in which the number of short lines rises beyond about 15 per cent.

The British tradition of typographic composition favours a standard depth of page throughout the book. The number of lines per page, however, varies from page to page at the beginning and end of chapters, and when there is any insertion of extra space between lines and any change of type-size. Passages to be set in smaller type, already separately assessed, can now be separately cast off and added in. When the character-content of the whole text in copy form is ready to be converted into an estimated number of printed pages, the three stages of casting off, introduced above, can be represented by simple formulae.

The character-content of the copy = the average number of typescript characters per line × the average number of lines per folio of copy × the number of folios.

The character-content of the printed page = the average number of typographic characters per line × the number of lines per page.

The minimum extent of the printed text in pages = the character-content of the copy ÷ the character-content of the printed page.

The extent is a minimum because these formulae make no allowance for the essential addition of items which can only be cast off by page. Such items make a page or part of a page regardless of their character-content, and they may indeed have no significant character-content at all. Illustrations, tables, preliminary pages other than such textual items as a preface, and the index are usually added as pages to the product of the formulae. The index cannot be cast off until it has been compiled, and the page increment for it may be from about 2 to 10 per cent of the number of printed text pages, according to the nature of the index. If the text has not been cast off by chapters, a page should be added for each chapter when each is to begin a new page, and may end part-way down its last page.

If the final total falls short of a multiple of 8 pages, anything other than a very accurate cast-off should be increased to that multiple. Other additions may be justifiable because it is better to expect too many pages than too few.

Any typographically simple book, such as a novel, which is to be printed in large numbers should whenever possible be planned to make an *even working*. In an edition so planned, every sheet which passes through the press has room for the same number of pages as the other sheets. An octavo book, for example, when printed on a quad perfector, has 64 pages printed on most of the sheets if not all. If its extent is 256 pages, it is printed on four such sheets. If its extent is 288 pages, the *oddment* of 32 pages will be more expensive page for page to print than the rest of the book. A book printed as 312 pages, on four sheets of 64 pages, one of 32, one of 16, and one of 8, is likely to cost more to print and bind than a longer book of 320 pages. More complex editions cannot often be planned or restricted in this way, and when the number to print is not a large one the effort is not economically necessary.

A cast-off is only an estimate of extent, and an error of less than 1 per cent would often be enough to introduce an oddment. When an exact extent is crucial, the book should be cast off in galley before it is made up into page. Printer or designer may still be able to adjust the original design to compensate for inaccuracy in the cast-off of the copy.

§ 18-2 Estimated expenditure

An estimate is a forecast of the cost of work or of material or of both; the cost may be anything from that of a single item — such as the paper for the text, or alterations to type — to that of an edition. The physical quality of any book is limited by the highest price at which it can be sold. The price depends very largely on the cost of production, and the maximum expenditure on production depends upon the price. This interdependence is balanced only when the ratio between price and cost has been carefully calculated, and when cost has been precisely and comprehensively estimated.

The comparative costs of work and materials have to be borne in mind when decisions about them are made. Estimated costs should be relied on for action only when based on prices quoted by suppliers for precisely specified work and materials. Complete copy for text and illustrations is part of this specification, without which a supplier's estimate is not contractual.

The cost of any item has to be assessed in two ways — as a total for the whole edition, and as a cost per copy. Costs are best assessed by comparison with variant forms of the same item, with the costs of other items within the same edition, and with the costs of the same item in other editions. The total cost of composition, for instance, is likely to be much the same when quoted for by a particular printer offering a particular composition system, whatever the style of setting; a substantial reduction in cost may be found only by seeking another composition system or another printer or both. But a reduction in the composition price may not be worth looking for when the number to be printed is great and the printer's presswork price is favourable, or when the quality of composition is likely to be reduced together with its cost.

An economically valid design for an edition is based on a specific number to print, since a variation of a thousand copies in the number might be enough to indicate the economic advantage of other plans. A design so based must still be proved by means of a comprehensive estimate. Every single event in the course of the edition's production, from the first communication with any supplier to delivery of the last copy to the publisher's warehouse, has to be foreseen, evaluated in terms of cost, and included in the publisher's estimate.

§ 18-3 Preparatory costs

Production costs may be divided into three classes, each having different characteristics. The first of these comprises most of the preparatory costs which have to be incurred before presswork can begin on a new edition. When completed, the work giving rise to such costs does not have to be repeated for a reprint. In an edition of a few thousand copies, this class of cost forms a high proportion of the total production cost of the first printing.

Fees of various kinds may have to be included in the production estimate. Which are to be included, and indeed whether any are to be included at all, is decided by the publisher. Translation from a foreign language may incur the biggest of these fees, and may cause some modification of design plans towards stringent economy. Editorial and sub-editorial fees tend to be less onerous. Text or illustration not originated by the author may give rise to copyright fees. Fees for commissioned illustration, book design, and jacket design are usually included in production estimates. The designer has to know which fees are to be shown in the estimate, and how much to allow for each.

Like presswork and binding, composition is in fact a series of processes, but is usually estimated for as a single sum. The composition price per page depends mainly on the number of characters per page and the degree of complexity. The price per thousand characters rises for foreign languages, particularly in foreign alphabets; for any element of handwork in metal make-up; for heavily accented languages such as classical Greek; for complicated typographic arrangement, as in mathematics; for a high proportion of special signs and letters, as in phonetics; and for anything else, including inadequately legible copy, that tends to retard the standard rapidity of keyboard operator and proofreader.

Each specimen page is likely to cost up to twice as much as the composition price per page, and should be included in the printer's total composition price: a minor reduction in price is then within reach when a specimen is unnecessary. Alterations to type should always be allowed for, since they will nearly always be made; the amount to allow is that allocated by contract to the publisher before the author begins to pay for the excess beyond it. The cost of any extra proofs not allowed for in the specification in the printer's estimate should appear in the publisher's estimate.

Authors of works of information sometimes provide their own diagrams, intended to be ready for camera, but not always good enough to contribute an appearance of authority to any publication. In accordance with the contract between author and publisher, the author may have to pay for re-drawing, but otherwise this is an additional item for the estimate. The reproduction of illustrations, up to the point of either providing a printing surface (such as letterpress blocks) or preparing one (for instance, the positive or negative film for offset half-tone), is a preparatory cost; any additional charge (printing down offset plates, for example) for the printing surface itself is better allocated to presswork. Different forms of illustration proof before offset printing give rise to different costs. The cost of alterations to illustration proofs can be assessed only by experience of handling similar illustrations; there are too many variables for theoretical calculation.

The cost of preparing the exterior images of the book is the last example of this class. Blocking dies, type-setting, and artwork for the cover or jacket, and reproduction in the sense defined above, are the principal items.

The second class of cost is that incurred only by preparation for a reprint.

There will usually be alterations to the title and its verso, and there may be other typographic amendments within the text, for instance to correct literals. Imposition may or may not be necessary, depending on the printer's usual procedure. If no other printing surface is available or can be prepared from existing printing material, two copies of the printed book will have to be used as camera copy for offset film. Cover or jacket may need modification to fit round a different bulk.

Like the items mentioned in § 18-4, these are the main sources of cost within their respective classes. When the plans for any edition are analysed for estimating purposes, other items are likely to appear.

§ 18-4 Manufacturing costs

The third class of costs is different in nature from the first two, defined in § 18-3. Manufacturing costs are on the whole proportionate to the numbers of books printed and bound: preparatory costs are on the whole proportionate to the number of pages, and as a total are affected only indirectly by the numbers to print and bind. Manufacturing costs are incurred by all new editions and all reprints.

The cost of paper, bought for any part of the book or for the main text, is directly proportionate to the number of copies printed. The quality of presswork and binding depends on the quality and suitability of the paper; the physical presentation of any edition can be debased by unduly cheap paper. The cost of paper per copy may be reduced by a printing order for a large number, as lower prices are available for larger tonnages, and paper bought on the reel for web-fed presses is less expensive than sheets. The amount of paper ordered will have to include whatever overs the printer will need; *overs* are additional sheets beyond the number of sheets per copy × the number of copies ordered. The printer cannot avoid spoiling a number of sheets in the course of make-ready, and more in the course of printing; and the binder cannot avoid spoiling more. A printer who supplies paper usually adds a materials surcharge to its price.

Presswork costs differ from other manufacturing costs in the initial cost. This consists mainly of make-ready, but may include off-machine processes and materials required each time the edition is prepared for printing. The initial cost is higher for the first sheet of a book to be printed than for subsequent sheets of the same book. The ratio between initial cost and running cost may have a significant effect on the cost per copy of a small edition or reprint. The cost of the actual printing, after make-ready, is directly proportionate to the number printed. Presswork charges generally are influenced by sheet size rather than by the number of pages. On the whole, a smaller press costs less per page to make ready, and more per page to run, than a larger one. Oddments cost extra per page. When the standard sheet of an edition is printed from formes of 32 pages, for example, an oddment of 16 pages is usually charged for as though it were 24 pages. Colour printing tends to cost more per colour than printing in black only.

Binding costs also include an initial charge for setting up machinery. This is small compared with that of make-ready in printing, but bears heavily on binding orders for 1,000 copies or less. Once the machinery is set up, the difference in cost per copy varies only marginally with variations in the number to be bound. Materials account for a high proportion of any binding price, and offer opportunities for ill-judged price reductions which may threaten the quality of the book. Handwork of any kind, applied to every copy bound, gives rise to substantial cost increases. One way of reducing the edition's total cost of production is to with-hold part of the edition from binding. This will increase the final cost per bound copy of the whole edition, but reduces risk and investment.

The cost of bulk delivery, whether of sheets to the binder, or of bound copies to the publisher's warehouse, should be allowed for; provisional delivery instructions therefore form part of the designer's specification when requesting a price.

The copy itself, together with design instructions, provides a basic specification for quotations from suppliers for preparatory costs. For manufacturing costs, the designer has to rely on comprehensive descriptions of the work to be done and the materials to be supplied to or by the supplier, whether or not each item is part of his design plan. The specification in each quotation from a supplier will also have to be scanned, in case an item proposed by the supplier in the absence of other instructions becomes contractual when the price is accepted.

§ 18-5 Price calculations

Each publisher has his own method of calculating an appropriate published price for an edition. A book designer who intends to make purposeful use of the publisher's resources to provide well-made books has to understand this method. Until he does, he cannot deduce from the published price a maximum permissible expenditure on production, nor from the expenditure he expects can he deduce an economically minimum price.

The tendency of publishers in Britain is to lead their competitive effort with price, as a trained right-handed boxer leads with his left hand in a fight. The tendency of a book designer should be to lead with quality, since nobody else in the organization may be particularly interested in doing so, and since somebody should. Unless the designer reconciles his own preference with that of the publisher, his own future and that of quality in the publisher's book production may be in danger. For this reconciliation, calculation provides a better base than appeals to the history of typography or to the instinct of the craftsman, since even a successful publisher may have little knowledge of either.

From among assorted methods of price calculation, two only are described here, as examples. The first is appropriate when the odds (in the sense of another kind of bookmaker) are on large numbers to print and an indefinite continuation of reprints. These odds may apply, for instance, to a successful list of school-books for primary education. Preparatory costs may then be allocated to a budget, controlled by a managing editor, which is taken into account in pricing calculations as part of his overheads which he expects to recover from gross profit (6, in table 18-5). The published price is proportionate to the manufacturing cost, and is influenced only indirectly by the preparatory costs of the edition itself. The editor is then in a position to extend his planned expenditure on the composition and illustration of the edition without driving up its published price. The effect of such a method on book design may be to combine in one edition a high quality of preparatory work with a lower grade of manufacture.

A second method is more common. It is appropriate when the odds are against any reprints, as they usually are; the publisher's aim is to profit from the sale of the first printing, whether or not reprints follow. The profit is not always monetary; the prime purpose of publication may be propaganda, or prestige, or culture, and monetary profit may have to be provided by other titles to pay for it. Preparatory and manufacturing costs, in this method, are combined in a direct relationship with published price.

Any relationship between production cost and published price is most easily defined in terms of the unit. The *unit cost* is that of a single copy of a book. It is equal to the total production cost applicable to price calculation, divided by the number of complete copies to be produced. The *unit price* may be defined as the retail price of one copy in the publisher's own country.

A potentially profitable ratio between the unit cost of production and the unit price can be ascertained by projecting in percentage terms the expenditure and income to be expected from production and sale of the whole of the first printing. To do this, the designer will have to find out what proportion of the books printed are likely to be sold at which book-trade discount, home and abroad; what the royalty payments to the author will be, as a percentage of the value of such sales at retail price; and what gross profit, also as a percentage, but of actual receipts from book sales, the publisher intends to recover. The designer will have to base his calculations on a specific number to print, as changes in the rate of royalty are usually dependent on the actual number of copies sold. Gross profit is the surplus of income from book sales remaining after production and royalties have been paid for. It is usually expressed as a percentage of receipts from book sales. It expresses the intent of the publisher with regard to the edition, and comprises a share of all his organization's expenditure other than production, together with a margin of net profit after that share has been earned. Income from subsidiary rights is often omitted from price calculations, but the publisher may alternatively choose to take it into account.

A ratio between the minimum economic unit price and the maximum economic unit cost can be calculated by a percentage formula, of which an example is given in table 18-5. The published price is represented by 100, but all the other figures are hypothetical.

In the formula's suppositious circumstances of discount, royalty, and gross profit, just over 20 per cent of the unit price is available for production. This ratio enables the designer to calculate from any unit price a production

TABLE 18-5. Ratio between unit price and unit cost

Except where otherwise stated, percentages are of unit price   %

(1)	Unit price				100.00
	Book trade discount and free copies				
	Home:	60% of edition at	35% discount	21.00	
	+ Export:	35% of edition at	50% discount	17.50	
	+ Review copies:	5% of edition at	100% discount	5.00	
(2)	= Average discount per copy			43.50	43.50
(3)	Average receipts per copy from book sales				56.50
	Author's royalty				
	Home:	50% of edition at	10%	5.00	
	+	10% of edition at	12.5%	1.25	
	+ Export:	35% of edition at	5%	1.75	
	+ Review copies:	5% of edition at	nil royalty	—	
(4)	= Average royalty per copy			8.00	8.00
(5)	Average receipts per copy (3) *less* average royalty per copy (4)				48.50
(6)	Gross profit: 50% of average receipts (3)				28.25
(7)	Available for production: unit cost				20.25

budget for the edition, or from an estimated unit cost to derive a unit price. The unit price in this example should be not less than five times the unit cost; the unit cost should be not more than one-fifth (20 per cent) of the unit price. When the published price has been provisionally decided, the estimate of unit cost brought up-to-date if necessary on the completion of preparatory work, and some information about probable sales received, the whole formula can be repeated in monetary terms instead of percentages, to verify the price calculations.

The effect of spending too much on production, or of fixing too low a price on the edition, may be seen from an assessment of the *break-even points*. These are stages in the continuing sale of an edition at which real expenditure such as total production cost, or notional expenditure such as overheads, will be recovered from receipts.

Total production cost = unit cost (7, in table 18-5) × the number to print, which for demonstration purposes might as well be 1000, so investment in production is 20.25 × 1000 = 20,250. When this figure is divided by the average receipts per copy less the average royalty per copy (5 in the table), 20,250 ÷ 48.50 = 418 copies which will have to be sold before receipts have reimbursed the publisher for his production investment and before they begin to make some contribution to the overheads or general costs of running the business. For any title sold in the economic circumstances postulated in table 18-5, then, there will be no profit of any kind from the sale of the first 42 per cent of the edition. And if the unit price were to be reduced below its economic minimum or the unit cost allowed to rise above its economic maximum to a point at which unit cost were 25 per cent of unit price instead of about 20 per cent, there would be no profit of any kind from the sale of the first 52 per cent of the edition, and the gross profit on sale of the whole edition would sink to 42 per cent. If 40 per cent of receipts were allocated to overhead including advertising, the net profit of 10 per cent expected from sale of the whole edition at the economic price would be reduced by this imbalance between price and cost from 10 to 2 per cent.

Having begun to make some contribution to the publisher's overheads on selling its first 42 per cent of copies printed at the economic price, the edition has next to complete that contribution and begin to provide some net profit. An overhead of 40 per cent of average receipts would be 22.6 at position (6) in table 18-5, so the total overhead allocated to an edition of 1000 copies would be 22,600. Added to the total production cost of 20,250 already calculated, this would represent a total investment in the edition of 42,850, which divided by the average receipts per copy less the average royalty per copy (5 in the table) would indicate that net profit would begin to be made after the sale of 884 copies or about 89 per cent of the edition.

These formulae represent an aspect of reality which book designers will do well to bear in mind when planning expenditure.

CHAPTER 19

Intention and result

T HE DESIGN OF AN EDITION consists of those plans which determine the form of the completed book. When there is a change of plan during production, the change becomes part of the design, and the altered part of the original plan becomes part of an earlier stage of the design. The successful implementation of design plans depends on co-ordination between suppliers, between materials, and between materials and suppliers, and at all stages on verification of the effectiveness of the design and of the workmanship applied to production. Co-ordination and verification are part of book design, just as reading proof is part of authorship. The edition itself, ready for dispatch from the bindery, represents completion of the design.

From the time he starts to form his ideas for the edition, until the bound books are ready for dispatch, the designer needs to validate some of his plans in various ways. This may be for his own purposes, as when he sketches the title-page to make sure exactly how it should be set. It may be for his publishing colleagues; for instance, jacket designers usually provide a coloured sketch of a proposed picture, to be considered by editor and sales manager. Detailed plans provide a specification for an estimate of production cost, which is usually prepared before expenditure begins, and which may indicate a need for design alterations; economic aspects of design were outlined in chapter 18. The plans become effective only when expressed clearly and comprehensively to all responsible for carrying them out. Throughout the edition's production, the results of the plans have to be compared with the plans themselves, to make sure the plans were valid and are being rightly used.

Some conventional methods of design control before and during production are described in this chapter. They are recommended because misunderstandings happen all too often in the complexity of book production, and mistakes can be expensive. Also tastes differ; everybody entitled to influence the style of production should have a chance to see for himself what the designer intends for the edition.

It is an axiom of book production that if anything is given a chance to go wrong, it will go wrong. The terms of a supplier's contract — and even verbal orders may prove to be contractual — will often protect him from having to pay for putting it right. Things are best prevented from going wrong by meticulous handling of the various specimens and proofs which pass to and fro during production. The usual procedure for text proofs is described in § 19-3, and other evidential forms of control have much the same requirements. By these means, the designer can maintain the integrity of his plans at every stage of production.

§ 19-1 Layout and marking up

A *layout* may be defined as a sketch of part of the book's typography, including some of the display, without which there is no need to lay out text lines. The dimensions of margins and the position of illustrations are also likely to be shown or at least worked out in layouts. The sketch may be little more than a scribble, in which the typographer tries out his ideas about the choice of type size and alphabet for different lines, and about their lateral and vertical placing. Another kind of layout may consist of a page or pages from a previous publication, with an instruction that it is to be followed for general style. Imprecise composition instructions such as these have their uses, but are not usually enough to result in a satisfactory proof.

A carefully drawn layout is more likely to be acknowledged by composition the designer will be able to accept. It should be accompanied by typed copy, in case any sketched letter is difficult to identify. The height and width of each drawn display letter should be traced or copied from a type specimen, or should at least be roughly equal to those of the printed letter; without type specimens, the width of the lines and the general appearance of the page will be guesswork rather than plan. Letter-spacing should be marked between letters at every occurrence, and if it is to be visually even, as it should be in display lines if possible, the necessary instructions should be written in the margin. Also in the margin, the type selected for each line or group of lines should be specified in writing. The margins and the edges of the leaf should be drawn in exact position. Spaces between display lines are best drawn to the nearest point, and the number of points required should be written opposite each in the margin. For hand make-up in photo-composition, headline position should be indicated by *row distance* (base-line to base-line, in points and fractions or millimetres).

Finished layouts are rarely necessary for this purpose, unless a client is to be impressed. Drawn in black or coloured ink with great care, with all pencilled guide-lines erased after use, they simulate the printed page. But the designer's time is not less valuable than that of a printer who could produce a better result by setting specimen pages.

The layout accompanies the copy to the supplier, and should always be present where it is needed. The printer should return it with the specimen setting or proof to which it applies, for comparison, but may need it back at a later stage. When illustrations are printed as plates, the binder may need a layout to indicate their margins. A layout for a die, returned by the engraver with a proof, may be necessary to show the binder the position of the lettering on the case. When possible, the designer will do well to photocopy his layouts, keep the original sketches, and send a copy to each supplier with a request that it should be retained.

Layouts are not always necessary, particularly to experienced designers. The copy can be marked up to indicate the designer's intentions, so long as he is confident of the outcome or at least determined to accept it. The copy for some of the preliminary pages may be easier to mark up if it is first re-typed in the general style of the printed page; the half-title, title, verso, and contents list are frequent candidates for this treatment. Excessive marking up of such items as tables may also be avoided by re-typing and marking up one of the tables as a style guide for the rest. Re-typed folios or passages always need to be carefully checked.

Marking up to indicate typographic style is clearest when based on standard proof correction marks such as those listed in British Standard 5261C:1976. Any instruction not represented by such a mark needs to be written clearly and horizontally in the margin, and ringed to indicate it is not part of the text for setting. An instruction which applies to the whole of the text copy should include some such word as *throughout*, and the folio on which it appears should be indicated on the first folio of the typescript. An instruction which applies to one folio only, such as *all centred*, *all letter-spaced* on a title-page in capitals, should include the words *this folio only*. Preliminary pages, because they differ from each other, usually need comprehensive marking up; the limited number of non-standard text items can be marked on first appearance, at least by the designer. Every folio of the copy should be examined, however briefly, for items which will have to be planned if the designer is to be satisfied with every page in proof.

§ 19-2 Specimen setting

A specimen setting is a composition sample prepared as part of the design validation of a specific edition; it is different from a type specimen, which is produced to demonstrate type-faces for use in any edition. A single-line specimen may be enough for a limited purpose, perhaps to show the alignment between a Greek and a roman type. A group of lines may be evidence enough of the average number of characters in a certain measure. One page may have its uses, but an opening of two facing pages will be better even for such plain composition as fiction. For informative text which includes such variant items as extracts, tables, sub-headings, and footnotes, four pages should be enough; but the cost of a four-page specimen may equal that of adding eight or more pages to the book.

The designer should ensure that the specimen will show a sample of the main text, and of any variant item of the kind mentioned above, together with an example of repetitive items such as chapter headings. The most difficult instance of each item may be the best to select. In planning the specimen, as elsewhere in his work, the designer should bear in mind the irregularity of copy; if the specimen chapter title makes two lines, but others make one, the printer will need instructions for both kinds.

Specimen pages in metal should be carefully pulled in proof from type on respectable paper, cut to exact size to show the correct margins, and precisely folded. In photo-composition, some kind of photographic proof is less expensive than a printed one, but the outline of the typographic image must be sharp, as evidence of its ultimate appearance on the printed page. Such a proof should still show the exact position of margins and cut edges, as an example of the eventual make-up and imposition. This kind of guide is likely to be more precise than a layout.

A specimen setting should be considered methodically, detail by detail, and the slightest divergence from the design should be marked. If it is not, it may appear throughout the book as a recurrent mistake, and the specimen might as well never have been submitted. Word-spacing throughout the page, and word-division at line ends, need particular care with unfamiliar composition systems. The position of display lines, spacing round sub-headings, and paragraph indents are all worth examining, as well as the more conspicuous features of any typography. Specimens sometimes reveal in metal matrices a state of wear which if ignored may have deteriorated further by the time they are used for the book.

Even before the designer sees the specimen pages, they may have served as a useful guide to the printer while *casting off* (§ 18-1). When the pages have been examined and approved, one set should be returned to the printer with all the marks on it, including any he has made himself. Other sets deserve to be kept by the designer, who is likely to find a collection of specimen pages useful in more ways than one, particularly in discussing future editions with typographically impercipient colleagues.

§ 19-3 Text proof and revise

Nearly all authors and publishers work at some distance from the factories where editions are made, and indeed from each other. Most book designers work in publishing firms, but freelance practice in planning typography and illustration is growing. Design plans and their effects have therefore to be communicated across various distances, and probably to strangers, in order to be effective. The control items listed in this chapter are part of a conventional method of transmitting and verifying the author's text, the editor's amendments, and the designer's plans. The usual procedure for handling all such items is another part of that method. Elements of the procedure outlined in this section are worth considering not only for text proofs but for other forms of specimen and sample.

When more than one version of copy is available for type-setting, the version from which type is to be set should be designated at the outset. All marks intended for the edition, whether editor's or designer's, should be made on that version. Until a corrected proof has been returned to the printer, this version of the copy is the only valid source of the correct text. The printer's amendments to the copy, and also to the designer's instructions written on it, should be marked in a distinctive colour, for checking by author, publisher, and designer; but the printer's queries should be entered on the marked proof.

The first stage of proof is sometimes called *proof* to distinguish it from a second stage known as *revise*, though a proof at any stage may also be known merely as a proof. When the printer sends the marked proof, he should send the copy with it. Wherever proof and copy go together, the designer should see them together, and should mark all design amendments on the marked proof.

When it has been comprehensively marked by all responsible for doing so, the marked proof replaces the copy as the source of the correct text. Any marks on the copy which the printer will need, for example as guidance to the placing of illustrations, will have to be transferred to the marked proof. The printer will not normally consult the copy again, so it need not be returned to him. Only the marked proof should be returned to the printer, whether for revise or press, as he should follow no other.

Proofs in galley are often ordered for illustrated books in which the designer intends to adapt illustration size to suit the page make-up. Galley proofs also emerge, as first output, from composition systems in which alterations, corrections and make-up into page are combined in a second output; this second proof is derived from an amended record of the original keyboarding. In some systems, the outline of the typographic image may be less exact in the first output than in the second.

When illustrations are sized and shown in proof before make-up begins, the printer should be able to show satisfactory proofs in page, even of heavily illustrated books. But if the make-up requirements are strict, and likely to be met only by editorial alterations to copy and by changes of illustration plan, the designer will probably prefer to work on galleys. Otherwise galleys may be an additional and perhaps unnecessary stage of proof, as editor and author will need to see proofs in page anyway, if only for

[345]

index compilation; and any designer who passes an edition for press at the galley stage is taking an unwarrantable risk.

When the designer has to make up an illustrated book from galley, the make-up of every page in his paste-up should follow the style of experienced compositors working in a first-class printing-office, and every item should be exact in dimensions and position. Otherwise the printer will have problems to solve in his own way, which may not be that intended by the designer. All proof correction marks should appear on the paste-up, as the printer tends to work from one proof at a time, and as the make-up may otherwise fail to make allowance for the effect of inserted or deleted lines. If the printer is to work from more than one set of proofs at any stage, he will need special instructions, and the limited purpose of each set must be clearly marked on it.

The sales promotion of editions has been well served by the *book proof*, as a first stage. Usually this kind of proof is first imposed, at least temporarily, and then pulled on both sides of sheets which are folded and bound up in paperback form. Pulled from metal type, this kind of proof is expensive but not prohibitively so. Photo-composition, which does not readily lend itself to economical book proofs, is compelling publishers to dispense with this promotion material. When book proofs are to be pulled, the designer may have to make sure that printing almost without make-ready will not cause the metal to show signs of wear before the edition itself is printed.

A second stage of proof, usually in page, is loosely known as a *revise*, a term applied — also loosely — to marking corrections on the proof and to carrying them out in type. The designer should examine any such proof page by page, however briefly, searching for anything which he knows, from experience of that kind of book, of that kind of printer, and of his own past design work, may go wrong. Page make-up, in particular, is not normally under the designer's control, and the printer's solution may not always be the same as the designer's.

A *press revise* may take various forms, but the printer usually regards it as a *house* proof, pulled for his own purposes only, after imposition and just before the formes go to press. If the designer or anybody else outside the printing-office is to see it at all, he should do so quickly, minimize the number of alterations, and restrain himself from inessential improvements. There is no time left for changes of design, unless the press time reserved by the printer and the publication date agreed by the publisher are to be relinquished.

However precise and comprehensive his instructions were, the designer will do well to check certain points throughout any specimen setting or text proof, even if he does it almost unconsciously. Skill in this kind of examination grows from never setting eyes on printed text without assessing its quality. The exact arrangement of display lines, any degradation in the typographic image which might appear on the printed sheet, letter-spacing, word-spacing, and word-division at line-ends can be studied or at least glanced at on every page. Except for a proof shown after final imposition, and then only if it is untrimmed, no proof is a guarantee of margins; and except for pages which include some item in a fixed position, such as a page number, no proof before imposition guarantees the position on the leaf of any item of text or illustration. When the designer changes the printer's make-up locally in proof, he should also mark the consequential alterations to adjacent pages, where lines may have to be carried over or taken in. One of these pages may indicate the reason for an apparently careless example of make-up; a minor defect could have been avoided only by tolerating a larger one.

Layout, specimen, proof, and sample generally, as a means of communicating the design plans for an edition and of verifying their implementation, are most effective when all evidence is clearly identified, carefully examined, and sent to the right place at the right time.

§ 19-4 Illustration proofs

The main purpose of most proofs in book production is the purpose of any other kind of proof — it is to prove something. No proof can prove everything about the reproduction of an illustration, because the presswork conditions of an edition differ from the proofing conditions. Any print is likely to differ, however minutely, from its proof; but the printer's aim is to match, as closely as he can, the quality of any proof which sets a quality standard. Some proofs are required to prove little more than the identity and dimensions of an illustration; they are still known as proofs.

A printer's proof is in effect the most limited of editions. A single copy may be enough; the average number of proofs pulled on most occasions is probably less than ten. Not all proofs are intended to prove anything; reproduction proofs pulled from type are often used only as camera copy.

Illustrations to be printed from letterpress blocks are so convenient at all stages of proof that designers may well wish at times that no other printing process had ever appeared in book production. The process-engraver pulls proofs which demonstrate the quality of reproduction, without extra charge. The printer makes up the blocks with text and captions, and pulls proofs of the whole page, showing each illustration in position. The printer's make-ready may be relied on for local adjustment of tonal balance in half-tones.

Pulling proofs from offset plates is usually too expensive for text and illustration to be shown together. Various forms of photo-proof replace the proofs which would otherwise emerge from the printing process itself, as in letterpress. These identify the illustration, show its dimensions, and can be used to indicate position, but they provide no guarantee of quality. The *photo-mechanical transfer* (PMT) is a photographic print made to scale in the camera without a separate negative; a half-tone screen may be used, and the result may be sharp enough to be mounted with text for line reproduction. Half-tones are then reproduced dot-for-dot, a method suited to the coarser screens. Each transfer requires a separate exposure. When more than one print is to be made from one original, bromide prints may be made from a process negative. The bromide print's faithfulness to the original may be less certain than that of the PMT, but such prints are commonly used as camera copy for line subjects. Positive images on film can be printed down on to positive-reversal sensitized paper to show a positive proof image. In order to prove identity, dimensions, and position only, by means of several proofs of each subject, photo-copies can be made from any of these photo-proofs, but no electrostatic photo-copy should be used as copy for a process camera when anything better can be found.

When illustration quality is critical, offset half-tones are often shown in *scatter proof* — an offset proof without text, usually in more or less random positions, crowded together to reduce the number of plates required. This expensive form of proof is rarely necessary for line work.

Colour proofs should be amended with caution. To match a flat colour exactly with another, when one is printed in proof conditions and the other is printed as part of an edition, is far from easy. Some tolerance of slight colour variation is advisable. Colour half-tone separations ordered by the publisher are likely to appear on the designer's desk in the form of *progressive proofs*. Essentially proofs of this kind show the step-by-step effect of successive colours, by means of several proofs of the same illustration. The first colour to be printed appears alone; then the first and second printed together, followed by the first three together; and finally the complete illustration in all four colours. There may also be a separate proof of each colour, register marks, and a strip of variegated panels for quality control. The whole proof is essential to the printer, as a check on colour intensity and balance. The designer should concern himself with the complete illustration only; amendments to the rest are better left to specialists. The proof must be examined in colourless light for any critical examination of colours, and comparison with the colour transparency or any other original must be by the same light. Alterations should be indicated precisely, and described in non-technical terms, without reference to individual trichromatic colours. Locally reducing the intensity of magenta, for instance, might distort the colour balance if no compensating adjustment were to be made in the other colours. The probable cost, in money and time, of minor alterations, together with the possibility that the illustration is already good enough for its purpose, should be borne in mind.

All illustration proofs should be critically examined. Unless the scale, dimensions, and rectangularity are visibly correct, for instance by appearing in alignment with the text lines, they should be checked. Scatter proofs of half-tones usually show more of the illustration than will be reproduced in the edition; the edges should be drawn in on the marked proof, even when this has already been done on the original. Slight irregularities sometimes appear at the edges of any half-tone. The inclusion of the whole subject, and the angle at which it is to be reproduced, should not be relied on without proof; the designer's own instructions may have been at fault, or may not have been clear. In half-tones the designer should make certain of the screen gauge and angle, and examine the dot pattern under a magnifying glass. When a half-tone original is reproduced by half-tone, some moiré pattern is inevitable, and will be visible to the naked eye. Details in highlights and shadows should be equally visible in proof, unless extreme contrast is preferred. In a proof intended to verify quality, the intensity, the balance of tones, the appearance of contrast, and above all a general resemblance between original and proof, should satisfy the designer on behalf of publisher and reader. The position of each illustration on the page should be checked, together with the presence of detail at any edges which are to bleed.

§ 19-5 Paper, presswork, binding, and jacket

Paper is best selected in one of three ways. When its specification is in all respects the same as that of a paper previously selected by the designer, the book in which it was used provides a sample which in effect is approved before the making is ordered. When paper of the required specification is already offered for sale, the merchant will be able to submit sample sheets for approval. When paper is to be made to a new specification, the required characteristics are likely to have been projected from those of different samples of paper. Whatever the method of selection or specification, the designer will do well to base it on sample sheets he can handle and examine, and send to the printing-office unless already familiar there. The binder will also need a *paper dummy*, a set of folded sheets based on the estimated number of pages in the book. This is for return in the form of a bound dummy, described below.

When the paper for the edition is delivered to the printer, he should submit an *out-turn* sheet to the designer. Not all printers examine paper in bulk before printing, or test even the out-turn sheet; this sheet, taken at random from the consignment, is often the only certain report of the paper's quality and condition.

Some assessment of the paper can be based on comparisons with samples from other makings of the same kind of paper, or with samples of other kinds of paper. Colour, opacity, surface, caliper, stiffness, and furnish can be observed while the designer handles the advance sample or the out-turn sheet, looks through it by transmitted light, and compares it with other samples. Weight, sheet dimensions, and rectangularity can be measured, though a single sheet will not provide a guarantee of the whole consignment. If the paper is delivered in good time, the printer will have an opportunity to obtain special ink if necessary, measure humidity, and generally prepare to make sure of a good result.

When all proofs and the out-turn sheet have been approved, the quality of presswork becomes the responsibility of the printer. The designer is not entitled to supervise presswork, since the printer must be allowed to control the costs of his production process. *Running sheets*— sample sheets of the edition sent to the publisher while printing continues — usually arrive too late to be of much use, and may anyway have been wet when folded for dispatch. The essential quality check at this stage is a complete set of sections, printed, folded, gathered, and cut, examined by the designer before binding begins. The evenness of impression throughout the book, and the general appearance of the presswork, cannot be effectively examined by any other means.

A *bound dummy* or sample binding is needed at an early stage, to demonstrate the bulk of the book, and the width of the spine. This should be prepared as soon as the number of pages is known and the paper selected. *Publicity proofs* of jacket or cover may be needed before the book is bound, and jacket design is therefore urgent. The dummy also serves to show the book's general presentation in terms of page size, bulk, and weight, in relation to its provisional published price; there may still be time to improve the market aspect of the designer's plans. The designer may also wish to make sure that the dummy matches his specification and the implicit requirements of the binding at all points, including the dimensions, endpaper quality and colour, squares, turn-in, rounding, tapes, and lining, discussed in chapter 16.

A *specimen case*, or sample case presented without the text pages, provides more convenient evidence of the quality of board and hollow (§ 16-7). This too should be checked for dimensions and turn-in.

An unblocked specimen case, however, is of limited use, particularly when it is made before the covering material intended for the edition is available. Even when one has been submitted, the designer should see a blocked specimen case. He may already have seen a proof of the die, but this cannot demonstrate the appearance of the blocked image. The blocked specimen case should be made with the exact materials intended for the edition, including the correct colour of the covering material and of the blocking foil. This kind of sample provides a final check on all aspects of the outer part of the binding, including the precise position of the lettering.

A picture jacket or cover is likely to be so carefully aimed at the market that the book designer is not authorized to decide what kind of design will be best or how it can best be produced. In these circumstances, a *rough* or sketch, in colour if colours are to be used, will be needed if publishing colleagues are to approve the plan before process artwork is begun. Deft and imaginative jacket designers sometimes provide several roughs for selection. When the designer submits roughs to other people, he is likely to hear a diversity of opinion. Comment may be worth passing on to a jacket designer, but instructions are unlikely to call forth his best work. Within the limits of his brief, he is better left to design as he thinks best.

Most editions are subscribed, or sold in advance of publication, to the book trade, mainly by means of jacket or cover 'proofs', which are in fact an advance publication of the outermost part of the book only. Publicity proofs are printed in colour on the paper chosen for the edition, and may even be varnished or laminated. These proofs are useful as a final check before jacket printing, but their main purpose is fulfilled only when they appear some weeks before publication. The necessity for this early proof indicates to the designer the priority he should accord to all stages of jacket design, starting with the paper dummy.

§ 19·6 The completed edition

When the work has been well planned and well carried out, the arrival of advance copies can be a delightful moment. Surely the least inspired of book designers must feel a quickening of enthusiasm and hope as he studies the result of his efforts. It is for this he has trained himself and practised; this is the fruit of his slowly gathered technical knowledge, of his experience and taste, of his ability to create and translate, and of the standards he has set himself to maintain.

Even before that one edition is completed, there are likely to have been others to prepare, and to return to the finished task is not always easy. Quality, however, can be maintained only by continuous effort, not by spasms. Subsequent binding orders of the first impression, reprints, and new editions should all match the original product. Alterations to the text or to the style of the book's production may have to be supervised, and advance copies of each new impression compared with a copy of the first.

To observe the book's endurance in use is less easy. Only the most conscientious designers seek out their books on the shelves of the public library, to study their condition after being read by a number of borrowers in the first year or two after publication. It is usually the binding that is most worth examination — the appearance of the covering material and the blocking, the strength of the back, and the tenacity of any tipped plates.

To be too much a technician and too little a layman is a fault in a book designer. Most of the qualities of good book production can be appreciated by any perceptive reader; a flat opening, a clean impression, and an obviously readable page are of more value than extreme subtleties in the choice of type. The designer's task on a book can best be completed long after publication, when he can draw it from a shelf, open it as though it were strange to him, and ask himself whether it was so produced that it would now tempt him to acquire and read it.

CHAPTER 20

The purposes of book design

T HE QUALITIES of a book's appearance and structure, which are determined by design, depend to some extent upon each other; a well-designed and well-printed page, for example, will not please the reader if it will not lie flat, or if after a little use it begins to deteriorate. Visual as well as structural qualities depend almost entirely on the adjustment of all variables to each other, rather than on successful attention concentrated only on a few items. A well-produced book is distinguished not by a single characteristic but by harmony between the various parts, and by a pattern imposed upon the whole. To design a single graceful chapter opening is not enough — the chapter headings throughout the book deserve equal care.

The processes of book design may be classified as editorial planning (in which the text may be locally re-arranged if necessary in preparation for composition, for the benefit of author and reader), visual planning (which determines the appearance of the printed image), and technical planning (which is concerned with the structure of the book and the methods of its manufacture). The editorial and visual aspects of the design derive most of their effectiveness from technical planning. Success in one process or in one aspect alone is never enough; failure at one is more than enough.

Clearly, then, the different processes of book design are too closely linked to be undertaken as entirely separate stages. Some specialization, of course, is common. The graphic or technical specialist — the illustrator, perhaps, or the binder — may achieve more in his own field than the all-rounder can hope to do. The utmost efforts of the most studious of typographers will hardly bring him level in knowledge of one particular aspect of his craft with the artist or technician, whose skill derives from concentration on that aspect alone. The book designer cannot be an expert in everything he needs to know, and should always value the opinions of others engaged in book production, of those who keep and distribute books, and of those who read them. Indeed, he should know much of these different points of view without having to inquire. But whatever influence or advice he accepts, and however far he delegates different parts of his task to different specialists, the designer who wishes to produce entirely successful books must seek to gain and to keep authority over all the planning and supervision of which book design is comprised, subject only to the reasonable requirements of his client or his employer.

The possession of technical knowledge is vital — it is the principal tool of the designer's craft, which shapes not only his books but his methods. It is not so much that he may find this knowledge useful, as that every problem he approaches must be seen in its light. The appearance and structure of the industrially printed book result from the techniques used in its manufacture; the book is not an art-form but an industrial product, shaped by an ancient kind of industrial design. The book designer relies equally on a technician's knowledge of a variety of processes and materials, though not necessarily of details of operational method; on a sub-editor's understanding of the needs of author and reader; on an industrial artist's creative ability in the arrangement of mechanically produced patterns; and on an unrelenting care and mastery of detail. In the words of a great book designer, 'The success of book printing lies in never for one instant relaxing in the inspection of details until the book is actually bound' (Rogers, 1950).

If these are the qualities which the designer needs, how are they to be put to use? The purposes of industrial book design spring naturally from those of the book itself, and from the nature of mass-production. The purposes of the book may be gathered into four groups.

A book is to be sold. The designer's task is not so much to settle the price as to make the best use of the permissible expenditure, planning the edition for economical production, and exploiting to the full the techniques and materials available. The book must attract the buyer, and be worth possessing as a physical object, not merely worth borrowing. Its price must be within the buyer's reach, and its appearance and construction should make the price a bargain. The requirements not only of the public but of booksellers and librarians must be allowed to influence its form.

A book is to be laid open, held, and carried. All but a few books are held while being read, and most books are carried about to some extent before and after reading. No book can be considered legible unless it lies flat when open; it should not have to be held open. The printed part of the pages at which the book is opened should be flat, not curving inwards towards the spine. Bulk should be proportionate to page size, as far as possible; a very squat, stout book is as inconvenient to hold as a very large thin book. Every book should be designed to withstand whatever handling it may receive, without unduly rapid deterioration.

A book is to be seen. If it fails to attract more than a glance, it may not be read at all. Then it must be capable of being read with ease, speed, and accuracy, by the reader for whom and in the conditions for which it is intended. This can be achieved only by the precise adjustment to each other of all the variables of the text page, and is a matter of paper and presswork as well as of the arrangement of printing images. Illustrations no less than composition need to be planned by the designer. The well-made book presents an appearance of pattern and purpose; all its parts are planned to suit each other. The designer must concern himself with the intellectual as with the optical process of reading, arranging the text and illustrations with their headings, captions, notes, reference systems, and other accessories in a clear and convenient manner.

A book is to be kept. After being read it is set aside, usually on a shelf, to be read again one day. The book should if possible be of a size to stand between ordinary bookshelves; particularly large books are apt to be a nuisance. Once it is on the shelf, the book should be able to stay there indefinitely without undue deterioration, retaining its qualities against its next use.

The advantages of mass-production techniques are speed and low cost rather than high quality. High quality, in book production at least, has to be imposed on the product by technical planning. The first purpose of this planning is to adapt the book's form to the methods and materials involved in its making, in the service of low price and high quality. It is also by technical planning that the structure of the book is designed to withstand the stresses of use. An even strength throughout is worth more than extra strength at one point, except that books intended for rough use may well be reinforced at points where the mechanically bound book is known to be weakest.

Quality has not only to be achieved, it must also be maintained. The designer's duty is to plan not simply a book but an edition, the last copy of which should be very nearly as good as the first. Indeed, he may have to plan a whole series of editions, and the likelihood of frequent reprints may well influence the design of the book. Quality in a book has little value without endurance in production, in use, and on the shelf.

This conclusion, and the chapters which precede it, may seem to emphasize the arid, utilitarian aspects of book design, and the service rendered by planning to industry and commerce. But the intention of this book is to describe some methods of book design, not to prove the value of the craft or to suggest how it may be enjoyed. The value of book design is derived from the value of books. Pleasure is most easily to be found in a craft by those who have mastered its methods. Skill in book design may begin with reading about it; mastery, and the pleasures that attend it, spring from practice, in the service of author and reader, and from work on book production rather than from theories. No reader should suppose that this book, or any number of books, can offer him all he needs to know. What has been said here should be no more than a beginning of his studies, particularly in technical matters; about these he should never cease to learn, with constant vigilance in his watch for technical possibilities and limitations in an area of unceasing change. Books need and deserve all the study and practice that can be afforded. The book designer who has ceased to learn, and to see the books he makes as they are seen by the amateur of books and reading, may have to expect little more achievement of himself.

Publications

I<small>N</small> the first two editions of this book, the lists of publications seemed long enough to provide for a life-time's reading in which there would be next to no time left for the practice of book design. Any book designer who neglects the literature of his craft is likely to restrict his own scope, but practice is a better form of study. Accordingly I have drastically abbreviated the list of publications I recommend for further reading, and I have combined it in a single list with the titles I have referred to in the preceding pages. These references, and the value of further reading in general, together with the possession and use of works of reference, I also recommend.

Aldis, Harry G. (1951) *The printed book*, 3rd edition, revised and brought up to date by John Carter and Brooke Crutchley. Cambridge University Press.

Anderson, M. D. (1971) *Book indexing* (Cambridge authors' and printers' guides). Cambridge University Press.

Bailey, Herbert S., Jr (1970) *The art and science of book publishing*. New York.

Beaujon, Paul [Beatrice Warde] (1926) 'The "Garamond" types: sixteenth and seventeenth century sources considered', *The fleuron*, V.

Berry, W. Turner, A. F. Johnson, & W. P Jaspert (1970) *An encyclopaedia of type faces*, 4th edition. Blandford Press.

Bland, David (1962) *The illustration of books*, 3rd edition (enlarged). Faber & Faber.

Bland, David (1969) *A history of book illustration: the illuminated manuscript and the printed book*, 2nd edition. Faber & Faber.

British Federation of Master Printers (1962) *Book impositions*. BFMP.

British Standard (1967) *Typeface nomenclature and classification*, BS 2961. British Standards Institution.

— (1970) *Specification for page sizes of books (metric units)*, BS 1413. British Standards Institution.

— (1971) *Specification for the title leaves of a book*, BS 4719. British Standards Institution.

— (1972) *Specification for metric typographic measurement*, BS 4786. British Standards Institution.

— (1975) *Guide to copy preparation and proof correction*, BS 5261. Part 1: *Recommendations for preparation of typescript copy for printing* (1975). Part 2: *Specifications for typographic requirements, marks for copy preparation and proof correction, proofing procedure* (1976). British Standards Institution.

— (1977) *Recommendations for the presentation of tables, graphs and charts*, draft for development 52. British Standards Institution.

British Standard (1978) *Citing publications by bibliographical references*, BS 5605. British Standards Institution.

Burbidge, P. G. (1952) *Notes and references* (Cambridge authors' and printers' guides). Cambridge University Press.

Burbidge, P. G. (1969) *Prelims and end pages* (Cambridge authors' and printers' guides). Cambridge University Press.

Butcher, Judith (1982) *Copy-editing: the Cambridge handbook*, 2nd edition. Cambridge University Press.

Carey, G. V. (1957) *Punctuation* (Cambridge authors' and printers' guides). Cambridge University Press.

Carter, Harry (1969) *A view of early typography up to about 1600*, the Lyell lectures 1968. Clarendon Press.

Cartwright, H.M. (1961) Ilford graphic arts manual. Volume 1: *Photo engraving* (1961). Volume 2: *Photolithography* (1966). Ilford.

Chaundy, T. W., P. R. Barrett, & Charles Batey (1954) *The printing of mathematics: aids for authors and editors, and rules for compositors and readers at the University Press, Oxford.* Oxford University Press.

Cheetham, Dennis, & Brian Grimbly (1964) 'Design analysis: typeface — Univers typeface, designer Adrian Frutiger, maker Deberny & Peignot, available in Britain from the Monotype Corporation and Lumitype', *Design* 186.

Chicago (1982) *The Chicago manual of style for authors, editors, and copywriters*, 13th edition, revised and expanded. University of Chicago Press.

Clapperton, Robert Henderson (1952) *Modern paper-making*. Blackwell.

Clutton, Sarah (1960) 'A grammar of type ornament: an analysis and classification of typographic border designs and their behaviour in use', *The Monotype recorder*, 152, 1.

Collins, F. Howard (1973) *Authors and printers dictionary*, see Oxford.

Coulson, Anthony J. (1979) *A bibliography of design in Britain 1851–1970*. Design council, London. [Book design, book illustration, book jackets, bookbinding, pages 117–135.]

Craig, James (1974) *Production for the graphic designer*. New York, and Pitman Press.

Crutchley, Brooke (1967) *Preparation of manuscripts and correction of proofs*, 4th edition (Cambridge authors' and printers' guides). Cambridge University Press.

Curwen, Harold (1966) *Processes of graphic reproduction in printing*, 4th edition, revised by Charles Mayo. Faber & Faber.

Davis, Alec (1975) *Graphics: design into production*. Faber & Faber.

Day, Kenneth, ed. (1966) *Book typography 1815–1965 in Europe and the United States of America*. Ernest Benn.

de la Mare, Richard (1936) *A publisher on book production*, the sixth Dent memorial lecture. Dent.

De Vinne, Theodore Low (1904) *Modern methods of book composition: a treatise on type-setting by hand and by machine and on the proper arrangement and imposition of pages* (The practice of typography). New York.

Dreyfus, John, ed. (1963) *Type specimen facsimiles: reproductions of fifteen type-specimen sheets issued between the sixteenth and eighteenth centuries*. London.

Dreyfus, John, ed. (1972) *Type specimen facsimiles II: reproductions of Christopher Plantin's Index sive specimen characterum 1567 & folio specimen of c. 1585 together with the Le Bé–Moretus specimen c. 1599*. London.

Gaskell, Philip (1974) *A new introduction to bibliography*, reprinted with corrections. Clarendon Press.

Glaister, Geoffrey Ashall (1979) *Glaister's glossary of the book: terms used in printing, bookbinding, and publishing, with notes on illuminated manuscripts and private presses*, 2nd edition, completely revised. Allen & Unwin.

Grannis, Chandler, ed. (1967) *What happens in book publishing*, 2nd edition. New York.

Gray, Nicolete (1976) *Nineteenth century ornamented typefaces*, with a chapter on 'Ornamented types in America' by Ray Nash. Faber & Faber.

Hart, Horace (1978) *Hart's rules for compositors and readers at the University Press, Oxford*, 38th edition. Oxford University Press.

Herdeg, Walter (1981) *Graphis diagrams: the graphic visualization of abstract data*. Zurich.

Holland, F. C. (1967) 'Typographical effects by cathode ray tube typesetting systems' in *Journal of typographical research* 1:1. Cleveland, Ohio.

Hunter, Dard (1947) *Papermaking, the history and technique of an ancient craft*. New York, reprinted 1978: London, Constable.

International Standardization Organization (1975) Draft proposal 61. ISO.

Jackson, Holbrook (1938) *The printing of books*. Cassell.

Jammes, André (1965) 'Académisme et typographie: the making of the romain du roi.' *Journal of the Printing Historical Society*, 1.

Jennett, Seán (1973) *The making of books*, 5th edition. Faber & Faber.

Johnson, A. F. (1937) *Specimen books of Lamesale and Gando*. Bibliographical Society.

Johnson, A. F. (1946) 'On re-reading Updike' in *Alphabet and Image 2*.

Johnson, A. F. (1966) *Type designs: their history and development*, revised edition. Deutsch.

Kindersley, David (1976) *Optical letter spacing for new printing systems*, 2nd revised edition. Wynkyn de Worde Society.

Lamb, Lynton (1962) *Drawing for illustration*. Oxford University Press.

Langdon-Davies, B. N. (1951) *The practice of bookselling, with some opinions on its nature, status, and future*. Phoenix House.

Lee, Marshall (1951) *Books for our time*. New York.

Lee, Marshall (1979) *Bookmaking: the illustrated guide to design/production/editing*, 2nd edition, completely revised and expanded. New York.

Lewis, John (1967) *The twentieth century book, its illustration and design*. Studio Vista.

Lockwood, Arthur (1969) *Diagrams: a visual survey of graphs, maps, charts and diagrams for the graphic designer*. New York.

Lowry, Martin (1979) *The world of Aldus Manutius: business and scholarship in Renaissance Venice*. Oxford, Basil Blackwell.

Mackays (1976) *Type for books: a designer's manual*, new edition. Bodley Head, for Mackays.

McLean, Ruari (1958) *Modern book design from William Morris to the present day*. Faber & Faber.

McLean, Ruari (1980) *The Thames and Hudson manual of typography*. Thames & Hudson.

McMurtrie, Douglas C. (1943) *The book: the story of printing and bookmaking*, 3rd edition. New York.

Mansbridge, Ronald (1980) 'My publishers are terrible . . . a catalogue of complaints from authors about their publishers — a list that includes too much copy editing, poor communications, lack of consultation, broken promises, mistakes in accounting, and general inefficiency' in *Scholarly publishing, a journal for authors and publishers* 11:2.

Morison, Stanley (1962) *On type designs past and present: a brief introduction*, revised edition. Benn.

Morison, Stanley (1967) *First principles of typography*, 2nd edition (Cambridge authors' and printers' guides). Cambridge University Press.

Morison, Stanley (1973) *A tally of types*, edited by Brooke Crutchley, with additions by several hands. Cambridge University Press.

Moxon, Joseph (1962) *Mechanick exercises on the whole art of printing* (1683–4), edited by Herbert Davis & Harry Carter, 2nd edition. Oxford University Press.

Norris, F. H. (1952) *Paper and paper making*. Oxford University Press.

Oxford University Press (1981) *The Oxford dictionary for writers and editors*, compiled by the *Oxford English dictionary* department. OUP. [Formerly F. Howard Collins: *Authors and printers dictionary*, 11th edition 1973].

Phillips, Arthur (1980) *Handbook of computer-aided composition*.

Rice, Stanley (1978) *Book design: text format models*. New York.

Rice, Stanley (1978) *Book design: systematic aspects*. New York.

Rogers, Bruce (1936) *An account of the making of the Oxford lectern Bible*. Monotype Corporation.

Rogers, Bruce (1943) *Paragraphs on printing elicited from Bruce Rogers in talks with James Hendrickson on the functions of the book designer*. New York: reprinted 1979 by Dover, New York: London, Constable.

Rogers, Bruce (1950) *Report on the typography of the Cambridge University Press, prepared in 1917 at the request of the syndics by Bruce Rogers and now printed in honour of his eightieth birthday*. Cambridge, privately printed.

Rosner, Charles (1952) 'The book jacket: first principles' in *The Penrose annual* 46.

Rosner, Charles (1954) *The growth of the book jacket*. Sylvan Press.

Ryder, John (1976) *Flowers & flourishes*, including a newly annotated edition of *A suite of fleurons*. Bodley Head, for Mackay.

Ryder, John (1976) *Printing for pleasure*, revised edition. Bodley Head.

Simon, Oliver (1963) *Introduction to typography*, 2nd edition edited by David Bland. Faber & Faber.

Spencer, Herbert (1969) *The visible word*, 2nd edition. Royal College of Art.

Steinberg, S. H. (1974) *Five hundred years of printing*, 3rd edition revised by James Moran. Penguin.

Straus, Victor (1967) *The printing industry: an introduction to its many branches, processes and products*. New York.

Taylor, F. A. (1962) *Colour technology for artists, craftsmen and designers*. Oxford University Press.

Tschichold, Jan (1967) *Asymmetric typography*. Faber & Faber.

Twyman, Michael (1970) *Printing 1770–1790: an illustrated history of its development and uses in England*. Eyre & Spottiswoode.

Updike, Daniel Berkeley (1924) *In the day's work*. Cambridge, Massachusetts.

Updike, Daniel Berkeley (1937) *Printing types, their history, forms, and use: a study in survivals*, 2nd edition. Harvard University Press: reprinted by Dover, New York: London, Constable: 1980.

Western Printing Services (1970) *The Western type book: analysed specimens of*

PUBLICATIONS

Monotype, Linotype, and Intertype faces suitable for bookwork and available at Western Printing Services, Bristol, new revised edition. Hamish Hamilton.

Westwood, John (1976) *Typing for print.* HMSO.

Williamson, Hugh (1981) *Photocomposition at the Alden Press, Oxford: a printer's type-specimen book.* Bodley Head for Alden Press.

Wilson, Adrian (1967) *The design of books.* New York.

Zachrisson, Bror (1965) *Studies in the legibility of printed text.* Stockholm.

Acknowledgements

The first edition of this book, published nine years after I started work on book production, could not have been compiled without a great deal of help. This for the most part was acknowledged in the first two editions. I hope the many friends and organizations who were generous with assistance in the 1950s will forgive me for not referring to them individually again; I am still grateful, and this edition is based on their support. One omission has to be made good. David Neale, then like myself a member of the staff of Oxford University Press, edited the typescript and marked the proofs of the first edition in his own time at my invitation and with a degree of skill and care I would have expected of nobody else. At that time members of staff in some eminent publishing houses were excluded by policy from printed acknowledgements by authors, however grateful.

The first twelve chapters of the third edition were read and marked in typescript by Anne Charvet, and the rest by Ena Sheen, at Oxford University Press. John Robson read the entire typescript and improved the first draft at many points. All or much of the typescripts of revised drafts was read and annotated by Harry Myers, Andrew Thompson, and John Trevitt. Draft chapters were read by Nicolas Barker, Michael Chater, Tim Chester, Philip Cohen, Richard Norton, and Anne Sworder. I am most grateful for all their vigilance, but I remain wholly responsible for whatever imperfections may have escaped them, particularly as much of the book has been redrafted since they saw it. I owe thanks too for other forms of help to Dennis Avon, Matthew Carter, John Dreyfus, Adrian Frutiger, the late John Jarrold CBE, K. C. S. Lea, James Mosley, Walter Tracy RDI, Professor Michael Twyman, and M. J. Walker.

The passages from Updike's *Printing types* (1937) on pages 4, 33, 84, and 149 are reprinted by permission of Harvard University Press, and from his *In the day's work* (1924) on page 96 by permission of William E. Rudge's Sons Inc.; that by Bruce Rogers (1936) by permission of the Monotype Corporation Ltd; those from Hart's *Rules* (1978) on pages 53 and 146 by permission of Oxford University Press; that from Steinberg (1974) by permission of Penguin Books; and those from Aldis (1961) and Burbidge (1969) by permission of Cambridge University Press. The Monotype Corporation also permitted me to refer to 'Monotype' machinery and type-faces without the quotes otherwise obligatory round a registered trade mark.

ACKNOWLEDGEMENTS

I was fortunate in having the diagrams drawn for the first edition by Ralph Mabey, and in having his drawings available for this edition; they are examples of the highest standard of quality in this kind of work. The wording on his diagrams is manuscript, and that on the additional diagrams is typographic. I acknowledge with thanks other help with illustrations from the Bocardo Press, Oxford, Linotype-Paul Ltd, the Monotype Corporation Ltd, the St Bride Printing Library, and the University Press, Oxford. Other illustrations, including examples of printing types, are reproduced, by permission where appropriate, from *Brendan Behan's New York*, Baskerville's *Book of Common Prayer* of 1761, *IBM economical composition at the Alden Press*, *The journal of typographic research* 1 : 1, 1967, *Linotype Matrix* 17 of 1953, Mackays (1976), Morison (1973), Phillips (1980), Rogers (1936), Ryder (1976), type specimens of the Monotype and Linotype companies, Updike (1937), Western Printing Services (1970), and Williamson (1981).

I compiled the index of the 1956 edition in order to find out what kind of index I wanted. I had commissioned indexes before then but I had never written one before and I have never written another. As soon as I started work on the index I found I was moving away from the conventions to which I was accustomed, towards something of an innovation. The glossary was combined with the index so that the reader would have to look up any word on one page only. The head-words or lemmata were set in italic to distinguish them from the rest of each entry. The colon, rather than the semi-colon which would have been corrrect, was used as a dividing point to reduce the number of commas. The use of ampersand for 'and' throughout the index reduced the number of characters per line in the narrow columns. The order of sub-entries was as far as possible the order of occurrence. The first part of a lemma was repeated rather than being replaced by a dash. These arrangements provoked favourable comment from some observant and knowledgeable people, and continue in this edition, apart from minor alterations. Tom Colverson, who reviewed the 1956 edition and who has since used the 1966 edition for teaching, very kindly offered to compile the index for this edition in the style previously established.

All authors are indebted to their printers, and I as much as any, though I think I have caused them less trouble than some; and I offer my thanks to all my friends at the Alden Press who have had a hand in this edition. Above all I am grateful to John Nicoll of Yale University Press; on the basis of the out-of-date draft of the third edition, he committed himself to the book, and encouraged me to undertake the necessary revision, at a time when I had begun to lose interest in so long and so demanding an effort without certainty of publication.

Oxford, 1983   H.W.

Index and glossary

[365]

quad, quadrat: in composition, see under *em* & *en*: *quad* – as paper size, four times the area of the broadside – 14–17 228–9 237

quarter-bound – bound with spine in one material & rest of cover in another – 312 332: *quarter cloth* 312

quarto – page area ¼ that of basic sheet – 13 15 18–19 95 112 229, use of 21–3, & two-column setting 18 108, & folding 320, demy 4to book 19 312, example 22–3

quoin – wedge for locking up type matter in chase – 54 232

quotation marks, quotations, quotes – punctuation convention for quotation marks – & typescript 6, single preferred 145

RAO – see also *A sizes (ISO)* – an ISO standard quad sheet, 860 × 1220 mm suitable to produce books of A4 (297 × 210 mm) & A5 (210 × 148 mm) formats – 16

reader of books 1 4 5 8 11, & format 13, & type size 19, & conventions of setting 48, & complex make-up 50, & footnotes 52, convenience of 55–6, & proportions of letters 88, & irregularities 89, & hyphenation 89, & good composition 90, & author's communication 91, & letter form 99, & legibility 101, & title-page 100, & illustrations 267 270–1, & end papers 272, & plates 273, & copyright 275, & paper 278 289, & thickness of books 287, & tipped-on plates 298, & register 306, & spine-blocking 316, & blurb 324, & production of quality 330 349 352 353 354

Réale – see *Old Style*

re-binding & librarians 21 263 351

recess processes – (intaglio) printing techniques such as photogravure in which ink is contained in recesses in printing plate or cylinder – 211 226–7

recommended minimum row distance – minimum vertical space between the baselines of adjacent rows (or lines) of type by which all characters including accented capitals can be separated from all others – example 34: also *body*

re-composition – operation in support of good page make-up – 51 270

rectangularity of paper 350

recto – right-hand page, conventionally odd-numbered – 25 166 169 171 173 175 261 270 271 274, used for title page & other main display pages in prelims, & for first page of main text 173

reduction of artwork for reproduction 205 257 258, & bromides 265, & legibility 274

reel – web of paper – 336

reference, books of 160–1 169 174 181: *reference signs* 44: *reference works & margins* 24

reflective range – range of light radiations reflected by a surface – 258–9

register – relationship of position between the images produced in the course of two or more printings on one sheet – 232, & colour 248 252–3, & paper on offset press 286, & guillotined paper 288: *register marks* on artwork overlay 260: *register (bookmark)* – coloured silk ribbon bonded between a book's spine linings, to serve as bookmark – 306

reglets – metal interlinear spaces of more than 3 points – 46

related alphabets 98: see also *series*

repeat half-title – recto page repeating bastard title & immediately preceding text – 169 180

representative, publisher's 319

reprint by offset 203 211, replacement for worn type 215, from film as reserve material 317 221 241–2, & re-imposition for 228 232, & paper for 287 320: *reprint by rotary letterpress* from curved duplicate plates 239, & paper bulk 320, & quality 351, & estimates & cost 335–6

reproduction (repro) 54 347: *reproduction, photographic* 203

repro proofs – pulled from metal type with care on carefully selected paper as *camera copy* – 54 103 206 215 235 238 257 275 347, & cost 335, for blocking dies 314: *repro sheets* – unbacked machine proofs pulled after make-ready, kept as camera copy for future reprints – 215

reserve material – film, plates, or typematter stored in preparation for reprinting – 55–6 95 215–7 222

resolution – frequency of output scan lines from CRT & digital composing systems – 80 94

retouch – adjust, by hand, tones and imperfections in camera copy or continuous tone positives to ensure good reproduction – 205 207, using *air-brush* 259

re-typing bad copy, to avoid excessive marking up 343

reversal, reverse – in process-engraving or in litho plate or gravure cylinder-making, turn from black to white or left to right by photo-mechanical means – 202 208, & type & 4-colour blocks or plates 244 322, & lettering 274, on jackets 322 325